D0987091

FORM 19

Science under Socialism

Science under Socialism

EAST GERMANY IN
COMPARATIVE PERSPECTIVE

EDITED BY
Kristie Macrakis
and Dieter Hoffmann

Harvard University Press

Cambridge, Massachusetts, and London, England | 1999

Library of Congress Cataloging-in-Publication Data

Science under socialism : East Germany in comparative perspective /
    edited by Kristie Macrakis and Dieter Hoffmann.
        p.   cm.
    Includes bibliographical references and index.
    ISBN 0-674-79477-X (alk. paper)
    1. Science—Germany (East)—History.   2. National Socialism and science—
Germany (East)—History.   I. Macrakis, Kristie.   II. Hoffmann, Dieter.
Q127.G3S357   1999
509.431—dc21   98-41275

Preface

Science under Socialism is appearing in print almost ten years after the fall of the Berlin Wall, in November 1989, and the collapse of the German Democratic Republic (GDR). Since that time much knowledge has been uncovered about the former GDR, and investigations based on recently opened archives have led to many insights into the inner workings of the system. Much work remains to be done, however, owing to the sheer volume of paper left behind and the challenge of reconstructing the appropriate interpretive framework. The magnitude of the task can be gauged by the fact that despite a deluge of publications in the 1990s in GDR general history, social science, and political science, the history of science, technology, and medicine, including science and technology policy, has received scant attention.

Writing about East German history can be controversial and complicated, especially since many of the historical actors are still alive. Whereas Kristie Macrakis wished to include a discussion of some of these problems in a prologue, Dieter Hoffmann did not; therefore we agreed that the prologue should be printed with the understanding that it is Macrakis's view and approach and not necessarily those of Hoffmann or the other contributors. Whereas Macrakis wished to discuss the creative process involved in an East-West group project and the controversy surrounding who should write East Germany history and how, Hoffmann wanted only the finished product presented.

Science under Socialism is being published in both an English- and a German-language edition. Macrakis has edited the English edition and Hoffmann primarily the German edition. We are listed as joint editors, however, because we have discussed or corresponded about most of the chapters. We both have different styles and approaches to editing a volume, and so the final products are different. For example, they begin differently: the English edition has an Introduction that provides an interpretive analysis of East German science, whereas the German edition has a Prologue that gives background information on the book and a short overview of the chapters.

Another major difference is the inclusion of four extra chapters in the German edition—three written by Germans and one jointly authored by an American and a German. Three of the four chapters fall under the category of "disciplines" and include a study of the failed East German aviation industry by Burghard Cisela. The second disciplinary chapter is by Friedrich Naumann, on the history of East German computing up to the 1970s. The third disciplinary chapter is a study of biologists' careers by Ekkehard Höxterman. The fourth chapter, written by Hoffmann and Mark Walker, is a biographical chapter on Friedrich Möglich, a scientist who lived and worked both in the National Socialist period and in the early GDR period until 1957. The German edition also has an extensive bibliography current up until the summer of 1997. Finally, the American edition contains a chapter by Gary Geipel on computers and politics under Erich Honecker that is not included in the German edition.

This book could not have been realized without the financial support of two institutions: the Alexander von Humboldt Foundation and Michigan State University. In 1993 the Humboldt Foundation awarded us a Transatlantic Cooperative Research Grant for three years to support various stages of the book, from research to the workshop and conference. The foundation allowed us total freedom in pursuit of our project goals. We are particularly grateful to Kurt-Jürgen Maaß and Heide Radlanski, who continually inquired about the project at Humboldt Foundation meetings and invited Macrakis to write about it for the *Humboldt Magazine*. Michigan State University generously matched those funds; there Macrakis is indebted to George Leroi, the dean of the College of Natural Science, who resourcefully found funding and was supportive of the research afterward. Macrakis is also indebted to the Institute for Advanced Study, Princeton, for providing her with a fellowship during the academic year 1993–94. Hoffmann's work on the project was facilitated by support from the Max Planck Institute for History of Science in Berlin.

Two student research assistants at Michigan State University, Andreas Boos and Erika Dirkse, aided Macrakis in her research and in preparing the book. We are grateful to both of them for their efficiency, organization, intelligence, and good-naturedness. Charles Molhoek prepared the index, and several students in Germany also made contributions. Finally, we thank the two recent reviewers of the manuscript for their careful reading and thoughtful comments; they will find many of their suggestions integrated into the book.

K.M./D.H.

Contents

Prologue: The Makings of *Science under Socialism*

Kristie Macrakis

Science under Socialism is a group project with an international cast of authors—six Americans, four former East Germans, and three former West Germans.[1] The idea for the project emerged shortly after the fall of the Berlin Wall, in 1989; it took shape in the wake of unification, in 1992, as my application for the Transatlantic Cooperative Research Program sponsored by the Alexander von Humboldt Foundation. Thus the book was conceived and written in a particular time and place. This circumstance not only gives the topic more significance, but also makes knowledge of the process of its creation especially important. Our intention was to produce a unified, multi-authored book with this mixed group through the steps of an initial workshop, a final conference, revision of book chapters, and publication.

We see this book as a first attempt to grapple with the history of the sciences in the former German Democratic Republic (GDR) in comparative perspective (National Socialist Germany, Eastern Europe and the Soviet Union, and West Germany).[2] Since the fall of the Wall many archives have opened up that had never before been available to the public or to scholars—for example, the archives of the Communist Party and the Ministry for State Security (MfS or Stasi). Since this project began, our group has been working in these archives, discovering material that sheds new light on the construction of a socialist state and the role of science and technology in it. Interviews have been conducted with, among others, the Politburo member Kurt Hager and the legendary spy chief Markus Wolf.

Our coverage includes general policy issues, institutions, disciplines, and biographies in their social and political context. We have tried to touch the major bases in East German science and technology but realize there is much that cannot be covered in a first effort. We hope this book will provide a historical framework and will spur others to study different aspects of the subject in greater detail.

The project officially began in the fall of 1993, when our group met in Berlin at the Berlin-Brandenburg Academy of Sciences for the preliminary workshop, which was part of an ongoing discussion in Germany on the East German past. At the time, discussions of the activities of the Ministry for State Security, especially the exposure of prominent personalities who were agents, dominated the press. I invited Werner Stiller, a defector from the MfS's science sector, as a surprise guest. Since then we have added a chapter on science and espionage in East Germany.

The final conference took place five years after unification at the World Fellowship Center in Conway, New Hampshire, host to numerous conferences on the GDR, between September 15 and 17, 1995. The setting for our second meeting was a far cry from the tumultuous Berlin of 1993. It was a good time to take stock of what we had learned about the former East Germany since unification.

Although we were a diverse group, we worked well together, despite differences in historiographical approach. In some ways I had hoped for more fiery debate between the East and West Germans about the East German past. Unlike some West Germans, those in our group offered a balanced, fair assessment of East Germany. It was clear that the participants' objective was to try to understand the past, not to condemn East Germans and their state. Since unification, others have also begun working on the contested history of the GDR, although not on the history of science. German historians have asked the questions, Who should write the history of East Germany? What should their politics and background be?

Historical writing on East Germany has emerged from state-directed projects, academic centers on the territory of the former GDR that grew out of unification, university group projects, and projects initiated by historians themselves.[3] Several of the state-directed projects, often called "truth commissions," were initiated not to gain an understanding of East German history, but rather to investigate crimes that may have been committed by the SED (Socialist Unity Party) and the Stasi. Two prominent examples are the so-called Inquiry Commission for Working through the History and Consequences of the SED Dictatorship and the Investigative Committee for Alexander Schalck-Golodkowski and GDR Assets, both sponsored by the German parliament. The Inquiry Commission was led by the former East German dissident priest Rainer Eppelman. Both commissions have issued massive reports—the Inquiry Commission's report has been published in eighteen volumes and runs 15,378 pages—heard hundreds of witnesses, gathered many documents from various archives, and had results reported in the media.[4] Some of the historical judgments of both these works are tainted by political motivations, but they still provide valuable documentation. Perhaps, as Timothy Garton Ash writes, the Inquiry Commission is for East Germany "what the record of the Nuremberg trials is for the student of the Third Reich."[5] Not surprisingly, a "Counter-Inquiry" group designed as an extra-parliamentary oppositional counterpart to the Inquiry Commission was formed by historians close to the PDS (Partei Demokratische Sozialismus, the successor to the SED).

Before the inception of state-sponsored reports, a discussion on who should write East German history emerged among East Germans themselves. Many of those deeply involved saw this historical discussion as a "dispute about cultural hegemony in Germany" that broke out in the last months of 1989 in East Germany between "the forces of the ancien régime and the new political forces of the GDR-Citizens Movement."[6] The conflict came to a head when two historians, Armin Mitter and Stefan Wolle, founders of the Association of Independent Historians and authors of the first books based on archival material, published an attack on the Potsdam Center for Contemporary History in the *Frankfurter Allgemeine Zeitung* because it housed many scholars from the ancien régime.[7] Much ink was spilled on this dispute, and American historians visiting the Potsdam Center also offered their comments.[8]

To an American the dispute seems odd; most of us would agree that anyone who wishes to research and write about history should be able to do so, with quality determining what gets published. But in Germany the dispute was complicated by the fact that there were very few professional jobs in history available. In those charged years following the fall of the Berlin Wall, the politically correct answer to the question who should receive those positions was dissidents or those similarly disadvantaged by the East German state.

As the biographies of the contributors to this book illustrate, the authors come from very different backgrounds. Some of the East Germans held important positions in the GDR. At least one was not a friend of the regime and was even under surveillance by the Stasi. My main concern was to find the best historians for the book, and so participants were not vetted to meet certain political criteria; rather, they were chosen for their ability to write an informative chapter that fit the structure and concept of the book.

As this is a book about East Germany, there was no question that East Germans would be invited to participate. The four East German contributors brought to the project not only their expertise, but also the intimate experience of life in East Germany. Three are historians of science and one was trained as a physicist, but is now working toward a degree in history. Two East Germans were at the Institute for the Study of the Theory and Organization of Science at the academy; although they worked on history of science, the institute also had the task of analyzing East German science. Three contributors are West Germans—one has written extensively on the history of German physics and the other two worked at the West German Institute for Science and Society in Erlangen, which studied the East German science system during Germany's time of division. Five of the American participants are historians of Germany or of German science, and one is a political scientist.

There is no doubt that there are differences in national and personal style in historical research and writing, and this project was no exception, especially given that it included former East and West Germans and Americans. It was my philosophy not to edit out differences in style and points of view. Several of the authors, especially the East Germans, consciously avoided a heavy interpretive overlay such as a Marxist analysis. The reason for this was not only that Marxism had been discredited, but also that the East Germans had spent twenty years listening to or reading Marxist interpretations and thought that any sort of explanation, whether Marxist or postmodernist or something else, would be ideological and would skew the historical events being described. At one of the conference dinners, I asked one of the authors whether he could add some historical interpretation to his account. He replied that, because of the way history had been used and abused in East Germany, he just wanted to state the facts precisely. In the end he did contextualize his story a little more, and it still remained a balanced and even-handed account.

At the other end of interpretive analysis is the contribution by the American Mitchell Ash, who directly addresses a question that has occupied historians and sociologists in science studies for several decades—namely, what kind of influence does the social context have on the *content* of science?

Although we never scheduled any historiographic sessions at our workshop or conference, collectively we all strove to tell the "truth about history."[9] Several scholars in our group have addressed current debates on historical relativism and postmodernism as they apply to their other work, but they did not explicitly bring these historiographic issues into their book chapters. Because most of us have exploited newly opened archives and woven that material together with other sources, our collective approach was to reconstruct the GDR past from the records it left behind—notably the records of the Socialist Unity Party, the Ministry for State Security, various commissions, and the ministries for science, as well as interviews with scientists. Thus to some extent, the history we have reconstructed here is an interpretation of the records. We are not then "neutral and passionless investigators" reconstructing the past exactly as it happened.[10] The way West Germany characterized East Germany after its demise would negate the existence of any "objective" history. Precisely because East Germany came to an inglorious end, historians often view the events teleologically—that is, analyzing them as an inexorable movement toward a negative outcome.

Although most historians avoid explicitly revealing their attitude toward a subject, sometimes a topic is so controversial that the historian feels compelled to do so. This editorializing was the case with at least one of the contributions to this book. As editor I felt I had to make a decision to edit the piece or to allow the author to leave in his opinions (provided that I could comment on the chapter in the Preface). The chapter in question is written by Eckart Förtsch, a former West German expert on science policy in the German Democratic Republic. Förtsch, who worked at the West German Institute for Science and Society, was one of the first people asked to take part in our project because of his background and knowledge. Although the majority of Förtsch's chapter chronicles a process of politicization and economization in science and intelligentsia policies from the 1940s to the 1980s, Förtsch is clearly disapproving of the developments. For example, he argues that "reminders that human beings and their fates were at stake . . . did not surface . . . The imposition of rigid demands for change on people and institutional structures ultimately destroyed some of them: that was the point . . . Claims that the destruction wrought by the politicization process was 'creative' strike me as dubious" (p. 34). After an exchange with Förtsch, I decided to leave the passages as he wrote them.

Since I have singled out a Western participant for commentary, to offer symmetry it is only fair to turn to the contributions of his East German counterpart, Hubert Laitko. Laitko was at the Academy of Science's Institute for the Theory, History, and Organization of Science, where he headed the history of science research group. He has contributed a chapter on science and higher education policy during the 1960s. Although he offers no overt commentary on the reforms he discusses, he does admit that "in 1995 it was still very difficult, if not impossible, for a German author to be objective about this topic . . . The very fact that the GDR as a state came to an end can seduce a historian into viewing its history teleologically" (p. 44). Nothing in Laitko's account lacks objectivity, however, and so it, too, remains as written. In his conference contribution, however, he diverged from the text of his chapter and offered the group members some personal insight into his life in the GDR and his role in policy making.

Laitko received his *Diplom* (the equivalent of an M.A. in the United States) in 1959 and then became an assistant in the Marxist-Leninist Basic Studies program. He received his doctorate in philosophical questions in the natural sciences in 1964 at Humboldt University. He was then active in the field of the theory of knowledge and was especially influenced by Derek de Solla Price and the science of science movement. By the late 1960s he was one of the founders of an institute for research on science at the Academy of Sciences. It is for this reason that he urged the group to see him as an "eyewitness" to history rather than as an "analyst."

The early post-unification atmosphere also shaped the book in another way. Originally, we planned to offer even more of a comparative perspective in *Science under Socialism* than we do now. We were very interested when Reinhard Siegmund-Schultze proposed to write a chapter on mathematics in Nazi

Germany and in socialist Germany, examining whether comparisons could be made between the two. Since unification, the GDR has often been characterized as totalitarian or as Germany's second dictatorship, primarily by West Germans but also by the British historian of East Germany Mary Fulbrook, and so we were intrigued by Siegmund-Schultze's proposal.[11]

Siegmund-Schultze presented his comparative paper on mathematics at the first workshop but changed the topic to the concept of antifascism for the conference and book after misinterpretation by a West German journalist and a sense that the political atmosphere was not right for his original subject. Rather than comparing the two periods of time, Siegmund-Schultze examined the legacy of National Socialism in East Germany and its effect on personnel issues. In his book chapter he argues that it is dangerous and misleading to make comparisons because of the "absence of a shared base of reference and comparison for the two systems of science." More interesting to Siegmund-Schultze is the "historical derivation of one system (East Germany) from the other (Nazi Germany)" (p. 79). Further, he editorializes on the unfair comparisons some Germans have made between mathematics in Nazi Germany and East Germany.

As these examples show, GDR history is "living history" in that those who experienced it can still provide invaluable commentary on what life was like under socialism, as well as take part in debates such as whether comparisons can be made between East Germany and Nazi Germany. But East German history is living history in another respect as well: it is very much a part of the present. That is, East Germans' past has partly determined their future in United Germany. After unification, society became polarized between those who were victims of the regime and those who were perpetrators, with victims theoretically benefiting and perpetrators being ostracized. This and other issues have led to increasing rather than decreasing tensions between former East and West Germans. As part of this coming to terms with the past, a more general debate has erupted over the issue of amnesty for people with questionable political pasts. These developments not only produced transitional problems in united Germany in the 1990s, but also were present as we wrote the chapters for this book.

Science under Socialism

Introduction:
Interpreting East German Science

Kristie Macrakis

In 1993, four years after the fall of the Berlin Wall, a German entrepreneur initiated plans to set up a theme park in Prenden, Germany (about twenty miles north of Berlin) in his former homeland, the German Democratic Republic (GDR).[1] Once there, nostalgic citizens could relive forty years of communism, and tourists, who had never had the opportunity before, could experience life in the former communist state, complete with guardtowers, border crossings, low prices with little to buy, military parades to attend, and, of course, spies everywhere. Visitors would be able to talk to waitresses, bartenders, and chambermaids and would risk being thrown into jail if they criticized the communist system or its leaders. In the future theme park, the only cars allowed will be Trabants (the Trabi); East German flags will flutter in the breeze; and banners exhorting people to follow "socialist commandments" will hang from the walls.[2]

Although science and technology were absent in these plans for the theme park, the state had high hopes for the role science and technology would play in building socialism on German soil and in contributing to a "developed socialist society." Not only that, but East German society, like other communist societies, had an optimistic faith and belief in science and its ability to solve social and economic problems. From the start, state policy supported and advanced science, scientists, and their institutions.

In a sense, the organization and practice of science can also be seen as a theme park. The scientific landscape is dotted with institutions where science is practiced or policy formulated. Scientists go about their work in those institutions and generate knowledge in their various fields. As a result, scientific disciplines are created. Scientists' biographies and career trajectories often mirror the societies within which they have lived. And the sociopolitical, economic, and national context shapes the landscape's contours. Although politics influences science in every modern industrialized country, the interaction is particularly magnified in countries in which the state dominates society.

1

This book examines the effects of an emerging socialist society on the development of the scientific enterprise in its various layers. It is therefore simultaneously about the dreams and aspirations of science policy makers as they attempted to build socialism on German soil and the reality of life for frustrated scientists and engineers living under "real-existing socialism."[3] The aim of the book is to gain a *historical* understanding of the development and course of the East German scientific enterprise in comparative perspective.

Major topics that run through the volume include the legacy of National Socialism, the exodus of scientists to the West, the establishment of and contradictions in socialist intelligentsia policies, the changing and exploitive relationship with the Soviet Union (including the extent to which the Soviet model was transferred to East Germany), the survival of a German scientific tradition, the role of the Communist Party and the Ministry for State Security, as well as comparisons with West Germany and the impact of a centrally planned economy.

Although the book is organized topically, the individual chapters are for the most part chronological. This raises the issue of periodization in the German Democratic Republic and whether a general, politically demarcated periodization applies to science and its organization.

GDR history is usually divided into four periods: since the country cannot be understood without looking at the occupation of the Eastern zone, the 1945–1949 occupation and division period is generally considered to be phase one, followed by the founding of the GDR in October 1949, shortly after the founding of the Federal Republic of Germany (FRG) in the West in May 1949. The second phase spans 1949–1961, when the Berlin Wall was built. Politically, however, one could see Walter Ulbricht's rule from 1949 to 1971 as the second period, followed by Erich Honecker's accession to power in 1971 until the revolution of 1989, when East Germany was on the road to unification with West Germany. Most of the authors in this book would find the general political periodization applicable to their topics; most of us have found that the major political events had repercussions for the scientific enterprise. However, other phases also emerge in the development of science and education policy.

Unlike the Disneyland theme park, historical periodization and analysis allow us to reconstruct the past and to organize it in ways the actors themselves, living in the moment, may not have recognized. While a theme park allows us to relive the past, it offers us only a static reconstruction. Historical reconstruction orders the past for us chronologically, thematically, and analytically; it allows us to see events with hindsight or a bird's-eye view.

Even though the book is organized topically, each chapter documents the interaction between the scientific enterprise and society through four decades of GDR-style socialism. The early years were influenced by the legacy of National Socialism (the occupation, 1945–1949, and the early 1950s), the middle phase by the societal goal of building socialism on German soil (the 1950s until the building of the wall); the third phase by a period of educational and economic reform influenced by notions of the scientific-technical revolution (1961–1971, the end of Ulbricht's rule); and the final phase was that of a "developed socialist society" exhibiting seeming stability and solidification of structures (1971–1989, the Honecker period). Throughout all these periods, the GDR maintained a friendly and supportive attitude toward science while undermining itself through social policies, economic deficiencies, and working conditions that were ultimately hostile toward achieving the highest quality science and technology.

While coming to terms with the Nazi past and officially rejecting it through the policy of antifascism, the Soviet Occupation Authorities and East Germany purged personnel and restructured universities. Although de-nazification had to be modified for the nation to survive, socialist transformation at the universities slowly succeeded over the decades, though it also led to a mass exodus of qualified personnel.

Nevertheless, planners in the 1950s sought to harness science and mobilize scientists to build socialism on German soil. Thus 1955 was a banner year, with the return of German scientists from Soviet captivity, major resolutions on scientific-technical progress, the lifting of the ban on nuclear research, the launching of research in aviation, and the mobilization of State Security to procure the results of Western science and technology for scientists at home. Modernization became a watchword. The 1950s also saw significant transformations at the Academy of Sciences, as it became the center of scientific research in the GDR. Projects like the Chemistry Program (1958) were developed to boost the economic strength of the country. By the end of the 1950s severe economic problems exacerbated the flow of scientists to the West. Leaders saw the building of the Wall as the only way to stem the flow and build socialism without "disturbances" from West Germany.

The 1950s can best be described as a decade of trying to build socialism amid frustrating personnel problems. As the GDR formulated a policy for the intelligentsia, it implemented one to retain the "old bourgeoisie," even as they fled to the West. By the time the Wall was built in 1961 and external disturbances were minimized, the GDR had most experimental variables at its disposal, although it had fallen behind the West in most fields. The decade saw reforms in the scientific-technical revolution—reforms in the economy, in education, and at the academy. The slogan "science as a productive force" became official state policy. Since Marx had left no instructions on how to implement his ideology in a modern, industrialized society, the idea behind the New Economic System (1963) was to combine a planned economy with some features of a market economy. It turned out to be Ulbricht's science policy finale, and by the time Honecker succeeded him, the reforms had failed.

Honecker's rule was also characterized by an interest not only in the "key technologies," such as computers, laser research, and biotechnology, but also in consumer goods. Honecker once again stressed the importance of attaining international levels in science and technology. By the 1970s and 1980s, not only had the GDR's vow to "surpass" West Germany fallen short, but its multibillion-mark investment in computers had most assuredly contributed to its economic decline.

While the GDR had always had problems procuring modern equipment and materials for scientific work, thus requiring scientists to build their own equipment (there was even an academy institute for instrument building), by the last years of the GDR, there was an urgent need for modern, expensive equipment.[4] Indeed, data indicate that 21 percent of East Germany's physical plant had industrial machines and equipment that were more than twenty years old, compared with only 5 percent of West Germany's physical plant.[5]

Thus, despite a highly educated supply of scientists and the steps it took to modernize, East Germany remained surprisingly old-fashioned in its infrastructure for science.

Some of the other questions we considered were, What were the intentions of the leaders regarding science? To what degree was the Soviet educational and scientific model transferred to East Germany, and to what extent did the German scientific tradition survive? In what precise ways did the new socioeconomic conditions and political context affect science? Was a political overlay simply placed over science and its institutions or did the politics affect the actual content of science? And how does the influence of political ideology in East Germany compare with other historical periods such as the Third Reich?

We were not interested in providing sociological arguments about which conditions or values are necessary for science to flourish and survive. Nor did we wish to follow the well-worn thesis of the adverse effects of authoritarian political ideology on science. Implicitly, when explanations are offered, for example, as to why East German science (like that of other East Bloc countries) began to lag behind the West in the 1960s, authors point to multiple factors, both systemic and contingent.

Nor was it the goal of the historians assembled to write this book to assess the quality of East German science, a task the West German Science Council undertook after the two Germanys merged. Others have completed scientometric studies using science citation index materials such as total publications of a country, citation rate per publication, most-cited institutions, and city publication output. Another method used for measuring the degree of scientific-technical excellence has been the tallying of patents.[6] Some interesting results emerge from these studies, although the scientometric studies and the Science Council's judgments do not always coincide. For example, the Science Council concluded that GDR research in agricultural science was seldom published in international journals, but in the publication and citation data revealed that the GDR had much higher activity than the FRG in dairy and animal science, soil science, and veterinary medicine.[7] Nevertheless, both the Science Council and the scientometric studies found that science and technology of a high quality existed in many fields in the GDR. Interestingly, a study of the Munich Patent Office shows that between 1950 and 1970 the GDR was awarded more patents than Austria—a German-speaking country of about the same size (the GDR was ranked seventh and Austria ninth); during the 1980s Japan emerged as a technological leader and went from eleventh place to sixth place, pushing the GDR down to ninth place.[8]

Some areas identified by the Science Council as having achieved "international levels" include mathematics, geology, cosmos research, and medicine. Research in astrophysics was considered to be better than in the former Federal Republic of Germany.[9] Other scientometric studies find high citation rates for chemistry, mathematics, biomedical research, and physics.[10] According to one source, the Science Council's results and the scientometric studies agree only on the excellence of three institutes in the GDR: the academy institutes for semiconductor physics, for nutrition, and for polymer chemistry.[11] In addition, the GDR had a good reputation for its work in optics, lasers, and electronics, much of which aided the Soviets' military research. The fact that the material assessing the quality of East German science and technology is contradictory and variable shows the problems inherent in trying to assess East German science quantitatively or qualitatively.

The book is organized topically according to the different levels of the modern scientific enterprise, ranging from large macroscopic *science policy* developments, to the central *institutions* where science was done, to the development of individual *disciplines* and professions, to biographical sketches of individual *scientists*. Many chapters concentrate on the first half of the GDR coinciding with the founding of the German Democratic Republic in 1949 to the end of Ulbricht's rule (1949–1971), while a few span the whole period of the GDR through the Honecker era (1971–1989). Some emphasize the 1950s and 1960s because of the dynamic activity of the early period, including the socialist construction phase of the 1950s, and the reform period of the 1960s. The Honecker period was more stable, lacking the urgency and turmoil that characterized the transformative phase, though the former ended with fireworks.

Part 1 of the book seeks to paint policy developments in the GDR in broad strokes. The first two chapters, by Eckart Förtsch and Hubert Laitko, respectively, describe and interpret the interaction between science and politics and the formulation of government policies toward science and scholarship in general. While Förtsch details how state institutions were developed to control and guide science and the economy, Laitko uncovers the structure of scientific organization in the GDR. Both contributions encompass state policy toward institutions such as universities and academies as well as toward science and scholarship, including the humanities and social sciences (as the German term *Wissenschaft* connotes). Förtsch covers the whole GDR period (while concentrating on the earlier period) and distinguishes three phases— an early period of cognitive and social autonomy for science, followed by a period of politicization, and finally economization—while Laitko focuses on the crucial years of the 1960s, a period of both economic and scientific reforms. With the so-called reform packet of the 1960s, the leaders of the Socialist Unity Party (*Sozialistische Einheitspartei Deutschland,* or SED) attempted to solve economic and political problems through reforms in education and science as well as through economic modernization.

Förtsch and Laitko, the two West-East counterparts, offer us different interpretations of science. While Förtsch depicts science in the GDR as "short-circuited modernization" in a closed system, Laitko describes it as a "failed laboratory experiment." Förtsch argues that because the scientific process was distorted, science never had any chance of succeeding in the GDR. Laitko believes that the Eastern system may have been reformable if GDR leaders had not had to resort to preserving their "power through pragmatic maneuvering devoid of any strategy." As the reforms came to an end, "improvements could be conceived only in opposition to the system."[12]

The third chapter in Part 1 deals with other general policy issues that are important for understanding the development of science in the GDR. In the "Shadow of National Socialism" Reinhard Siegmund-Schultze examines the legacy of National Socialism in shaping personnel policies and its effects on the lives of several scientists, primarily mathematicians, during the 1950s. He analyzes the SED's "antifascist" campaign and its ideological function, and reveals the way it was used to transform politically the professoriate. He concludes by pointing out that a negative German tradition connects the National Socialist period with the GDR: the absence of political responsibility among scientists.

Chapter 4 is really a bridge between policy issues and the institutional section of the book. Like Förtsch and Laitko, I presents a political history of science while exposing how one of the state's main institutions operated. The Ministry for State Security (MfS or Stasi) controlled, guided, and supported science and the economy. Whereas many treatments of the Stasi have been sensational, with a heavy focus on the informants (IMs), this chapter approaches the topic from a different angle: it examines the issue of scientific-technical espionage as it applies to technology transfer from the West to the East. Although the covert technology transfer itself was efficient, the GDR's scientific and economic system was unable to absorb sufficiently the equipment and techniques transferred, despite the fact that the Ministry for State Security worked hand in glove with the Party and government ministries responsible for shaping science policy.

The second part of the book focuses on three major institutions involved in pursuing science and scholarship—the universities, the Academy of Sciences, and the Leopoldina, a learned society. My rationale for constructing the book this way was that institutions played a key role in implementing the socialist transformation of East Germany. This part of the book looks at the extent of the SED's success in this endeavor. The Academy was central to scientific research in East Germany, and the universities were centers for educating the new socialist intelligentsia.

In Chapter 5 John Connelly explores the foundations of university formation in Central Eastern Europe in comparative perspective. He argues that it was during the formative years, 1945–1955, that essential structures were transferred from the Soviet Union and implemented diversely in East Germany, the Czech Lands, and Poland. These formative years determined future developments. Compared with Poland and the Czech Lands, East Germany implemented the Soviet model more thoroughly and the Party had greater control of higher education; this was especially the case with the cadre policies through which universities became centers for the socialist elite.

Since there was an emphasis on heavy industrial growth, and since communist ideology glorified science and technology, technical subjects were emphasized. In the GDR in 1953 there was a 463 percent increase in the number of students studying technical subjects, while the philosophy, arts, and languages student body increased by 112 percent.

The Academy of Sciences of the GDR was the most important and largest center for scientific research in East Germany. Originally a learned society— the Prussian Academy of Sciences—it was transformed during the GDR period into a sprawling research center. Peter Nötzoldt traces the steps in this transformation from the occupation period through the academy reform in 1968.

Nötzoldt argues that before the Wall was built, a dialogue existed between the SED and the scientists, and the scientists' input was important. As in other areas, the old bourgeois scientists were encouraged to contribute despite their lack of party credentials. After the Wall was built, the development of the academy was "shaped by forces outside of science."

A combination of SED success in personnel policies and the academy's expansion into a research institution made the academy "socialist." Although the Learned Society and research in the humanities survived to a certain extent, the main emphasis was on research institutes in the natural sciences. Once the Central Committee apparatus and the party members in the academy leadership had consolidated their influence, they proclaimed the building of a socialist academy, which meant abolishing remnants of the all-German character of the academy, giving the state more influence in setting the goals of research projects, and developing party leadership; traditional ties of research to the West were to be cut back and cooperation with the East increased.

In contrast to the academy, the Leopoldina—the German Academy of Nature Researchers—enjoyed more freedom: in Chapter 7 I show that attempts at socialist transformation by the Party and state were met with stiff opposition. In many ways, the Leopoldina was one of those "niches" in GDR society that could remain free of communist influence. This phenomenon— and the extent to which a myth was cultivated—is examined by analyzing the relationship between the Leopoldina and the Party and state. The Leopoldina's credo that it was an all-German institution believing in the unity of science in a time of national division (the "clip" or "clamp" theory) clashed with East Germany's power triangle in scientific and educational matters— the Party, state, and Ministry for State Security.

Part 3 focuses on scientific disciplines and professions and outlines the situation of the chemical industry, engineers, nuclear physics, computers, and biomedicine.[13] Although the contributions do not explicitly evaluate the quality of East German science, they do examine issues such as the structure and scale of research installations, comparing them with international trends, especially in West Germany or the Soviet Union. All of them point to economic (the deficient economy), social (the cadre class policies), and political (a centralist bureaucracy) impediments to achieving goals in the various disciplines and professions.

Part 3 opens with a chapter by Dolores Augustine, in which she provides a social history of the engineering profession. Technological advance had been at the heart of the socialist ideology, yet the ideal was most often not realized, and this is especially apparent in the case of engineers. In the Ulbricht era, technology was proclaimed to be "the engine of socialist change," yet the

GDR's economic deficiencies and class quotas impeded the implementation of this ideology. In the next chapter Ray Stokes writes about the important chemical industry in the GDR. Excellence in chemistry was an inherited mixed blessing for the GDR. Stokes analyzes how and why the GDR chemical industry fell behind its West German counterpart even though it received strong support from the state with the inception of the Chemistry Program in 1958.

After the ban on nuclear research was lifted in East and West Germany, and after scientists returned from captivity in the Soviet Union, the GDR, like other industrial countries, offered handsome state support for nuclear science and technology. But unlike its West German counterpart, East Germany abruptly curtailed this funding in the early 1960s. Burghard Weiss traces the evolution of nuclear science and technology in the GDR compared with West German developments. He also shows how the Soviet Union was a wanting partner for the GDR because military interests and secrecy did not allow for openness and the free flow of information necessary for scientific cooperation.

The problem of the Soviets' secrecy and restrictive information policy also affected the development of an aviation industry in East Germany. During the late 1950s the GDR sought to develop its own aviation industry after the return of German specialists in the mid-1950s. Surprisingly, the project was dropped by the SED in the early 1960s, ostensibly for economic reasons, but scholars have speculated that the Soviets also contributed to its end.[14]

The story of computers is a dramatic example of East Germany's intense effort to overcome technological backwardness. Gary Geipel analyzes this drive in the context of East German technology policy and the impact of the international environment on that policy. Technology had become the "holy grail" of economic international competition and was the life blood of a modern state and world economic prowess. The microchip became a status symbol of considerable importance.

Many of the disciplinary chapters refer to the development of the GDR's major state-funded programs—for instance, the Chemistry Program of 1958, nuclear research and technology, the aborted 1950s aviation industry, the Energy Program (only touched upon), and the Microelectronics Program (discussed in both the Geipel chapter and my chapter on espionage). There was a tendency on the part of the centralist state to overconcentrate and overinvest limited resources in some areas at the expense of others. Although total investment figures for science policy are not available, we do know, for example, that more than fourteen billion marks (35 percent of total industrial input) was invested in the so-called Microelectronics Program launched in 1977. Only the Politburo thought that this illusionary program had any chance at success; this meant that other areas in science and technology suffered. Even though they had been progressing well, several of the programs—the aviation industry and nuclear research and technology—were aborted or curtailed soon after the state pumped in millions of marks.

State funding of scientific research is certainly not unique to centralized or communist states. Since World War II, for example, the United States has had many major state-funded national-political projects like the NASA space program and the Strategic Defense Initiative. The main difference is that this was just part of America's public complement to much larger private research activities. America had the personnel, the resources, and the raw materials to carry them out. In the case of computers, the GDR was striving for a goal that was limited by a lack of material resources, blocked trading possibilities, and an insufficiently modern infrastructure for science and technology. Moreover, the successful computer industry had originated in the West, where innovations in computer technology came initially from government-funded and military projects but then expanded into a market-based economy.

In Chapter 12, Rainer Hohlfeld places the development of genetics and biomedicine at two privileged academy institutes in social and political context. He looks at issues such as the extent to which scientists were allowed a degree of autonomy and the extent of state control. He argues that the social and political overlay on science did not threaten its cognitive autonomy. Economic deficiencies and political cadre policies limited scientists' access to first-rate equipment and hindered international travel. Nevertheless, he concludes that biomedical research (molecular biology as paradigm) and genetics (Mendel-Morgan tradition) at those institutes did not diverge from research in Western industrialized nations. In another area—that of open public discussion on genetic engineering during the 1970s—Hohlfeld found no difference between East Germany and the West.

Finally, the two chapters of Part IV chronicle the lives of several important scientists in the GDR, demonstrating how the legacy of National Socialism affected scientists in the 1950s. Robert Havemann and Kurt Gottschaldt dealt with the two opposing ideologies they experienced—National Socialism and communism—in different ways. Dieter Hoffmann narrates the story of Robert Havemann's shift from a believing Stalinist to one of the GDR's most well-known dissidents. What makes Havemann's story tragic is that he initially embraced communism and benefited from the state's interest in promoting science and technology, but he then began to experience a bureaucratic, autocratic form of communism—what he called the "bureau of eternal truths."

By contrast, Mitchell Ash examines the life of a psychologist whose career spanned four different political regimes in Germany. Unlike Havemann,

Gottschaldt demonstrated great skill at adapting his science to the changed sociopolitical contexts in which he worked, and was able to take advantage of resources from the state to do this. While Havemann had suffered under the Nazis, Gottschaldt had opportunistically accommodated them.

Issues and Interpretations

The Comparative Perspective

Although this book focuses on the development of the scientific enterprise in East Germany, the authors were encouraged to compare developments in the GDR with the structures and influences during National Socialist Germany, in other East Bloc countries and the Soviet Union, or in West Germany. John Connelly organized his entire chapter around a comparison of university policies in East Germany with those of other East Bloc countries, while Weiss and Stokes make direct comparisons with developments in West Germany. Other chapters, such as the one on the Leopoldina (Chapter 7), discuss similarities and differences between the GDR and earlier totalitarian dictatorships in Germany.

Since East Germany was a child of the Cold War and was therefore influenced by the legacy of National Socialism, shaped by the Soviet Union and by its place in a divided Germany, it cannot be studied in isolation. However, we were well aware of the dangers involved in making comparisons that may be perceived as politically motivated. For example, at times, general commentators or general political historians have tried to equate East Germany with Nazi Germany or compare East Germany with West Germany in order to show that the West's democratic forms were superior to the "totalitarian dictatorship in the East."[15] Similarly, comparing science under capitalism and communism can lead to technological determinism or to triumphalist history. In this book the intentions and rationale for using comparisons as an analytic tool are totally different. Indeed, our interest in making a comparison between East and West Germany lies in the experimental laboratory conditions available. Both Germanys shared a common past, language, and culture, yet the division brought about two separate states with opposing political, economic, and social systems. How did the democratic market system compare with the communist planned economy in countries that started with similar scientific capabilities and traditions? Although this was the initial question some of us asked, the historical realities changed the emphasis of our research.

Two historical forces strongly influenced laboratory conditions in the GDR. East Germany always had the centrifugal force of West Germany pulling at it. Starting in the 1950s, scientists fled to the West; by the 1980s East

Germany had become increasingly dependent on West German hard currency, subsidies, and imports, as the work of Alexander Schalck-Golodokowski, the head of the Commercial Coordination Unit in the Ministry for Foreign Trade, shows. The other historical force shaping East Germany was the Soviet Union. East Germany was in essence an artificial state born of the Cold War, and as such it lacked indigenous legitimacy and the support of the population. In the world of science, East German–style communism was therefore a synthesis of a strong German scientific tradition and Soviet ideology, structures, and influence.

East Germany often compared itself with West Germany. As Laitko aptly describes, West Germany became "a moving target"—not only in the sense of being an enemy but also because of the rival competition between the two countries. As many of the authors found, East German scientists and leaders continually complained of falling behind the West or West Germany. For example, in 1967 Ulbricht was already complaining that the GDR was ten years behind the West in computer technology.

East Germany itself aspired not only to "catch up" to West Germany, but also to "surpass it" in science and technology. A slogan made popular by Ulbricht in the late 1960s and early 1970s was "überholen ohne einzuholen"—surpass West Germany without trying to catch up to it by pursuing the same course of development. Seemingly an oxymoron, this phrase is a variation of the phrase adopted from the Soviet Union and used in the mid-1950s in East Germany: "einholen und überholen"—catch up to and surpass the West. Erich Apel, the GDR's economics minister, proclaimed at the Third Party Conference in 1956 that "we want to, can, and must catch up with and surpass Western capitalistic countries in a short time in technology."[16] Given that East and West Germany seemed to start on the same footing in the 1950s, the GDR did not think it unrealistic that it might surpass West Germany in science and technology and thus prove the superiority of communism.

But the two countries were not in a similar position in the 1950s. West Germany was almost twice as large as East Germany and had three times the population and five times the industrial output. Economically, this asymmetry was aggravated by Soviet reparations and the absence of Marshall Plan aid in the GDR. While West Germany achieved an "economic" miracle in the 1950s with the support of the wealthy United States, the East Germans made only modest gains at rebuilding their own economy. Thus the competition was uneven from the start.[17]

But East German leaders at the time did not want to accept that they had started off at a disadvantage because they lacked the equivalent of a Marshall Plan to pump up an infrastructure weakened by the war and Soviet reparations. Moreover, in the 1950s they were not yet able to foresee the extent of Soviet exploitation. Although efforts were made in the 1960s to reform the economic system, the planned economy proved to be inefficient, as well as detrimental to science and especially technology. By the 1960s and 1970s, catching up to and surpassing West Germany on all fronts was no longer possible, and Ulbricht therefore posited finding an alternate technological path to supremacy.[18]

A more general, structural comparison between East Germany, on the one hand, and Nazi Germany and the Soviet Union, on the other, has the potential to offer theoretically significant results. In an excellent introductory text entitled *Totalitarian Science and Technology*, Paul Josephson has made interesting comparisons between science and technology in Nazi Germany and the Soviet Union using examples from biology, physics, and technology.[19] Interestingly, his descriptions of many of the features of both those regimes do not always extend themselves to East Germany! For example, there was no equivalent in the GDR to racial hygiene and Lysenkoism in biology, to the ideologization of physics, or to gigantomania in technology. This is in part due to the historical factors outlined above—the centrifugal force of West Germany and the nature of East German communism. In addition, although East Germany shared the left wing ideology of transforming society along class lines, it was not a powerful, large country with hegemonic tendencies, as were Nazi Germany and the Soviet Union. As will be outlined below, the transformist vision manifested itself most clearly in the cadre policies for scientists and at the level of institutions. Transforming institutions like the academy along socialist lines influenced the structure and priorities of scientific research. Influencing personnel through the cadre and class policies had a similar effect.

Another feature East Germany shared in common with Nazi Germany and the Soviet Union was its strong centralist bureaucracy. This may be a characteristic of dictatorships, for, as the sociologist of science Robert Merton wrote, "dictatorship organizes [and] centralizes."[20] The East German bureaucratic structures in general, and those governing science in particular, were strikingly similar to structures in the Soviet Union. The Central Committee of the Socialist Unity Party was ultimately responsible for setting policy and took the lead in the socialist transformation of science and its institutions by controlling ideology as well as personnel appointments. Even if competing centers of power emerged—like the Ministry for State Security or Schalck's empire—they were ultimately subordinated to the "leading role of the Party."

From a study of the voluminous paperwork left behind by the Socialist Unity Party, it appears as though East Germany had even more centralist bureaucratic control than Nazi Germany. This may be due, in part, to the longevity of the GDR—forty years compared with Nazi Germany's twelve—and to its centrally planned economy.

The comparative perspective was also useful in addressing issues such as the extent to which the East German scientific system was modeled on that in the Soviet Union, comparisons with other East Bloc countries, the reasons East Germany developed differently from West Germany, and whether and in what way there were similarities and differences in the relationship between science and politics in National Socialist Germany.

Soviet Structures and Influence

Many Soviet structures and methods of control were transferred to East Germany, especially during the occupation period and the 1950s. The Soviet example taught East German leaders how to "secure and maintain power, create central systems for planning and control, and politicize the social environment and social structures." By the 1950s, "politicians ordered each discipline to adopt the methods and organization of Soviet science as a model."[21] The Soviet experience was also transferred to East Germany through personnel—that is, through the large group of German scientists who returned from their stint in the Soviet Union in the mid-1950s. A number of these individuals became influential policy makers and scientists, including Peter Adolf Thiessen (the chair of the Research Council from 1957 to 1965), Max Volmer, Max Steenbeck, Nicolas Riehl, and Manfred von Ardenne.

In contrast to Czechoslovakia and Poland, East Germany readily adopted the Soviet experience in university policies; the Soviet models were implemented more fully in East Germany because, as a result of the occupation, Soviet advisers were on hand to supervise the changes.[22] A similar argument can be made for other institutions as well. For example, a major impetus for transforming the academy from a learned society to a major research center came from the Soviets, whose own academy was organized in that fashion.

A different pattern emerged at the Leopoldina—an anomaly in the institutional landscape. The German scientific tradition survived there, in part, because the old bourgeois scientists were resistant to change along socialist lines. Although the battle between the scientists and the Party and state lasted several decades, the government found that the Leopoldina fit into its new internationalist policies of the 1970s.

In the area of covert technology transfer from the West, not only were Soviet structures and methodology—one thinks of successful Soviet espionage with the atomic bomb—adopted by the Ministry for State Security during the 1950s, but East Germany served as a continual conduit of information from the West to the Soviet Union, especially in the areas of nuclear physics and technology, computers, and military technology. The Soviet conception of the "technical intelligentsia," however, could not be fully implemented in East Germany because of the soil within which it was planted. The historical realities of East Germany and the engineers' adherence to a professional ethos and engineering mentality impeded this.[23]

The Soviet influence and senior-junior partner relationship with the GDR were considerable in the various fields outlined in this book, especially in the earlier periods of the Soviet Occupation and the 1950s. After the Wall was built, the GDR was forced to integrate itself more fully with the COMECON countries, but by the late 1970s and 1980s it had reestablished ties with West Germany through Alexander Schalck's Commercial Coordination Unit (*Bereich Kommerzielle Koordinierung,* or KoKo) in the Foreign Trade Ministry and the scientific treaty of cooperation, which was finally signed in 1987.

The Soviets' scientific relationship with the GDR was exploitive and uncooperative. The Soviet Union, for example, was not helpful in facilitating the technological transition from coal-based chemistry to petrochemicals in the GDR. In my chapter on espionage I reveal that, despite official cooperative agreements between the Soviet Union and the GDR in the area of computers, cooperation had failed on the 1-megabit chip because of the Soviet Union's "unwillingness" to work with GDR researchers and "insufficient conditions" in the GDR. In the late 1950s scientists complained to state planners about the Soviets' lack of cooperation in the field of nuclear physics; they failed to provide necessary materials like uranium metal, as well as access to socialist institutes or training facilities for reactor technicians, and instituted a restrictive information policy whereby the GDR did not receive any information about the Soviet reactor program, which was only revealed at the Third Geneva Conference in 1964. Instead of cooperation the Soviet Union tended toward "control and hegemony."[24] In some areas the GDR was heavily dependent on the Soviet Union; for example, with the curtailment of research on nuclear science and technology in East Germany in the 1960s, the GDR had to import nuclear power plants from the Soviet Union.

Other East Bloc Countries

The Soviet model for scientific organization was also implemented in other East Bloc countries, including Poland, Hungary, Czechoslovakia, Bulgaria, and Romania. The Academy of Sciences became the center of scientific research while the universities stressed teaching over research. Obviously, each country's history, scientific traditions, and national culture influenced the way science was organized, how it developed, and what it achieved. Just as the

Soviet educational system was variously implemented, so too was the scientific system. The greatest difference between the GDR and other East Bloc countries was that the Soviet tie to East Germany was stronger. By the 1980s, however, cooperation with the West was stronger in the GDR than in other East Bloc countries, with the exception of Hungary. Bulgaria and Romania were less developed countries even before the Soviet expansion, and so cannot be expected to be on par with the other East Bloc countries. Quantitative and qualitative analysis has placed the GDR at the head of the pack in terms of scientific achievement, although Hungary, Czechoslovakia, and Poland were strong in other fields.[25]

There is no doubt that collectively the Soviet scientific system with its bureaucratic centralist control, cadre policies, and planned economy impeded the development of science in these countries.[26] But it would be simplistic to indict the planned economy or the communist system more generally in comparing the achievements of Western science and technology with those in the Soviet Bloc countries. For a variety of reasons, the United States took off during the postwar period, dominating world-class scientific achievements as measured in external recognition such as Nobel Prizes, attraction of working conditions for scientists, or even the export (legal or illegal) of technology.[27] It does appear that a functioning and strong economy provides a more supportive environment for modern science, and especially technology, than does a weak economy. Economic shortages and the limitations of a planned economy certainly retarded scientific developments in the East Bloc; indeed, equipment was often outdated, preventing many of the East Bloc countries from achieving scientific and technological modernization.

A deterministic argument could conclude that the capitalist system won the technological battle hands down. The Cold War rhetoric about the superiority of the capitalist system compared with the communist system for the advancement of science and technology seemed to have been vindicated by the fall of the Berlin Wall. Since most commentators agree that capitalism provided a more conducive environment for science and technology in general, it makes no sense to list triumphantly the achievements of capitalism versus communism.

Clearly, more comparative work regarding scientific developments in other East Bloc countries needs to be done. For example, why was it that Lysenkoism did not take hold in East Germany, but did have influence in Czechoslovakia? Was it because there was no major agricultural crisis in East Germany, or because scientists resisted the influence?[28] Hohlfeld argues that Lysenkoism made slight inroads at the universities in the 1950s but none at all at research institutes because, in part, the respected geneticist Hans Stubbe and the plant geneticist Kurt Mothes resisted such efforts. Both Mothes and

Stubbe had the ear of Kurt Hager at the SED's Department of Science, whose policy in the 1950s was to support bourgeois scientists—the so-called old intelligentsia—because East Germany needed them. In 1962 the Central Committee's Department of Science initiated plans to develop three centers for "Mitschurin" biology in Greifswald, Leipzig, and Dummersdorf with emphasis on plant breeding, biology, and philosophy and animal breeding, respectively.[29] These plans did not seem to have had much influence on biological research in the GDR.

National Socialist Germany and East Germany

If we step back from the narratives of the various chapters in this book and reflect on general similarities in the relationship between science and the social, political, and economic structures of National Socialist Germany and East Germany, we do find differences and similarities, some of which are inherent in the interactions between science and politics in all states. The similarities in the way in which the scientific enterprise was affected by politics in the transition period—the first few years after the seizure of power—is striking.[30]

The major difference between the ideology of National Socialism and Marxist-Leninist ideology was that the former was *initially* ambivalent about science and scientists (some would argue hostile), whereas the latter glorified science and supported it from the start. This is not to say that science did not survive and often thrive in Nazi Germany.[31]

When Jews and politically undesirable scientists were dismissed from their jobs during the early years of Nazi Germany, it was because of social and racial policies, *not* because the Nazis had special plans for science and scientists. In fact, when Max Planck gained an audience with Adolf Hitler to complain of the loss of scientists and the effect it would have on German science, Hitler is alleged to have said, "We'll have to do without science for a few years."[32] It was only later, during the war, that leaders in National Socialist Germany recognized the importance of science and technology for the war effort; V-2 rockets are one well-known example. Thus ideology was modified or ignored because of the demands of technological warfare.

Since East Germany had a supportive and friendly attitude toward science and scientists from the start, the country's personnel policies differed dramatically from those of National Socialist Germany. The state attempted to *keep* scientists in East Germany even if they did not embrace the Party's class-based ideology. This was in part due to the philosophy that political transitions take time, and therefore one had to be patient in winning over the bourgeois scientists. The fact that in the 1950s the country faced a mass exodus of scientists also modified early attempts at transforming the professoriate in the GDR. In contrast to the situation in National Socialist Germany, the exodus of scientists was voluntary and was primarily a result of economic hardship and dislike of the new ideology. With the passage of time and the recruitment of the so-called new intelligentsia—scientists educated under socialism, SED members, and those coming from working-class or peasant families—the community of scientists would be transformed. And so it was. By the beginning of the Honecker era in 1971, 84.39 percent of all professors were members of the SED, compared with 28.43 percent in 1954. It is not surprising that more professors in the humanities and the social sciences were members of the SED and came from working-class or peasant families than professors in the natural sciences and engineering because of the close connection between ideology and the soft sciences. Figures for 1965 indicate that just 25.34 percent of professors in the mathematics and natural sciences faculty were SED members, whereas 66.45 percent of professors of philosophy and 85.03 percent of professors of Marxism-Leninism and law were SED members.[33]

The SED's effort to increase party membership at the academy was already successful by the mid-1950s; by 1956, 17 percent of academy members were also party members. By 1967, 18 percent of all scientists and 54 percent of all scholars in the humanities were SED members, and 32 percent of all regular academy members were in the Party.[34]

Those in power in both Nazi Germany and East Germany shared the goal of transforming society and the scientific community in order to align the population with the reigning ideology—what has been called *Gleichschaltung* (or putting into line) for National Socialist Germany. A common feature of major new movements or revolutions is the attempt to establish their vision in society. Some have characterized this transformative vision as a feature of totalitarianism because of the often rapid and forceful attempt to spread the ideology to all sectors of society.[35]

National Socialist Germany and East Germany shared several transformative efforts in relation to science and its institutions. Most important was the effort of party ideologues shortly after the seizure of power to transform scientific and educational institutions along socialist or National Socialist ideals, and to replace personnel. Once these changes were made, the state's disciplinary interests were easier to implement. Ideologues seized power at institutions, installed their own people (usually party members), arranged for party members to control the administration, and worked to increase party membership in general.

In the 1950s, Western Cold War commentators saw the transformation at universities in East Germany as a "sovietization," a "storming" of the "fortress of science."[36] Party bureaucrats themselves explained that their changes at the academy were transforming it into a "socialist institution."

In East Germany, ideologues began to make disciplinary inroads into the social sciences and humanities as early as the occupation period, while the natural sciences seemed unaffected. The humanities and social sciences underwent an ideological review in 1947, leading to a revamping of curricula in some disciplines. New institutions or departments at universities were created, including the Institute for Dialectical Materialism in Jena (founded in 1946), and new social science divisions were set up at the old universities in Leipzig, Jena, and Rostock (1947).[37] The Central Committee's Institute for the Science of Society *(Gesellschaftswissenschaft)*, the influential and important training ground for party cadres, was not founded until 1951. At this early stage, leaders were merely attempting to establish a competing philosophy among the disciplines—there was, as yet, no major attempt to establish the Communist Party's goals. Major institutional changes at the university and the academies had to be made before party ideology could influence the sciences themselves.

Although these changes were forced on them, some opportunistic scientists quickly adapted themselves and their science to the new ideology. For example, Kurt Gottschaldt's psychology began to emphasize class issues. The social sciences and humanities had a tendency to adapt and align themselves with the cognitive component of the ideology itself, whereas the natural sciences enjoyed support because the ideology was friendly toward science and its potential role in communist society. While the political ideology directly affected the cognitive part of the social sciences and humanities, it tended to instrumentalize the natural sciences and engineering.

Another important feature shared by Nazi Germany and East Germany was the emphasis on autarky, or self-sufficiency in the economy and technology and/or in procuring raw materials. The National Socialists had announced a policy of autarky with the four-year plan in 1936 in part because there was a shortage of foreign exchange to purchase raw materials from abroad. The GDR chemical industry tried to become self-sufficient by using the raw materials at hand, namely, coal. Stokes sees this philosophy as a continuity with Nazi Germany and earlier periods. While West Germans had abandoned this philosophy by the late 1950s, East Germans intensified their commitment to autarky. Like Nazi Germany, East Germany also had a shortage of foreign exchange of non-convertible currency.

The GDR aspired to build up an indigenous computer industry not dependent on the West, especially West Germany. A sense of political isolation and this fear of dependence led to the pursuit of autarky in this field as well: "Autarky for the GDR meant autarky in the mass production of IT [information technology] components and systems that its leaders felt were essential to the East German economy."[38] The Western embargo, as Chapter 4 shows, also played a role in the desire for an indigenous computer industry.

Another element both countries shared in common was isolation from the international community of scientists. In National Socialist Germany, scientists became isolated from their colleagues owing to the nature of Nazi rule, especially during the war. The East German scientific community became irrevocably isolated from the Western community of science once the Wall was built in 1961. All-German professional organizations gradually disappeared (with the exception of the Leopoldina), and it became increasingly difficult to travel to the West; this changed later with the travel cadre policy, but even then only a few, vetted scientists were allowed to take part in the international community of science. In Chapter 11 Geipel even argues that East Germany was isolated from the socialist community economically because the Council for Mutual Economic Assistance (CMEA) failed to deliver.

Continuity/Discontinuity with National Socialist Germany

More than structural similarities and differences connect National Socialist Germany to East Germany. As Siegmund-Schultze shows in his contribution, East German communist ideology saw itself as diametrically opposed to National Socialism, and thus developed a policy of "antifascism." This official rejection of the Nazi past, however, was modified by the realities of the scientific community in the 1950s. Because East Germany faced a shortage of scientists in the 1950s as a result of exodus to the West, it modified many of its most cherished principles. This legacy differentiated East Germany from other East Bloc countries.

Siegmund-Schultze's chapter, appropriately entitled "The Shadow of National Socialism," offers an exposé of the contradictions and myths inherent in the notion of "anti-fascism." In theory, the goal of the GDR was to construct a radically new socialist society; it thus sought to reject completely the ideological system that preceded it. The irony, however, was that in many ways East Germany was built and legitimized through its relationship with Nazi Germany. Thus, though the GDR did everything it could to demarcate itself from both National Socialist Germany and West Germany, it could not escape the shadow of National Socialism.

Many SED leaders had antifascist pasts, and therefore were on the right side when the Soviets occupied Germany and acquired the Eastern zone. National Socialist Germany left behind many scientists who either came of age or were educated between 1933 and 1945 or who were NSDAP (National Socialist Party) members. Although initial de-nazification proceedings attempted to purge East and West Germany of "nazis," many were reinstated a few years later because of the need for scientifically qualified personnel. In the end, the SED made a practical separation between professional competence and political activities during the early 1950s, allowing scientists who had been in the NSDAP to take up professorships in East Germany.

The issue of continuity and discontinuity surfaces most clearly in individual biographies. The two careers outlined in this book are a study of contrasts. Robert Havemann—a communist scientist who lived and worked under Nazism and socialism in East Germany—was, in the end, at odds with both regimes. In his first experience with what he perceived as tyranny, he suffered because he was a communist and a resistance fighter. In his second experience with a rigid and bureaucratic form of government, he suffered because he did not buy that regime's brand of communism, choosing instead to become one of its best-known dissidents. By contrast, Kurt Gottschaldt, rather than resisting, exemplified a more typical scientist as he adapted to the regimes he encountered.

Mitchell Ash takes the continuity/discontinuity theme further when he argues that Gottschaldt had to work very hard at "constructing continuities" in his work as he adapted to changing political regimes. Gottschaldt could not continue his work under the GDR without reestablishing his career. This meant submitting to the de-nazification process (though he was not an NSDAP member, he collaborated with the regime), establishing himself in a new professorship, and building an entirely new institute for psychology near the university. The emphasis of his scientific work shifted with the change in regimes; while under Nazism he drew strong hereditarian conclusions from his research on twins, in the GDR he stressed the "social superstructure."

The Quest for Scientific-Technical Modernization

"Modernity," "modernism," and "modernization" are notoriously slippery and vague terms, but they have nevertheless become popular concepts in the academic literature on a variety of topics, including art and the history of science. Several of the authors in this book—most notably Eckart Förtsch and Rainer Hohlfeld—use these terms in their discussions of the GDR, but intentionally do not treat the concepts in a systematic or lengthy way. Central to all concepts of modernization is the idea that scientific-technological progress and economic modernization are linked with science and technology, spurring on economic innovations in a modern industrialized society.[39]

It is not surprising that the term "modernization" is used in relation to the GDR, given that the country aspired not only to achieve scientific-technical "modernization"—in the sense of current, state-of-the-art, or new equipment and technology—but also attempted to use science and technology for economic modernization. Indeed, the 1960s reform period can be characterized as an "elusive search" for a "socialist modernity" using new economic and scientific reforms.[40] In retrospect, it seems that the economic reforms, had they succeeded, could have provided the GDR with the strong, shortage-free economy necessary for the twentieth-century modern scientific-technical enterprise.

For our purposes, the simple definition of the term "modern" as described above is more useful than a vague and abstract theoretical discussion that is difficult to relate to historical reality. Some of the conclusions of the book demonstrate that there was a constant tension between the ideals of the new socialist state and the realities of life in the occupation zone and in "real-existing socialism" in the German Democratic Republic. Most generally, this manifested itself in the tension between the proclaimed goal of achieving scientific-technical modernization and international excellence, and the glaring reality of an antiquated infrastructure for science and technology. Many of the contributions in this volume show that the GDR aspired to achieve scientific-technical prowess but continually fell short of that goal. It is precisely this tension between the ideological aspirations to achieve scientific-technical and economic modernization and the reality of an outdated infrastructure and a lack of modern equipment that make the case of the GDR so tragic.

While communist ideology is by definition "progressive" and "radical" because it is new, forward-looking, and not reactionary, when applied to a modern industrialized consumer state, it seemed to predate it; that is, it was a nineteenth-century ideology implanted in the twentieth century. This "premodern" political ideology and economic experimentation impeded the GDR's proclaimed efforts at modernization, and thus the process short-circuited.

The GDR's search for a socialist modernity through the reform package of the 1960s had presented an opportunity for the country to transform and update its premodern ideology in order to compete with West Germany once the borders were closed. Whatever the reasons for the failure of the reforms, by the fall of the Berlin Wall science and technology had not acquired the economic base necessary for achieving the GDR's goal of surpassing West Germany.

I

Policy

1

Science, Higher Education, and Technology Policy

Eckart Förtsch

The relationship between politics and science in the German Democratic Republic was complex and changing. The East German scientific enterprise was established after the Second World War and was, in its early years, largely autonomous with regard to both content and its connections with the rest of society. The ensuing era was characterized by a planned policy of politicization—that is, the manipulation of the sciences by politicians for political ends. As political leaders strove to establish "socialist methods of production," they realized that science could be utilized to help them achieve certain economic goals. Once the political leadership had proclaimed that science and technology played "a key role" in advancing the GDR, close links were forged between politics and science, which came to see themselves as interdependent, though they did not, in fact, enjoy equal rights.

While this historical process was marked by many experiments and numerous conflicts, three general features can be identified:

1. the development and institutionalization of policies to govern science and technology, with the goal of integrating both completely into the communist system;
2. the creation of a research policy establishing priorities, strategies, and use of resources, which determined how scientific data would be produced and used. However, a gap remained between the production of new knowledge and its application, for which the political system bore responsibility; and
3. the vision of science and technology as tools for modernizing society.[1] Although they were employed as such, the process of modernization in the GDR was characterized by a peculiar paradox: certain features of a premodern society were retained, such as a binding ideology, imposition of authority from above, and political control of most areas. As a result, the modernizers found themselves in a dilemma.

I will return to this last point in my conclusion. The bulk of this chapter, however, will be devoted to a discussion of interactions among the various parts of the system and how they influenced the dominant attitudes, modes of functioning, and structures of the different institutions and bureaucracies. I will place more emphasis on some aspects of this topic than others, as some of my fellow contributors also deal with events and stages in the development of science and technology policy (see the Introduction and Chapters 2 and 6).

Scientific Research during the Occupation

Scientific research had suffered considerably at the hands of the National Socialist government, which imposed fascist ideology on a number of disciplines and eliminated Jewish scientists from the profession. Further damage and widespread destruction of libraries, buildings, and equipment occurred during the war. The Soviet authorities seized control of existing research institutions in their zone of occupation and banned Germans from conducting research in fields such as nuclear physics and aeronautical engineering.[2] They also transferred some equipment and a number of scientists and technicians to the USSR, and science lost further personnel through voluntary departures and the de-nazification process.

As there had been virtually no Marxist tradition in prewar German science, political leaders in the Soviet-occupied zone had only a few "scientific cadres" at their disposal whom they considered reliable; they also lacked a rigorous and well-defined policy on science. They tended to oscillate between an ideologically founded faith in science itself as a progressive force and distrust of its practitioners as members of the bourgeoisie. For the first few years of communist rule some East German scientists worked in a relative vacuum, without directives, assignments to specific tasks, or clear government mandates. Neither political nor industrial leaders pressed for specific experiments or data, with the exception of orders from the military occupation for specific research in the fields of aviation, medicine, and health, and commissions to various East German research institutions from the Soviet bureaucracy. The first catalogue of research and development projects having a bearing on the economy, contained in the first two-year plan in 1948, was never followed up or implemented as concrete proposals, and remained entirely ineffective. Science, and research in particular, remained outside the focus of political interest for leaders in the Eastern zone, who were concentrating on transforming East German society and its government along Marxist lines. As the Introduction shows, however, the Soviets were very interested in exploiting German scientists in the Soviet Union and immediately recognized the importance of the academy as a center of science.

A framework for setting and managing policy in the sciences, higher education, and technology was still in a primitive stage of development. The responsible agencies within the government included the Central Cultural Committee of the East German Communist Party, the Central Committee on Universities, the German Central Administration for Public Education, the planning department of the German Economic Commission, and the ministries of education in the various states; the major journals devoted to the subject were *Einheit* (*Unity*, founded in 1946) and *Sowjetwissenschaft* (*Soviet Science*, founded in 1948). All strove to fit science into the general context of economic planning, education, and culture, but none provided more than the most basic orientation. It soon emerged, however, that certain structures within the Communist Party could be adapted without great difficulty to provide direct central planning and oversight.

At least four developments had a bearing on science policy during the late 1940s; three pertained to personnel policies, and the fourth pertained to ideology. First, communist leaders imposed new regulations outlining political and social criteria for admission to institutions of higher learning. The new admissions policy aimed both to end discrimination against formerly disadvantaged classes of society by altering traditional requirements, and to create a "new intelligentsia" while ensuring high levels of performance in future academic research in general and in science in particular.

Second, the "old intelligentsia" was granted material privileges, a measure that could be justified both politically (as a means of forging new alliances) and economically (preventing further "brain drain"). Those creating this policy of supporting the old bourgeoisie defended the controversial approach in terms of future science policy as a way to create acceptance of the Communist Party presence in the field of scientific research. Science policy in this context took the form of active support and funding as old institutions were reopened, new ones founded, and a resource allocations plan drawn up (see the Cultural Decree of March 31, 1949). As yet, however, no strict guidelines were imposed.

Third, the state implemented a new personnel policy in teaching and research. Its initial aims were the removal of scholars with incriminating Nazi pasts, provision of pensions for those who agreed to leave, and hiring of antifascist scholars and scientists prepared to cooperate with the new regime. For the time being work in the natural sciences remained unaffected. The humanities and social sciences, however, underwent an ideological review in 1947, leading to the revamping of curricula in some disciplines.

Finally, in the late 1940s the state introduced Marxist ideology into scholarly institutions. New departments and procedures were set up for this purpose, either run directly by the Communist Party or having close links to it. These included the Institute for Dialectical Materialism in Jena (founded in 1946), the new social science divisions at the old universities, for example, in

Leipzig, Jena, and Rostock (1947), the working group of Marxist Scholars in Leipzig (1948), official party conferences at all universities (beginning in 1947), and occasional conferences on Marxist theory. In this phase the leadership's chief aim was to establish Marxist doctrine as capable of competing with other theories, methods, and philosophies in the various academic disciplines; as yet no claims were raised about the superiority of Marxist-Leninist ideology, and no requirements imposed to follow its methodology.

During the phase of institutional rebuilding, there was no discussion or imposition of a leading role for the Communist Party in science or higher education. This was partly because the institutions of the scientific and academic communities were more highly developed and better established than those of the new policy makers. This asymmetry helped the scientists and scholars, with their older and better-organized system, to get the upper hand at first. They succeeded in securing funding and support without the imposition of restrictive guidelines, even though policy changes were being rigidly enforced at that time in other areas, as property was being expropriated and nationalized and the legal and educational systems completely restructured.

There are several reasons that political control of science lagged behind that of other areas, resulting in sought-after autonomy. Many problems of the most fundamental sort urgently needed solving; the population needed adequate food, clothing, and shelter; public health had to be maintained, and the devastated cities rebuilt. There was no time to come up with a separate policy for academia, and intellectual deficits were ignored under the pressure of other, practical deficits. The political system was concentrating its resources on securing its own hold on power, effecting a social revolution, and improving capacity levels in a restructured economy (including high-volume production requiring no great amount of research).[3] Furthermore, the political leadership had no persuasive alternatives to traditional modes of thought and social structures within the scientific community. Party ideology was still noticeably ineffective in articulating and defending its position vis-à-vis established scientific standards, and the proponents of the new world view— trained inadequately or by the wrong methods—found themselves no match for direct competition with those trained in the "hard sciences."

The Shift to Politicization

Political decisions made between 1948 and 1952 determined the character of the next phase. With the transformation of the East German Communist Party into a "party of the new type," Soviet policies came to serve increasingly as models in the development of the new communist government. As time went by, this trend spread beyond immediate party circles, affecting domestic policy making and creating tension between the political leadership on the one hand and the old intelligentsia and scientific and technological elites on the other. The newly founded East German state created new government agencies whose job it was to devise central plans for economic and social development. The task of "building a socialist society," announced in 1952, went even further, targeting social groups and individuals with the intention, now clearly articulated and implemented, of gaining control over science and research, and ultimately directing them toward specific political and economic tasks. From now on policy makers had to prove themselves capable of defining problems for research and devising strategies for their solution. This goal could not be achieved merely by expanding some officials' areas of responsibility and increasing others' specialization; rather, it required the political leadership to review, update, and rearticulate the sum total of scientific knowledge and identify the areas to which top priority should be given.

In retrospect, it appears that the leadership's approach to this task was determined by four chief factors: first, Marxist-Leninist-Stalinist ideology—that is, a knowledge of the aims of history and laws of social process, of revolutionary change, of the characteristics of the model of socialism they desired to implement, and the modes of dialectical orientation in the world—shaped their vision. Second, the experience of government accumulated in the USSR—an understanding of how to secure and maintain power, create central systems for planning and control, and politicize the social environment and social structures—helped party leaders as they seized power in the GDR. Third, a selective knowledge of history took the "progressive and humanist" aspects of cultural tradition as models for a particular interpretation of history, in order to legitimize the state's actions. Finally, the leadership's outlook was based largely on past experiences of discrimination, class warfare, persecution, and denial of its own errors, resulting in polarized patterns of perception ("us" vs. "them").

Although these factors may have sufficed to provide an ideological foundation for political goals and programs, they were inadequate for modernizing an entire society. For this reason the political leaders made the production and distribution of new knowledge and data a top priority. At the same time, however, they sought to gain more control by obscuring the distinction between scientific criteria and ideology, interfering in the research process, and insisting that experimental data be passed through political filters. Such a science policy led the leaders into a blind alley from which they later proved unable to escape: their "information base for modernization" never produced any results remotely resembling the claims they had advanced for it.

Expectations and demands changed in this phase, as the political leaders and their ideological agencies articulated their outsiders' view of what science is and does. New maxims included the notion that scientific research should be integrated into the general struggle between socialism and capitalism. Any remaining attention paid to the pan-German science of the past was linked to the demand that science in the GDR must "serve as a model for scientists in Western Germany."[4]

A political commentator in *Einheit* argued that methods of scientific reasoning must be linked to political orientation. This began negatively, with criticism of scientists' "objectivism" as too ideologically neutral. Two methodological directives were issued with the intention of altering scientists' methods of operation: politicians ordered each discipline to take Soviet science as a model and to use dialectical materialism as a general methodology.[5]

Political leaders thought that new rules must be provided for organization and communication in the various scientific communities, with the goal of altering existing structures and assumptions. Some of the key terms used in this effort were "practical orientation" (as opposed to the pursuit of knowledge *per se*), "collective effort" (as opposed to individual research, which was more difficult to oversee and control), "interdisciplinary focus" (as opposed to autonomy for each field), "all-roundedness" (instead of working on problems in segments), and "encouragement of debate" (as a vehicle to insert Marxist opinions).

Kurt Hager, the SED's chief ideologue, argued that four main categories and aims should be established for scientific research, all related to practical problems in the GDR. First, the political leadership expected scientists to provide expert advice and consultation on the new questions of economic planning and control. Second, they were to produce technology relevant to the interests of domestic and foreign markets. Third, priority was to be given to research in the sectors of health and education. And finally, scientists were to contribute to the "awakening of a new socialist consciousness."[6] All this was in the "service of building socialism."

This phase of politicization was characterized not only by the promulgation of such slogans, but also by their enforcement from above as administrative policy. On an intellectual level this was reflected in the far greater emphasis placed on systemic categories of Marxist-Leninist thinking (including power, the class struggle, and forces of productivity) than on issues affecting daily life (such as emancipation, overcoming alienation, ideological criticism, and so on). On the organizational level, this control from above was reflected in the successful creation within scientific and academic institutions of political organs designed to steer science—and research in particular—in specific directions.

Agencies were established within the Party and the government for the purpose of supervision. Foremost among them were two subdivisions within the Central Committee of the Communist Party, the Department of Science and Universities (created in 1952, renamed Department of the Sciences in 1957), and the Central Committee Working Group for Research and Technical Development (elevated to the status of a department in 1967). Their tasks were, or evolved into, developing a science and higher education policy (including specific plans, investment of resources, and hiring); implementing party decrees in scientific institutions, chiefly through the channels of the party organizations already established there; overseeing and controlling the administration of these institutions across the entire spectrum from universities and the Academy of Sciences to academic publishers and journals; and placing the appropriate cadres in leadership positions throughout the field and supervising their activities.[7] The authority of these party departments extended to the state bureaucracies responsible for science and research.

State bureaucracies were created to direct research in the natural sciences, technology, and industry: the Secretariat for Higher Education in 1951, and the Central Office for Research and Technology within the State Planning Commission, the Office for Nuclear Research and Technology under the Council of Ministers, and the State Secretariat for Research and Technology in late 1961. The most important body at the intersection of politics, the economy, and the scientific community was the Research Council, founded in 1957 and composed of leading scholars from all fields of science and technology. This group was assigned the tasks of developing long-term scientific programs, setting research priorities, reviewing economic policies, and deciding which specific lines of research deserved funding, staffing, and technical support.

It should be noted at this point that the persons to whom these tasks were delegated had to be not only willing but able. Since special qualifications and experience for such political intervention were required, key positions were increasingly filled from the mid-1950s on by scientists who had spent time in the USSR: Peter Adolf Thiessen (the chairman of the Research Council from 1957 to 1965), Max Volmer (the president of the Academy of Sciences from 1956 to 1958), Max Steenbeck (the vice-president of the Academy of Sciences, a member of the Scientific Council for the Peaceful Use of Atomic Energy, and the chairman of the Research Council after 1965), and Manfred von Ardenne (the founder and director of the Ardenne Institute in Dresden and the president of the Society for Medical Electronics). They and others functioned not only as scientists, managers of scientific institutions, and advisers, but also as exemplars of the kind of "positive development" desired by the framers of policy.

A chief characteristic of this phase of politicization was the expansion of the science and applied-science sectors, a necessary development if the envisioned goals were to be met. Numerous new institutions were created, particularly technical universities, institutes to teach the new economic policies, and teacher-training colleges. Between 1945 and 1954 institutions of these types increased from a total of six to forty-six. New research institutions were founded, such as the Academy of Building Sciences in 1950 and the Academy of Agricultural Sciences in 1951, and new departments were added to existing ones. The German Academy of Sciences in particular was expanded to include many new research facilities. Although the Party had already established its major training institutes in the first postwar phase (such as the Party Academy and the Institute for Marxist-Leninist Studies), one of its most important organizations for the training of party cadres was created in 1951 in the form of the Central Committee's Institute for Social Sciences. The fundamental reorganization of the sciences as envisioned by the leaders of the GDR also required new communications media, including publishers such as the Deutscher Verlag der Wissenschaften (German Scientific Publishers), founded in 1953, university journals founded from 1952 on, and separate East German professional journals founded from 1951 on. In the last instance the Central Committee's Department of Science and Universities played a key role in promoting the new publications, as well as in determining the membership of editorial boards and publishing strategies. In this phase the political leadership insisted on the foundation of new GDR organizations to replace and compete with the existing prewar pan-German organizations; these included the Physical Society of the GDR and the German Historical Association, founded in 1952 and 1958, respectively.[8]

As a whole, the connections between politics and science in this phase unfolded as a pattern of increasing intervention and control. While the instruments used for this purpose may have been of doubtful validity, they cut a wide and deep swath across the scientific establishment.

The problems scientists were to investigate were determined by economic planning; research plans outlined topics, priorities, and the criteria for relevancy. Although administrators experimented during this phase, the principle of planned research was adhered to throughout. For example, the Research and Technology Plan of 1958 listed approximately six hundred research topics. As it happened, one topic had passed through nine different agencies for approval, without any consultation taking place among them.[9]

Leading party institutions issued decrees concerning not only general ideological and strategic issues, but also specific tasks for particular disciplines. The first type were intended primarily to furnish party members with information about goals and methods, but at the same time to create a common language for communication among all members of the scholarly and scientific community. Decrees of the second type allowed the Party to steer and shape the various disciplines directly. A striking example of the latter is the

Politburo decree on research and teaching in the field of history (1955).[10] It established standards for theories and methods and presented political goals as scholarly ones (stressing terms such as "achieve," "uncover," "convey," "strengthen," "equip with," "prove," and "fill with pride"). It further prescribed topics for study, defined the results in advance (in phrases like "the main lessons of history are . . .," or "these great battles demonstrate . . ."), elevated other party decrees to the status of standard reference works, and created organizational and structural regulations.

The political monopoly on training, recruitment, and placement of personnel was used after the 1951–52 reform of higher education to favor the reigning ideology; a slogan of the 1950s was "storming the fortress of science [*Wissenschaft*]." In addition, the central political administration made decisions regarding publications, the creation and staffing of scientific journals and associations, the allocation of funds, and the founding and organization of new institutions. From this point on science policy included and regulated the political make-up of academic faculties.

In institutions directly under its control, the Party devised and employed authoritarian methods to create the desired transformations; these included authoritarian directives in political and scientific journals, conferences on theory and faculty training seminars, guest lectures by political figures, and study groups on dialectical materialism for faculty members.

Various government ministries created bilateral commissions, such as "scientific advisory councils," to ensure the transfer of information between government and the sciences and the continuation of political influence on the different scientific communities. Beginning in 1950, the political leadership also created the first regulated contacts between scholarly institutions and industry, such as "friendship contracts" and partnerships between universities and factories.

Party organizations within scientific institutions came to play an increasingly important role. Although their task was to organize discussions and other presentations on the political view of problem areas and strategies for dealing with them, they developed increasingly into organs of internal control. At the same time they also provided the political leadership with information on the views and attitudes of scholars and researchers.

One important feature of the politicization effort was the leaders' attempt to influence scientific and scholarly investigations, directly or indirectly. They justified their actions with arguments about the need to involve scholars and scientists in the larger process of reshaping politics and society and in the class struggle inside and outside East Germany. For this purpose political leaders intervened in the premises of scientific investigation (that is, criteria for establishing truth, and methodological standards), explaining their own ideology as a theory of science and initiating basic philosophical debates on such topics as the concept of matter, logic, and Hegel. They intervened further in the development of theory and in methods for disseminating scientific information, by granting privileged status or imposing sanctions on certain avenues of inquiry, categories, hypotheses, data, and results (for instance, the Lysenko debate). Failure to cooperate with the authorities resulted in withdrawal of teaching privileges, dismissals, and arrests for faculty members, and investigations and expulsions in the case of students.[11] Leaders sought to block specific fields of research (for example, sociology and empirical social research in the 1950s) through personnel appointments, publication, and promotion policies.

Although a certain amount of self-critical discussion took place within government circles about the politicization of higher education, the objections tended to remain on the level of means and methods, and the fundamental objectives went unchallenged. Official criticism and self-censure following the twentieth Party Congress in the Soviet Union targeted the style of the campaign but not its content or aims.[12] Reminders that human beings and their fates were at stake as well did not surface until a considerable time later, in the literary works of GDR authors.[13] The imposition of rigid demands for change on people and institutional structures ultimately destroyed some of them: that was the point.

Claims that the destruction wrought by the politicization process was "creative" strike me as dubious. The largely one-way traffic along the routes connecting politics and science had the intended effect of forcing science and scholarship out of its ivory tower or, to use the image of the time, its fortress, and of opening it up to outside influence. However, the resulting costs were high, and never quite paid off: the belief that they had to oversee and control every area of life overwhelmed political leaders and rendered them ineffective; various crises of leadership spilled over into and affected scientific and scholarly pursuits. To give just one common example, government agencies dealing with export trade hindered the use by, or transfer of scientific data to, production facilities by establishing prices for goods developed through intensive research, or prevented decisions from being reached. The spectrum of goods affected ranged from refrigeration cars for machines to skin cream for babies.[14] The colonization of research had the not uncommon effect of making output equal to input, without the production of any added value by the scientific process. Goals and values were neglected in favor of the means.

Emphasis on the Economy

By 1960 the collectivization of agriculture was completed, and the remaining private businesses had been drawn into the planned economy through a requirement that they accept partial state ownership. Obviously, the building of the Wall in 1961 had repercussions for science and its personnel. Once the Wall was built, the government could deploy the work force as it saw fit and impose conditions at will. In addition, the GDR was increasingly integrated into the political, economic, military, and scientific structures of the socialist world. Its cultural and academic ties to the Federal Republic were loosened, and scholars were forced to resign from pan-German professional associations and give up their related activities. (Despite the official "policy of demarcation" established in the late 1960s, however, some islands of cooperation survived, notably the Leopoldina.)[15]

The "victory of socialist methods of production" made economic reforms possible, although here again they were politically determined and experimental. And so the aim to be achieved with the slogan "economic leverage"—motivating production managers and employees to work more efficiently and introducing scientific and technological innovations into the production process—remained a mere episode. Reforms in the educational sector had a more lasting effect, such as the extension of compulsory school attendance through the tenth grade and the creation of polytechnic schools.

The party program of 1963 introduced a new policy that treated science and technology as a "productive force" and a factor of equal rank with labor, land, and capital resources. According to this policy, which had been stimulated by Gerhard Kosel's influential 1958 book *Science as a Productive Force,* priority in research and technology development was to be given to the needs of the economy, and an attempt was made to steer and control science with economic mechanisms. In this phase the strategies of research commissioned by industrial concerns and state and social institutions, and of project-linked funding (for projects developed in cooperation with industry) were introduced.

In the course of the campaign to direct science and technology toward economic ends, new policy agencies were created and existing ones restructured. The State Secretariat for Research and Technology was upgraded to a full Ministry for Science and Technology in 1967. As a flanking measure the Central Committee Working Group was raised to the status of Department of Research and Technical Development. From then on, as the minutes of the agencies' advisory sessions show, their activities expanded beyond merely preparing materials for policy decisions, issuing directives, and supervising research.[16] They now began to work more in the style of scientists themselves, studying Western literature on brainstorming and corporate management (at first for the purpose of educating themselves) and working out definitions ("What is big science?" "What is top performance?" "What is process engineering?" and the like) in order to provide a rationale for science policy, at least after the fact.

The scientific system was also adapted to match the guiding maxim of this phase. In one reform the German Academy of Sciences—renamed the Academy of Sciences of the GDR in 1972—was granted the status of a large-scale research institution in the area of natural sciences and technology, medicine, mathematics, and social sciences.[17] Another university reform reorganized degree programs and added to the curricula; institutes and faculties that had until then been left relatively autonomous were organized into divisions and placed under the supervision of university administrators. From the late 1960s on, industrial producers and scientists were supposed to be linked together in major research centers or research associations, in order to realize the aim of subordinating research to the interests of industrial productivity.

The research priorities envisioned by the economic and scientific policy planners were now concentrated on modernization and the creation of "perspectives for future development." For programs in the sciences and technology this meant a variety of things: a top priority was to close the productivity gap with the West through mechanization, automation, transition to industrial production in large series, reduction of planning and development time for new products and manufacturing procedures, and increased efficiency. Researchers in the field of chemistry, a focus of state investment since the inception of the Chemistry Program in 1958 (see Chapter 9), were to develop technologies for utilizing native brown coal and petrochemical products. In addition to mechanical engineering, optics, and the electrical industry, microelectronics was funded as a field of research and production as early as 1962. To solve the especially urgent problem of energy in the GDR, scientists were to develop new energy sources and new methods for converting, transferring, and saving energy. Basic research was concentrated in areas such as atomic energy, semiconductors, development of new materials, electrochemicals, photochemicals and photobiology, genetics, and health.

From this point on, these priorities remained relatively constant. Structural experiments such as the establishment of large-scale research centers or the economic reforms under the slogan "New Economic System (1963)," by contrast, ended at the beginning of the next phase. (See Chapter 2.)[18]

Modernization in the Honecker Era, 1972–1989

In the final phase of development, science and technology were envisioned as playing a key role in modernizing the GDR. With this concept as a basis, political control and use of the economy as an instrument of power had to be balanced against conditions under which scientific productivity and technical innovation are possible. It is not overstating the case to say that in order to achieve this, science and technology policy was organized virtually as an oligopoly.

To make this arrangement work and keep it functioning, a hierarchy was established. At the highest level were the central political agencies, which acted with a high degree of autonomy, never consulting the general public in their decision making. Thus one possibility for providing advice and the chance for improvement was lacking. The political leadership attempted to separate decision making from accountability (including accountability for scientific standards and technical criteria). It also attempted to differentiate its level of involvement, working with a lesser degree of control in the natural sciences and the technology field, for instance, and more in the social sciences.[19]

The second organizational level was intended as the site for discussion and negotiation of compromises when various groups clashed, to guarantee efficiency and smooth functioning. In many cases, however, political and bureaucratic interests continued to win out over scientific and economic arguments.

At all levels, any action taken in the field of science or technology policy had to unite contradictory aims (for example, keeping up with international progress and standards while simultaneously serving the ad hoc interests of GDR industries; satisfying both long-term planning goals and short-term expectations; forging ahead with modernization while maintaining old equipment in good repair).

The politically dictated structure made separate, hierarchically organized "columns" inevitable in research. Despite the effort and expense invested in planning and coordination, the organization of science in the GDR favored a segmented approach to problem solving, fields of research, sequences of scholarly work, and innovation.

Science and technology policy had to deal with four "columnar" sectors of research. Outside the universities, the Academy of Sciences, with its approximately twenty-four thousand employees by the end of the GDR, was the largest research organization. Along with some basic research it pursued applied research commissioned by either the state or industry, as did—with some differences—the Academy of Agricultural Sciences and the Academy of Building Sciences. The share of research undertaken at the country's approximately fifty universities and institutions of higher education was smaller, but at institutes with outstanding reputations it kept pace with international developments. The biggest contingent was industrial research, with about 85,000 employees. The intention was for industrial research and development in the combines to cover the entire spectrum of "branch-specific basic research" up to development, but in practice construction and development dominated. Social sciences research and teaching in the fields most closely related to politics were carried on in the major party organizations—the Party Academy, the Academy for Social Sciences, the Central Institute for Socialist Economic Management, and the Institute for Marxism-Leninism; in this area the Central Committee institutes also served as something like an ideological police force.

The science policy agencies (particularly the Central Committee Departments for the Sciences and for Research and Technical Development, the Ministry for Science and Technology, the Ministry for Higher and Vocational Education, and the State Planning Commission) continued to exist. At least in one part of this sector, however, a shift of perspective is recognizable, and perhaps also a shift in function. In the final years of the GDR, the Central Committee's Department of the Sciences appeared to act not only as a transmitter of policy in one direction down the hierarchical ranks, but also as a kind of "early warning system" in the interests of science in communications traveling the other way. The files of the "Hager Office" (the head of the Department of Science and Universities) contain, along with numerous other warnings and demands issued over the years by the leadership of the Academy of Sciences, concerns that basic research was being cut off, requests for support to import equipment from the West, complaints about reduction of funding from the State Planning Commission, and the warning that if its inadequate equipment were not replaced, the Academy of Sciences would be unable to present an increasing percentage of its experimental data to the international scientific community.[20] The Central Committee Department of the Sciences took up the academy's cause and argued on its behalf. A message of August 23, 1988, intended for Honecker, for example, contains a list of urgent problems in scientific development in the GDR that reads like the critical analyses made after the collapse: deficits in strategy; the danger that GDR science "will be relegated to the second rank"; "too little pressure on industry" to develop and accept innovations; "mediocre levels of performance" and "deformation of scientific potential"; and increasing difficulties in recruiting young talent "in view of the declining attractiveness of participation in scientific research as it is now conducted."

The proposals—better equipment for research; creation of a special new state plan for basic research; transfer of "more responsibility for research" to the Academy of Sciences and scientists themselves—were never implemented, of course. Yet I see in the open, matter-of-fact description of existing problems a transformation in some agencies' view of their task, especially given the high degree of stability in key positions of leadership. Johannes Hörnig and Hermann Pöschel, the heads of the two relevant Central Committee departments, were in their posts from beginning to end, just as Kurt Hager was.

Planning was the most important vehicle of science policy in this phase. The agencies responsible for planning, including the presidium of the Academy of Sciences, the Research Council, and the industrial combine directors, in addition to above-mentioned bodies, made up the oligopoly. The plans themselves consisted chiefly of the state plan for science and technology, the science and technology plans of the combines and production units, and the plans of the major scientific organizations; they covered research and development tasks, criteria for the relevance of science and technology, technical and economic parameters, tasks involved in marketing, distribution of personnel, and funding. The specifications of complex, high-priority projects were laid down in "commissions from the state." The plans were detailed and covered a broad spectrum. For example, the state plan for science and technology of 1989 encompassed many areas of research, from manufacturing processes for children's clothing to developing 4-megabit memory capacity, from cosmetics to removal of sulphur from flue gas, from medicines to biological insecticides.[21]

Academies and universities did have some freedom to select their own research topics outside these plans, and they made use of it. Nevertheless, central plans (to which the oligopoly contributed) and their specifics for individual institutions determined relations among the different parts of the system until the end. Compilation and oversight of the resulting data increased the bureaucratic aspects of conducting scientific research and led to erroneous evaluations: toward the end it had become common to identity about half the research and development projects as belonging to "the top international level," although this was in fact true for only a small fraction of them.

Ties between research and industry were regulated by "industrial contracts" in which the goals and conditions of cooperative endeavors were laid down in detail. Combines functioned as sponsors and funders in such cases; indeed, industry was one of the "main sources" of funding for research and development, along with the government, which funded basic scientific research and research in the social sciences. The official slogan "Ökonomisierung der Forschung" ("make research serve the economy") was designed to increase industry's direct influence on science and technology and harness them in its service. At the same time the contracts were supposed to ensure high quality in industrial research, make production more efficient ("more scientific"), and awaken mutual understanding among science, technology, and industry.

Although the political administration may have hoped to reduce its burden by handing some responsibility for planning over to the economic sector, the relations between science policy and the scientific institutions themselves remained politically regulated to a large degree. This was true of recruiting, conflict resolution, the mechanisms of control, incentives, and conditions under which communication took place. On the one hand the political leadership provided scientists with status and both material and nonmaterial privileges in order to stimulate productivity (and certainly also to reward good behavior). Yet on the other hand it was also responsible for their isolation from the international community. Close cooperative ties existed with "the countries of our socialist brothers," but all international contacts had to pass through the filter of official government contracts. Inner-German treaties on culture, the environment, and especially the agreement on scientific and technical cooperation led to more openness. Still, political control of all decisions relating to international cooperation and contacts was exercised in a restrictive fashion (as in the complex bureaucratic process of qualifying for membership in a "travel cadre").[22]

A rigid policy of secrecy contributed further to the difficulties of communication with foreign scientists, and also to the flow of information within the country itself (see Chapters 4 and 10 for more details). A variety of regulations impinged on the science and technology sector. They included the political leadership's right to monitor "outsiders'" access to institutes and industrial sites; the classification of both "treasonable transfer of information" (including information not classified as secret) and "illegal contacts" (including the transfer abroad of any information or manuscripts "likely to harm the interests of the German Democratic Republic") as criminal acts; and threat of criminal prosecution for unauthorized revelation of secret economic, technical, or scientific facts and information on research and development data, as well as new technologies or processes (including instruction booklets, "know-how," and products such as reagents). The Council of Ministers' decree of 1962 classified all data on the environment as state secrets.

There is thus a certain paradox in the fact that despite all such restrictions, the global economy and international standards in science and technology continued to serve as orientation points for research and development, as if this were the most natural thing in the world. The political system made achievement of its general goals dependent on "key technologies" and "matching top international levels"; the goals included becoming competitive in world markets; continuing to compete with capitalist West Germany; and ensuring political and social stability at home by providing better consumer goods, services, and social policy to its own citizens.

The resulting priorities for science and technology policy were thus determined by the fields of application. The five-year plan that began in 1986 and prescribed the "main directions" in which research was to proceed had the force of law. Roughly speaking, it divided research into three categories:

1. high-tech projects such as microelectronics, laser technology, computer-aided production and manufacture, automation technology, development of new materials, and biotechnology (including gene technology for medicine, food production, agriculture, and process technology);
2. normal science projects such as energy research, health research and medical technology, and electronic consumer goods; and
3. projects specific to the GDR on developing substitute materials and procedures, catch-up development, and the use and conversion of brown coal in the energy and chemical sectors.

Even in 1986, and certainly after the collapse of the system, observers noted that in individual areas scientists in the GDR had produced high-level results measured by international standards, despite the prevalence of conditions hostile to research. Some of these fields are materials science and materials research; solid-state physics, semiconductors, astrophysics, and plasma physics; optics and laser research; molecular biology and plant genetics; polymer and colloid chemistry; mathematics; nutrition science, geoscience, and also some areas in the humanities.[23]

Conclusion: The Paradox of Modernization in the GDR

Much or indeed most of the results of this research were stockpiled; that is, the innovators had no power to apply their discoveries. At least two necessary conditions were lacking.[24] First, those selecting the results of new research for implementation did not use economic criteria: no pressure for innovation was allowed in the marketplace, and no competition existed between manufacturers. Second, the social and cultural prerequisites for innovation were underdeveloped: many areas of the scientific enterprise lacked autonomy, and motivation, ambition, and a strong work ethic were frequently misdirected. Additional factors were the inherent conservatism of the existing structures and a political system that continued to lose flexibility, so that by the end it was completely paralyzed. Because the structures and procedures that would have corrected these tendencies had not survived in the GDR, the political and economic deficits, errors, and failures were passed on to central parts of the system and perpetuated.

It was not just technical knowledge and know-how that went unused as a result of this constellation of circumstances; indeed, those with a good understanding of policy, who were capable of analyzing crises and proposing solutions and improvements, both in the general opposition to the Party and to some degree within the scientific establishment itself, were stymied and unable to make themselves heard.[25] This short-sightedness was the most characteristic feature of the attempt to modernize the GDR. On the one hand the political system demanded, furthered, and oversaw an economic, social, and cultural transformation in which values, norms, structures, and relationships were to be adapted to the dynamics of modern science and technology. It proclaimed "scientific and technological progress" as the watchword of social development, and propagated its advancement, without limits or alternatives. It also adapted itself to fit this plan, in the oligopoly and in its planning agencies. On the other hand, however, the political establishment continued to see itself as the supreme authority and to act accordingly, imposing limits on activities and creative processes. Whenever it saw its authority threatened (and that was most of the time), it interfered in and restricted efforts to modernize. In the end, crises arose in so many areas—economic performance, technological development, planning, and the legitimacy of the government—that the system simply imploded. Thus the premodern ideological elements developed in the 1950s hindered attempts at modernization.

According to Jürgen Habermas, modernization must include all segments of society and the relationship among them (for example, implementation of different rationalities, differentiation and institutionalization of steering mechanisms, and colonization of the world). The GDR's modernization short-circuited (especially with regard to industry and the infrastructure, for example); because of the premodern nature of its ideology, the GDR could not follow a normal modernization process. And this is where the paradox lies.

A second dilemma is worth mentioning, though it would require a separate chapter to explore: things would have gone better for the GDR if it had formulated more successful innovation strategies, even though innovation is only one part of modernization. The role that the GDR played in the East Bloc and vis-à-vis the Soviet Union, and its contest with the FRG (where the GDR political leaders really thought they could surpass the FRG), hampered the GDR in a variety of ways. The GDR could not, for example, buy innovation (because of the embargo) and then selectively apply it to different branches. The GDR had high aspirations for science and technology but achieved very little. There are two reasons for this failure: the short-circuited modernization strategy, that is, the inability and unwillingness of politicians to organize themselves according to modern, rational principles; and the over-stretched innovation strategy, that is, the formulation of goals for which the political system provided little or no resources.

Within the short period between the collapse of the East German system and the end of the GDR as a separate country, there were a number of noteworthy attempts to reform science and technology policy. Basic criteria such as autonomous research institutes with their own independent administration, academic freedom in scholarly and scientific institutions, competition and openness, hiring based on competence in the field, structures legitimized by the democratic process, and inclusion of employees in the decision-making process, were supposed to determine the new climate of research, with the declared goal of reaching and maintaining international standards. The motors behind this process were to be new organizations (such as councils of representatives from different institutions) made up of scientists themselves, chiefly in the Academy of Sciences, as well as new regulations for universities developed independent of the state (for example, the envisioned reform of university governing bodies in Leipzig).

Reform was hindered by a lack of experience, time, financial resources, and leadership in the academic sector, as well as by a lack of support both within the country and in West Germany. As a result, the dismantled science system was not so much restructured from within as replaced from outside. The question of whether this led to neglect of opportunities to reorganize the existing science apparatus would require a separate study.

—Translated by Deborah Lucas Schneider

2

The Reform Package of the 1960s: The Policy Finale of the Ulbricht Era

Hubert Laitko

The 1960s were a period of profound change in the institutional basis of science in East Germany. Leading West German observers of the scene wrote in 1977 that "a phase of intense activity in the creation of organizational structures [occurred] . . . during those years, and there was a shift to strong economic growth; this boosted the demand for usable research results. In the end, this economic trend forced the political leadership to fundamentally rethink its ideas about science and research policy."[1] They went on to note that by then a "phase of relative stability" had set in, which made it possible "to take stock of the organization of research in the GDR."[2]

The 1960s saw the emergence of much of the institutional framework of science and its relationships to other sectors of society, and this entire infrastructure remained in place until the collapse of the GDR. It also formed the social basis for the accomplishments and deficiencies of scientific work in East Germany. An analysis of this period is, therefore, of central importance in understanding the history of science in East Germany.

In 1995 it was still very difficult, if not impossible, for a German author to be objective about this topic. The very fact that the GDR as a state came to an end can seduce a historian into viewing its history teleologically. There is a widespread consensus that the main reason for the collapse of the GDR was a tendency fundamental to its social order: all of social life was deterministically guided by one center of power—a tendency that runs counter to the complexity and innovative energy of modern societies. This is why most historians and writers focus on one strand in the history of the GDR: the progressive refinement of centralist control. They trace it in all its ramifications and examine in great detail its most extreme manifestations.

The excessive use of the concept of totalitarianism, and the readiness to label East German society as a whole as "Stalinist," have meant a return, to some extent, to the polemics of the 1950s. (Incidentally, many Western authors were once willing to concede that the Soviet Communist Party and its "junior partner," the GDR, had been "de-Stalinized after the Twentieth and especially the Twenty-third Party Congress.")[3] In the 1970s and especially the 1980s, "East European Studies" *(Ostforschung)*, "East German Studies," and "Comparative German Studies" had moved far beyond Stalinist paradigms. One remarkable example was the standard work *A Comparison of Education and Upbringing in the Federal Republic of Germany and in the German Democratic Republic,* which was produced at the request of the Federal Ministry for German Affairs and published by a scientific commission headed by Oskar Anweiler.[4] Anweiler noted "that macromodels of this type [totalitarianism vs. industrial society] can adequately grasp only certain tasks and functions of education and upbringing in society, while neglecting other aspects."[5] For this reason comparative research on education "has increasingly avoided deriving its questions and criteria directly from such macromodels. A much more fruitful approach when comparing systems, in particular, is to proceed from specific functions of institutionalized education and upbringing and to determine their role and expression in different social systems."[6] I have kept Anweiler's methodological warnings in mind while writing this chapter. I have tried to be guided neither explicitly nor implicitly by a specific macromodel of social theory. At times, however, I have used terms and concepts that belong methodologically to different models in drawing a picture that is integrated only on the historical level.

In the 1960s the concept of totalitarianism, which aimed at a radical juxtaposition of social systems, increasingly disappeared from discussion about socialist societies.[7] It was replaced by the concept of industrial society, which stressed similarities and stimulated questions about possible convergence between the two systems.[8] This development was not primarily a fruit of political détente, but above all the result of a maturing worldwide transformation of the technological basis of human life. This transformation was beginning to alter the conditions of economics and politics on a level that transcended individual systems. At the time, various labels were tried out to describe this phenomenon: second industrial revolution, technological revolution, scientific-technological revolution.[9] The opposing societies that were locked in confrontation around 1960 were thereby forced into competition with each other. The coexistence of confrontation and competition became the hallmark of world society in the 1960s. In the socialist countries, ambivalent conduct toward other states had its counterpart domestically in the coexistence of centralism and innovation. By undertaking reforms, the GDR accepted the challenge to compete. Those reforms were an attempt at controlled social innovation at home, or, to put it more paradoxically, an attempt at programming the unprogrammable. The tension inherent in this endeavor explains the frequent vacillations in the process, which were experienced in day-to-day life as an irrational back and forth between a "liberalization" and a "tightening up" of the political and ideological course.

The 1960s brought to a close the first half of the historical life of the GDR. The potential of this period was not realized, in large part because of political interference. As Hermann Weber has written, the deposition of Walter Ulbricht in May 1971 initiated "one of the most profound turning points in the development of the GDR."[10] If one accepts Weber's view, it seems obvious that we should interpret the 1960s not simply as a step on the road to the end, but also as one possible development abandoned as a result of political decisions. Of interest are both the power structures that were subsequently elaborated and the developmental potential that was generated under the authoritative influence of these structures.

The Political Context of the Reforms

The late 1950s and 1960s saw the culmination of the Cold War. During this time the East-West confrontation escalated to the brink of a third world war (Cuban crisis, war in Vietnam), and then de-escalated (treaty on the nonproliferation of nuclear weapons). The two German states were fully integrated into opposing economic and military blocs in spite of the sovereignty they had acquired following the withdrawal of the occupying powers. The ideological confrontation creates the impression that West Germany and East Germany were mirror images. The symmetry of opposing systems was upset at one critical point, however: beginning in the early postwar period, the economic productivity of the Federal Republic, and with it the material living standard of its population, was always much higher than in the GDR. Although the reasons for this are very complex, there is no question that the enormous burden of reparations that was placed on the GDR seriously compromised the country's starting position.[11] The economic gap triggered an exodus from East to West that continually drained the GDR of large contingents of well-trained workers and specialists, who were very hard to replace. On the surface, the East German leadership could counteract these losses in two ways, both equally dysfunctional: enact measures aimed at disciplining the population more tightly, or enact measures aimed at raising the living standard generally (through price reductions and the like) or selectively (by granting privileges to segments of the intelligentsia, for example). The former only served to increase social discontent and swell the stream of emigrants instead of stopping it. The latter did encourage people to accommodate to the system, but at the expense of the economy's already weak investment capacity.

In 1960–1961 the exodus grew to such numbers—especially in the wake of the forced collectivization of agriculture—that the GDR found itself in an existential crisis. At the same time the stream of emigrants was also a credibility crisis for the SED. At the Fifth Party Congress in 1958, Ulbricht had announced that by 1961 East Germany would surpass West Germany in per capita consumption of the most important foodstuffs and consumer goods.[12] The reality of 1961 showed that this had been a rash and groundless promise.

Under these circumstances the leadership of the SED opted to take a desperate step—physically closing off the GDR from the West by building the Wall in 1961. From one moment to the next, the Party and state leadership of East Germany had virtually all controllable social variables at its disposal. In a sense, the Wall created an experimental terrain in which a socialist society could prove what it was capable of accomplishing. There is not doubt that the closing off of the GDR and the borderline military regime that emerged were unacceptable from the standpoint of human rights. Still, had it been possible to generate—under laboratory conditions, as it were—forces of economic development superior to those operative in the Federal Republic, this step would have made sense in terms of practical politics. Success would have relegated the Wall to a relatively brief episode in the history of the GDR; indeed, its continued existence would have lost any political meaning with the disappearance of the real reason for its erection (as opposed to the alleged ideological reason). The government's efforts were not without success, but the results were not good enough, for the Federal Republic continued to grow at a faster rate.

The leadership of the SED had to learn that a change in property relationships alone does not lead to greater efficiency, unless it is possible, on the basis of this change, to institutionalize regulations that promote long-term economic growth. Moreover, these regulations had to reflect the restrictive conditions that existed in East Germany (such as a lack of raw materials, a relatively weak capital basis, and an antiquated infrastructure), and create a situation where yields greatly exceeded costs. What was needed were *nontraditional solutions* that were no longer evident from Marxist doctrine. The system made room for the kind of creativity necessary to develop such solutions, and opened a way for East Germans—especially creative intellectuals who saw how fascinating these challenges could be—to identify with it.

The only resource that met all these requirements was the intellectual potential of the population. This potential was less limited than any other resource in East Germany. Its disadvantage was that it was difficult to mobilize: long periods were needed to develop it and complex structures to use it effectively.

In the 1960s governments all over the world were searching for structures suitable for mobilizing intellectual potential; "big science" and "large-scale research" were prominent slogans of the time. Here the strategic needs of the SED leadership overlapped with a global trend, and the government consciously entered East Germany into the race to take advantage of it.[13] In fact, the prospects of centralized systems did not seem bad at all. Compared with other systems, they could, in theory, have greater institutional competence to engage in long-term planning and construct large, complex structures from heterogeneous elements (such as projecting, planning, research, development, testing, serial production, and sales), since there was no need for protracted negotiations among many forms of ownership. Moreover, there were indications that praxis was beginning to live up to the expectations of theory. The West was still suffering from the "Sputnik shock." Sputnik and subsequent products of Soviet space technology were high-tech achievements, which could not be coerced by administrative regimentation or ideological slogans but required flawless systems of production. If long-term, stable planning and the erection and maintenance of complex structures had been the only things that mattered, centralism could actually have proved superior to other systems. However, since it was at least equally important to take quick advantage of unexpected opportunities, to show flexibility in rearranging existing structures or dissolving them and making their constituent elements autonomous, centralism lost out in the long run.

The SED always thought it important to couch the ideological guidance of society in the form of "slogans." No other slogan expressed the spirit of this period as concisely as Ulbricht's "überholen ohne einzuholen" (surpassing West Germany without trying to catch up to it by pursuing the same course of development).[14] East Germany would not follow the Western countries on the paths of technological-economic innovation they had so successfully embarked upon. Rather, it would pursue its own unique paths. Historians of science and technology will have to determine whether the elements necessary for such a course of action were actually in place in the 1960s. What is clear is that East Germany did not find such an alternative path to pursue.

Ulbricht saw the independent technological development of the GDR as a means to another end: that of surpassing the economic standard of the Federal Republic. The Federal Republic was East Germany's fate: the GDR was fixated on its Western neighbor not only negatively, in the form of physical separation, but also positively, relating to it as a moving target. As a result, East Germany thought of itself as a growth-oriented industrial society.[15] Ideology aside, this was a moment when the two systems converged. Following a long period of confrontational paralysis, the Federal Republic, using the new *Ostpolitik* of the social-liberal government, seized this moment to put East Germany on the defensive regarding its policy on Germany and to initiate the normalization of relations between the two states on its own terms.[16] One of the implicit preconditions for the Basic Treaty of 1972, the first major step on the long and difficult road out of the confrontation shaped by the Cold War, was this objective moment of convergence created by East Germany's attempt to use science to modernize itself as an industrial society.

The Elements of the Reform Package

A whole series of reforms was crowded into a period of barely a decade, and chronology alone creates a coherent process. These reforms were integrated by the overarching intention of energizing economic development. Each reform had its roots in the preceding reform; now, however, they were deeply interconnected by the shared desire to mobilize the country's intellectual potential as the crucial factor in developing the economy. The following brief review of the most important reforms is guided by Eckart Förtsch's observation that "science, as a social instrument for perceiving and defining problems and attempting to resolve them, assumes an important place . . . in a socialist industrial society. Two consequences flow from this: the control and guidance of science becomes an increasingly important part of politics; [and] the important role of science places limits on arbitrary experiments in science policy, chiefly because of the dynamic inherent in science itself."[17]

Economic Reforms

Shortly after the building of the Wall, the government launched "economic experiments" in selected Associations of Publicly Owned Enterprises (*Vereinigungen Volkseigener Betriebe,* or VVBs) and in individual enterprises. The goal of these experiments was to find more effective forms of management and in the process try out different regulations and structures. This approach signaled, at first on a limited basis, the transition to an experimental view of the new ideology as a social undertaking. G. Lauterbach has noted that "Marxist social doctrine offered no functional solutions in connection with problems of control in socialist societies."[18] By admitting that it was necessary to experiment in order to find solutions to problems that had emerged, the government of the GDR conceded that the solutions could be neither found in nor "derived from" traditional doctrine. The economic experiments were raising concrete economic questions about the effectiveness of different ways of distributing authority, different arrangements of "economic levers," and so on.[19]

The trend toward identifying problems reflected both the increasing dependence of industrial society on science and the beginning generational succession in the SED.[20] The top echelons of the party hierarchy were made up primarily of "professional revolutionaries" who had no academic training, had never had personal contacts with the academic world, and for the most part instinctively distrusted intellectuals. The next generation of SED functionaries, who were moving into lower and middle positions in the hierarchy in the early 1960s, and in some cases already into higher positions, generally had university degrees or at least professional school diplomas. These functionaries introduced rational features of the scientific method into political work. They were more likely than their predecessors to understand that a society that was becoming increasingly complex could not be directed solely with ideological appeals and political orders, but needed leadership that was largely professional and qualified. Economic reform was their first big test.

There is no doubt that Ulbricht himself belonged to the circle of "professional revolutionaries." His political behavior was plainly marked by arbitrariness, high-handedness, and Stalinist traits. But he also had an unerring sense that in the 1960s one had to side with science and the young generation of its representatives. This set him apart from the usual apparatchiks and made him the patron of reform. It was apparently also his power and authority that prevented public revelation of the conflicts between the old guard that was still occupying the center of power and the rising new elite.

In 1963, the Council of Ministers of the GDR passed the "Guidelines for the New Economic System of Planning and Managing the People's Economy," the outlines of which contained all essential ideas of the New Economic System.[21] Two main weaknesses were identified in the previous economic management: instead of adopting a long-term perspective, managers were almost exclusively concerned with annual planning (even though long-term plans formally existed much earlier); and instead of looking at the big picture, they focused on individual enterprises.[22] The VVBs, in which nationalized industry was organized at the time, were essentially overarching administrations for groups of similar enterprises. The main objective of reform was to transform these associations from administrative entities into economic organizations operating with a view toward profit. The new tasks, it was said, demanded "a VVB organized like a socialist company and working in accordance with economic accounting."[23]

As part of the economic experiments, the VVBs switched over to economic accounting and were given funds that the managing directors could use at their discretion. These funds were intended to allow the VVBs to respond flexibly to unexpected situations and to devise a long-term technological strategy. Furthermore, the funds would be replenished by the profits of the VVBs, whereas previously the lion's share of expenses for research, development, and standardization had been financed from the government budget separate from the VVB's income accounting.[24] The purpose of these changes was to interest the economic units in innovation, and to ensure that "research and development, projecting, production, and sales" formed a coherent whole within them.[25] The economy itself was to create a surging need for innovative solutions, and the educational and research institutions of the country were to be empowered and motivated to respond by supplying both personnel with sufficiently modern qualifications and useful research results.

As Lauterbach has written, the economic reform in 1963 brought with it a "change in the style of research policy in East Germany, in that an attempt was now made to plan science and production as a single unit over the long term."[26] With the exception of a number of top companies that were intensely involved in scientific research, such as Carl Zeiss Jena, East Germany's industrial research and development resources were chronically insufficient. By contrast, the capacity of academic research outside of the universities had increased substantially in the 1950s. The number of staff members at the German Academy of Sciences in Berlin rose from 1,029 in 1950 to 9,000 in 1961. Its budget increased during the same period from 167 million marks to 1,596 million marks, which meant that by the end of the 1950s the academy was receiving about 10 percent of all the money spent on research and technology in East Germany.[27] Economic reform "initiated a policy whose goal it was . . . to create extensive research resources within the economy itself as suitable partners for academy research." Although funds for the academy also increased in absolute terms in the 1960s, its share of East Germany's research potential declined to about 7.5 percent during this period.[28] This reversal of growth ratios was aimed primarily at the establishment of large research centers for "structurally important" areas of industry, areas in which planners were striving for a breakthrough to complex automation on the highest scientific and technological level. This strategy was seriously pursued for only a few years, however, and that was not enough to strengthen industrial research decisively.[29] On average only 5 percent of the employees in East Germany's centrally directed industry were working in research and development in 1972.[30] Yet we must not underestimate the achievements of science and technology during this period of euphoria, difficulties notwithstanding: according to A. Bauerkämper, B. Ciesla, and J. Roesler, "Measured by the number of patents and industrial products newly brought to production, there was a remarkable density of innovation in the field of technology in the 1960s"—much greater than before or after.[31] Roesler has analyzed the situation in detail through the case study of numerical control mechanisms in machine-tool building in East Germany.[32]

Nevertheless, the reform efforts failed to implant an innovative orientation in the East German economy, one that was sustained by economic processes and did not need to be pushed forward administratively. Throughout the period of reform there was constant experimentation. Regulations were continually changed in response to short-term failures, which meant that no stable work regimen could take shape. The second phase of the New Economic System was implemented between 1965 and 1967 under different conditions. In 1967 Ulbricht presented the "Economic System of Socialism" at the Seventh Party Congress of the SED, once again with modified regulations, and a basic regulation in 1968 gave this system its shape for the years 1969–1970.[33]

Another negative factor was the maximization of the division of labor in the domestic economy. Under the motto of "concentration," the collectives were shaped into domestic monopolies, with economic competition among them virtually eliminated. The homogenization of business structures also proved to be disadvantageous. Collectives took the place of independent enterprises that varied greatly in size; in particular, there was nothing comparable to the innovative, small high-tech companies working with "venture capital" in Western countries.

Far more significant than all these factors was the extremely overheated pace of reform, which placed an excessive strain on the East Germany economy. The tremendous tensions of the reform process led to an accumulation of economic difficulties in 1969–1970. These difficulties discredited the idea of reform itself. They made it easier for the pragmatic traditionalists around Erich Honecker to bring about a change in power and retract the innovative elements of reform, though the institutional structures created by the reform were for the most part preserved.

A continuously innovative economy requires intensive interaction with flourishing basic research at universities, colleges, and other research institutions. Strengthening the economic orientation of higher education and the academy—a basic motif of reform in these areas—was therefore an indispensable condition for the success of economic reform. To ensure that such cooperation yields advantages for all sides, there must be well-developed research capacities within industry itself, extending all the way to basic research specific to industry in those branches most actively involved with science. Since this was rarely the case in East Germany, coercing universities and academy institutes into carrying out projects from industry meant forcing them to take on tasks that were much better suited to industrial research and development institutions. This situation was created by the "decree on the task-specific financing of scientific-technological projects," passed in 1968 by the executive council of the Council of Ministers.[34] According to the decree, research at the academy and at university institutions was essentially possible only if it was tied to a "contract" (Auftraggeber), and it was financed only on this basis. Collectives or the Ministry for Science and Technology (set up in

1967) generally played the role of the social principal for tasks related to the economy, proving that a basically constructive idea of reform had been radicalized to the point of absurdity. Under these conditions, industry was largely freed from the obligation to build up its own basic research, and it also held back in developing the capacities of applied research. Research at the academy and the universities, meanwhile, was a result of these excessive ties to industry, a situation that was partly remedied after 1971.

School Reform

In the general school system, the most important reform decisions had already been made in the 1950s. The 1960s saw the completion of these reforms but not the introduction of any fundamental changes. Anweiler has spoken of an "experimenting phase" in the East German school system, which lasted throughout the 1950s and was brought to a close with the Law on the Integrated Socialist Educational System of February 25, 1965.[35] Reform of the school system proved the starting point for subsequent university reform.

Unlike science, national education, as the most important instrument for socializing future generations and the only area that was open to direct influence by the state, had traditionally drawn the attention of Marxist parties. The SED's educational policy shows more of a continuous elaboration of original intentions than the kind of change pursued by economic reform. In keeping with the prevailing notion that society could be deterministically guided, the common thread that runs through school policy was the strong emphasis on "socialist education." The purpose of such an education was to imbue students with a thorough commitment to the socialist vision of society. The educational system itself was to be centrally controlled, structured along simple lines, and conceived as a single entity from preschool education to the universities—an "integrated system." The transformation occurred step by step. The mandatory number of school years for all children was gradually raised to ten.[36] As a result, all that was left of the Gymnasium of the German school tradition were the two years of the "Extended Upper School" (erweiterte Oberschule), which were added on after the tenth grade and lead to the Abitur, the equivalent of the American high school diploma.

The dominant substantive trend of reform was the progressive orientation of the general school toward the economy's need for trained personnel. The mathematical, scientific, and technical level of education was substantially raised. In 1959 the name "General Polytechnic Secondary School" was introduced for the school that ended with the tenth grade; the polytechnic curriculum was thereby made obligatory for all boys and girls for both theoretical knowledge and practical training. Initially the school carried the polytechnical principle to extremes. For example, in the early 1960s the Extended Upper Schools gave students complete vocational training in addition to preparing them for the *Abitur*. Beginning in 1964, basic vocational training in chemistry, metallurgy, electronics, machine building, energy, transportation, agriculture, and construction was gradually introduced in the ninth and tenth grades of the General Polytechnic Secondary Schools.[37] This strong vocational focus was retracted in the late 1960s in favor of a general technical orientation. Of particular interest is the fact that the focus on technology was established on a proficient mathematical and scientific foundation: at the end of 1962, the Council of Ministers passed a detailed resolution concerning the development of the mathematics curriculum.[38]

The competition between East and West was very pronounced when it came to the school system; the term "educational race" was used explicitly in the literature.[39] All measures of educational policy in East Germany must be seen from the perspective of gaining, defending, and elaborating positions in this race. According to Arthur Hearnden, since World War II one can observe a growing similarity in the factors forcing changes in the educational systems in highly developed countries. Moreover, these factors are applicable to both communist and non-communist societies "beyond the fundamental ideological division."[40] He goes on to argue that the development in the Federal Republic, still relatively sluggish in the 1950s, accelerated considerably in the 1960s, with the result that both the FRG and the GDR concentrated on fully utilizing their pool of talent.[41] According to Hearnden, in the 1950s and early 1960s the GDR was quite a bit ahead of the FRG in the percentage of students who had passed the *Abitur*. This difference in favor of the GDR reached its highpoint in 1963. A year later, Georg Picht published his alarmist tract *The German Education Catastrophe* in West Germany, triggering a heated debate over educational policy.[42] Immediately afterward, the balance shifted in favor of the FRG, and West Germany maintained its superior position during the subsequent years.

The two German states initially took contrary paths in seeking to utilize their pool of talent: while the prevailing idea in East Germany was the Comprehensive School (Einheitsschule), in West Germany it was a system with several types of schools, an idea that was reinforced by the federal structure of the state. Hearnden formulated his comparison very cautiously in 1973: "The FRG is one of the few countries that have resisted the trend toward a uniform school system; the preference was to more or less leave the selective Gymnasium intact, and the other schools were raised to a level that conformed as much as possible to the middle and lower grades of the Gymnasium." As far as the training of highly qualified workers was concerned, this policy "must be considered as successful as, if not more successful than, the opposite course taken by the GDR, that of dismantling the Gymnasium and creating a completely new, horizontally structured system."[43]

For the second half of the 1960s, however, Hearnden noted "a diminishing difference in educational policy between the two German states. Social and economic pressures gradually led to the realization in the FRG that the development of the educational system had to be planned at the federal level."[44] In East Germany, meanwhile, there emerged a clear need for a diversification of educational pathways that would break through the principle of the Comprehensive School. Characteristically enough, it was primarily demands from the economic sector that prompted the state to take action and led to the establishment of specialized schools and classes.[45]

To be sure, this convergence of the school systems in the two German states stayed within narrow limits. The principle of social equality, which underlay the idea of the Comprehensive School, kept the tendency toward diversification in check in East Germany. To the extent that this tendency was able to assert itself, it was expressed solely in a horizontal diversification according to fields, not in a vertical differentiation according to achievement levels. The development of achievement elites was checked by the structure and operation of the school system in the GDR. Elites that did emerge tended to be a secondary effect of horizontal diversification. The social system of the Federal Republic, by contrast, promoted vertical differentiation; however, in areas such as sports, music, and mathematics, where the GDR set up specific systems for selecting and promoting talent from an early age, comprehensive centralism certainly also demonstrated advantages.

In 1980, Wolfgang Bergsdorf and Uwe Göbel summarized a sober assessment of what had been achieved: "The SED has succeeded in largely harmonizing the needs of the economy with what the educational system produces." The results were "notable, but certainly not outstanding compared with other countries or with the Federal Republic."[46]

University Reform

The Third University Reform began around the time the restructuring of the general school system was more or less completed (the First and Second University Reforms were implemented in 1945–1946 and 1951, respectively). Its starting point was the "Principles for the Further Development of Teaching and Research at the Universities of the GDR," which were drafted under the aegis of the State Secretariat for Universities and Technical Schools and presented at the Eleventh Meeting of the Central Committee of the SED in December 1965.[47] After party leaders had approved them, these principles were opened to public discussion. At the time many East Germans were engaged in a very lively debate about the impending restructuring of universities and colleges. In a peculiar way, the interests of scientists and students, on the one hand, and of the party leadership, on the other, intersected in this debate. The university was at that time the target of criticism in both capitalist and socialist countries of Europe. Its institutional principles, which harked back to the time of "little science," no longer seemed able to meet the challenges of the scientific-technological age. Not long after, the student movement, which reached its climax in 1968, especially in France and Germany, demonstrated that the critique of outdated structures could easily turn into attacks against the university establishment and eventually against the existing social order. The SED noted another phenomenon as a danger signal to the political balance of power: the rising opposition in neighboring socialist countries—inside and outside the governing communist parties—to the central guidance of society. This opposition was carried largely by intellectuals and reached its highpoint in the violent suppression of the Prague Spring by the armies of the Warsaw Pact. In this potential crisis, it was a shrewd move on the part of the SED to initiate university reform from above and keep the critically inclined scientific intelligentsia busy with reform projects.

University reform was principally a component of the overarching strategy of increasing the economic productivity of the GDR by mobilizing modern economic growth factors. At the same time, however, the extensive debates had the unmistakable function of providing a safety valve and promoting identification with the East German system. The debates continued after the "Principles" were adopted at the Fourth University Conference in February 1967:[48] in January 1969, the Staatsrat of the GDR had before it a proposal entitled "The Continuation of the Third University Reform and the Development of the Universities to 1975."[49] This, too, was opened to public discussion, and 2,575 suggested changes and supplements to this text alone were received from individuals and institutions by the end of February.[50] Many creative ideas came to light in these discussions; the text that was actually implemented, however, was a relatively crude version that was most in keeping with the centralist pattern.

At the heart of university reform was the replacement of the outdated infrastructure of institutes and faculties with a new system that organized them into sections. In these sections, a number of disciplines were combined under the leadership of a director, and teaching and research were to be organized across disciplines. The remnants of corporate self-government were abolished, and the traditional faculties were turned into advisory bodies who reported to "single administrative heads" with full responsibility. This perfected the centralization of the administration; because the number of sections was much smaller than the previous number of independent institutes, it became much easier to get an overview of the structures. But it would be wrong to interpret the university reform as merely an instrument to further consolidate political power. Whereas the First and Second University Reforms had pursued largely political goals, in the early 1960s, according to Lauterbach, "university policy, too, was approached from a different perspective. Science was increasingly understood as a production site of innovations that needed to be used economically and socially."[51] This would account for the considerable attention the Third University Reform received in West Germany.[52]

The sections, first formed in the summer of 1966 and fully operational as of the spring of 1968, followed the ideal of concentrating capacities in central areas and projects. In addition, they were supposed to facilitate interdisciplinary work, though they were much less successful in this regard. The small university institutes that had previously been dominant gave way to medium-sized and large working teams in which professors, and especially the holders of chairs, were in the minority, and in which many collaborators of different ages and academic degrees worked together. These structures were advantageous for complex research projects and for university education that was closely involved with research.[53] University education was given a building block structure: basic, professional, and specialized education for all, and research studies leading to a doctorate for those with special talent for or interest in science. The organization into sections was meant to help make university education both research-oriented and practical.

In areas where the sections reflected the complexity of the field and the tasks at hand, they certainly represented progressive institutional forms. Frequently, however, the sections were nothing more than the joining together by bureaucratic fiat of institutes that had not developed any organic ties to one another. Small fields that did not fit into the section scheme had a difficult time of it, especially if they could not demonstrate any "utility to the national economy," in which case they were often "structured out of existence" ("wegprofiliert"). In an era dominated by enthusiasm for large structures, there was little understanding of the irreplaceable role that "small" fields and peripheral research played in maintaining the integrity of science and its capacity to develop. The result of the university reform was not clear from the perspective of science: on the one hand, it promoted growth in fields that benefited from structural changes; on the other hand, its structural "clean-up" damaged the flexibility of East Germany's scientific system and its ability to develop. What manifested itself here, as in other parts of the reform package, was the tendency of East Germany's leadership to implement standardized solutions and to homogenize structures, instead of making available a repertoire of institutional forms for different and changing tasks and circumstances.

The reform pushed very strongly for the link between the university and the economy. Specialized fields in science and technology were prevailed upon to make their research capacities available largely for projects from the economic sector. Research contracts (which specified certain topics) were frequently embedded within complex contracts between universities and businesses and state-owned enterprises or collectives. These contracts defined a whole network of relations: from cooperation in formulating projections and programs for scientific-technological progress, to research projects, the exchange of personnel, advanced training, hands-on training for students, all the way to the supply of apparatuses and equipment. The first model for such an agreement was the complex contract on scientific and technological cooperation concluded in January 1964 between the Institute of Technology in Dresden and the state-owned Kali firm; many similar contracts followed.[54] This kind of cooperation could be stimulating and useful for the universities. Given the chronic weakness of industry's research and development resources, however, this set-up developed in a direction that was disadvantageous to "basic" research, which is indispensable for a balanced development of scientific disciplines. The outcome was not that industry became more dynamic, which was the goal, but that the university system became more pragmatic.

Academy Reform

In 1960, the German Academy of Sciences in Berlin had by far the most extensive and productive resources in basic research outside of the universities. Chemical and physical disciplines, in particular, were strongly represented at the academy. These disciplines formed the scientific bedrock for what were projected to be the main directions of the scientific-technological revolution. Utilizing these resources to generate the basic innovations required by a rapidly advancing industry must have seemed the crowning achievement of the reform. This was not the first time the academy was confronted with the demand that it become active on behalf of industry. Ever since the founding of the Research Society of Scientific, Technological, and Medical Institutes of the German Academy of Sciences in Berlin in 1957, the academy had been moving steadily in that direction. Peter Nötzoldt, who has examined the entire process, is right in arguing that as far as the academy was concerned, this reform seemed "more like the conclusion of a previous, very dynamic process than the beginning of a new one."[55] Rudolf Landrock has given a detailed account of the reform, four characteristics of which deserve further attention here.[56]

First, the reform brought to a conclusion a process that had been under way since the early 1950s: elements of corporate self-government by the scholars and scientists were displaced by a hierarchy of single government administrators. This hierarchy was centrally organized and integrated seamlessly into the pervasive structure of state administrative relations. At the same time, this administrative pyramid was linked up with the party pyramid at critical points: for example, the first secretary of the SED district administration at the academy became an official member of its executive committee. The restructuring was to a considerable degree connected with the replacement of the leadership. Many scientists who had been in the "second tier" at the academy prior to the reform were now made directors of institutes or were given other prominent functions. Hermann Klare, an expert in the chemistry of fibers and an experienced industrial chemist who had previously been vice-president and chairman of the Research Society, was appointed president of the academy to make it unmistakably clear that the institution was to regard itself henceforth as a research organization that was predominantly tied to industry and whose chief orientation was scientific-technological. This meant the final abandonment of the postwar approach whereby the academy, constituted and supplemented by the election of members, was given access to research institutes. A very interesting variation on the institutionalization of research was thus finally put to rest. The institutes and the academy remained connected on the level of personnel. Given the complete integration of the former into the state administrative pyramid, however, they now had only a coordinating function.[57]

Second, compared with the situation prior to the reform, a much greater number of medium-sized, large, and very large institutes (Central Institutes) were set up. Depending on the disciplines to which they belonged, these institutes in turn were combined into research areas. Two motivations were behind the establishment of the larger institutes: first, to make the organizational structure of the academy more transparent and increase the resources available to the leadership; second, to raise the capacities so high above a critical threshold—which was different for each individual discipline—that practical needs could be efficiently met without excessively curtailing the possibility of autonomous, knowledge-oriented basic research. This restructuring was supposed to ensure the systematic integration of basic research and applied research in a single institution. This was unquestionably a novel approach, especially given the tension between these two kinds of research, and the strong tendency toward institutional separation to divide work along these lines. However, mergers and concentration also led to a reduction or cessation of research that was considered less important to society; at this time we have no inventory of the losses in the wake of the academy reform. One obvious disadvantage was that the reform further reduced the already limited variety of institutional forms at the academy. Whereas the Max Planck Society in West Germany is characterized by a great diversity of institutional types, in part because of the considerable autonomy it accords to its institute directors, the academy institutes had a standardized unitary structure foisted on them. This structure essentially remained in place until 1989, though a certain degree of secondary differentiation gradually occurred again within its framework.

Third, the academy reform brought closure to the long process of directing academic research toward practical results, especially those applicable to industry. This was its true goal, whereas the streamlining of the administrative mechanism and the concentration of capacities were means to achieve it. With this set-up, the academy could have become an important center of technical-technological innovation, without having to endanger knowledge-oriented basic research in a situation where personnel were being continuously increased. The overdrawn directive of 1968, which stated that the academy could pursue research exclusively in response to outside requests, could not be implemented. The Research Regulation of 1972, which set that share of research at 50 percent and accorded the president of the academy the role of "social contract-maker" for projects in basic research, left the academy appropriate latitude in making its own decisions about research. The ways in which basic and applied research were integrated into the large academy institutes, the transmission chains that were put in place, the considerable expansion of the capacity to produce technical prototypes, combined with the art of improvisation characteristic of East Germany—all this could have allowed the academy to become the center of innovation for industry. That it did not become such a center, or at least took only the first steps in that direction, had to do least of all with the academy itself.

Fourth, the academy incorporated into the buildup of its research potential the principle of embodying the disciplinary structure of science as a whole. Its institutes were for the most part defined by discipline. The motivation behind the establishment of new institutes was frequently to broaden the disciplinary spectrum represented by the academy. In connection with the academy reform, leading academics repeatedly emphasized that the academy, in view of the fact that the universities were heavily absorbed by teaching, was the true *universitas litterarum*. In this respect, the institutional principle of the Academy of Sciences differed fundamentally from the tradition of the Kaiser Wilhelm Society, which was continued in the Max Planck Society in West Germany. The academy reform was to consolidate the *universitas litterarum* principle and simultaneously promote unconventional interdisciplinary work. The first goal was achieved with the establishment of the research areas, which to a certain degree reflected the disciplinary structure of science as a whole.

Interdisciplinary work was facilitated only within the large institutes, how-ever; generally, only related disciplines were grouped together in these insti-tutes, so that unconventional combinations did not come about easily. In this essential area, the reform did not deliver.

Conclusion

Changes in the organization of science occurred in all industrialized countries after the Second World War. This was in line with the transition from the era of "little science" to that of "big science" described by D. J. de Solla Price. This historical trend was independent of the peculiarities of the political and economic conditions in the various countries, though it is clear that it was invariably influenced by them. In the socialist countries of Eastern Europe, changes in the organization of individual systems of social action occurred within the framework of centralized control by a single party.

Once the phase of postwar reconstruction had been more or less completed in East Germany, it became necessary to lay the institutional groundwork for a substantial increase in the number of university graduates and for coopera-tive research on a large scale, especially in scientific-technological areas. For particular historical reasons, all these reforms were packed into a very short period of time, essentially the second half of the 1960s.

The profound economic and social crisis of the GDR at the end of the 1950s, which posed an acute threat to the political power of the SED, was forcibly halted by the building of the Wall. However, Ulbricht and a number of younger science experts, on whose expertise he primarily relied, had grown convinced that only a substantial boost in economic performance could ulti-mately stabilize conditions in East Germany. Economic reform was the prel-ude to and basis of reforms carried out during this period in all areas of society.[58]

Reforms in education and science, in particular, were directly derived from economic reform; indeed, they assumed the form they did because their ultimate point of reference was economic reform, even though they also had roots independent of the economy. Given the grievous shortage in East Ger-many of material resources (raw materials as well as modern production facili-ties), reformers in the SED considered reform in science and education to be of great political significance. This resource deficit was to be made up quickly by a strong increase in highly qualified scientific and technical personnel and an expansion of facilities for research and development. This would allow the economy to become more efficient than it was in the Federal Republic, and in so doing slowly close the gap in performance.

The progressive abolition of private property in the 1950s meant that the earlier economic mechanism in the GDR had been largely nullified. It was replaced by a system in which economic planning and control (as a subfunction of the SED's political leadership of society) and the consumer needs of East German society were starkly juxtaposed. Independent economic actors with their own economic interests, who could have provided the economy with an internal dynamic and a mechanism of self-regulation, now existed only in rudimentary form; as a result, markets as a medium of self-regulation were correspondingly feeble.

The main objective of the economic reform was to shape the economic authority of enterprises in a way that would allow them to behave once again as relatively autonomous subjects, enter into market relationships, and develop their own economic interests. The goal was to create an economic mechanism that was partially independent of the political leadership. From the perspective of political centralism, this meant giving up direct guidance of the economy and establishing economic parameters as a means of maintaining indirect control. The economic historian Roesler has called this an economic "paradigm shift."[59] Reforms in education and science were intended to support this paradigm shift and make it irreversible.

Until the mid-1960s the New Economic System in the GDR was in an experimental stage. This was before the reforms in science had begun. Even at this undeveloped stage, the new system had contributed to a noticeable stabilization of the economy and slowed the pace at which East Germany was falling behind West Germany.[60] However, modest stabilization did not seem sufficient; rather, the goal was to bring about a sweeping structural change in industry through the accelerated development of modern branches. In particular, numerous ambitious automation projects that would be scientifically served by new installations for large-scale research were planned. The feeble economic situation from which these efforts started created new imbalances, which manifested themselves in the second half of the 1960s in rising economic tensions and worsening supply problems.

These developments made it easier for the strong antireformist circles in the SED leadership to obstruct the economic reforms—which they did covertly hitherto because of Ulbricht's personal power—more openly and aggressively, and to support their position with populist arguments. The experience of the Prague Spring of 1968 played a decisive role in their opposition: traditionalists within the SED Politburo saw the situation in Czechoslovakia as proof that any deviation from strict centralism—no matter in what area— undermined the political power of the Party. Ulbricht's growing isolation within the Politburo, which eventually led to his downfall, was, as far as we now know, furthered by conspiratorial contacts between Honecker and his confidantes and Leonid Brezhnev in Moscow. The leadership of the Soviet

Communist Party saw in Ulbricht's course of reform the danger that the GDR could break away from unconditional subordination to the Soviet model of society.[61]

The economic paradigm shift was completely reversed under Honecker. The ambivalent nature of the university and academy reforms resulted largely from the fact that they got under way only after the restorationist forces in the SED began undermining economic reform. Because of this, the science reforms took on certain schematic traits: while they did lead to a considerable expansion of educational and scientific capacities in the GDR, they emphasized above all those features of structural change that were compatible with centralized social reform. There is no question that they had some lasting rationalizing effects. However, the main idea of the reformers, to turn science into the generator of innovations urgently needed by the economy and thereby make it highly creative, was not implemented beyond preliminary steps. Even though the science reforms did result in a restructuring of the scientific enterprise, and were not reversed during the Honecker era, it is accurate to regard them as a partial failure. The reason lies largely with the politically driven reversal of economic reform and the restoration of a planned economy. Ultimately, "intense creative action within the framework of a strategic concept" was exchanged for the preservation of power through pragmatic maneuvering devoid of any strategy whatsoever.[62] The opportunity to initiate and implement substantial reforms in the social system as a means of transforming society was thus missed in the GDR. Henceforth, reforms seeking to be more than minor improvements could be conceived only in opposition to the system.

—Translated by Thomas Dunlap

3

The Shadow of National Socialism

Reinhard Siegmund-Schultze

The division of Germany into two opposing social and scientific systems was the result of the defeat and breakup of National Socialist Germany, and the shadow of that dictatorship fell on both. The politicians and historians of the Western zones (after 1949 the Federal Republic) looked back chiefly to the history of the Weimar Republic. Their view of the past was not uncritical, for the Weimar Republic had caused its own demise in 1933.[1] But they had at least some positive lines of continuity in German history to connect to the conception they had of themselves and their state. By contrast, the political leadership of East Germany had the far more radical goal of constructing a completely new society on German soil. Consequently, the leadership's relationship with Germany's political past was almost entirely negative. The history of Germany after 1949 can therefore not be written without taking into account "three contemporary histories"—that of National Socialist Germany (and its prehistory), the Federal Republic, and the German Democratic Republic.[2]

This chapter deals chiefly with *one* side of this historical and historiographical triangle: the after effects of National Socialism on East Germany, specifically on the development of East German science to about 1961.[3] I am principally concerned with the ideological function of the officially proclaimed notion of antifascism. I will examine the psychological and social influences of the Nazi era and will determine how predisposed older scientists were to engaging the new ruling ideology of Marxism-Leninism or accepting it to some degree or another. At the conclusion of the chapter I will begin to examine possible "comparisons" between the National Socialist and socialist systems of science, a question that has been raised repeatedly in recent years, for obvious political reasons. As I am more concerned here with the political issues than with developments specific to the field of mathematics, my findings will have general applicability.[4]

To be sure, it will become clear in the course of this chapter that another side of the historical "triangle"—the relationship between West Germany and East Germany—cannot be ignored. Given the specific concerns of this chapter, the relationship of the two German states to their respective victorious powers, a relationship that was also relevant for the history of science, especially in East Germany, can be largely excluded. The antifascism officially proclaimed in East Germany was on its most consequential and historically important level an ideology of demarcation from the Federal Republic. Given certain uncritical attitudes in the Federal Republic toward the Nazi past, this ideology of differentiation was not *only* demagoguery, but it did produce some absurd results, for example, the designation of the Berlin Wall in 1961 as an "antifascist barrier." On the other hand, Christoph Klessmann emphasizes that the history of science in West Germany, as well, cannot be written without a discussion of its deliberate demarcation from East German science.[5]

The Common Point of Departure in East and West

Needless to say, the Eastern and Western occupation zones after 1945 had to work with the people, resources, and structures that still existed in their respective territories. Other chapters in this book highlight the tremendous losses resulting from emigration, particularly in basic sciences like mathematics, and the subsequent absence of nearly an entire generation of capable young scientists in these fields—realities that affected both parts of Germany. To that extent, the "after effects" of National Socialism in East and West were initially similar. However, the exodus of German scientists to the Western part of the country, which had already begun during the final months of the war and was encouraged by widely held anticommunist and anti-Soviet attitudes, created considerable differences, especially with respect to the materials supply and early lack of manpower.

Certain changes that had been made during the Nazi period in the organization and structure of the sciences were retained as long as they were not politically compromised through their direct association with the Nazi regime. For example, one of the innovations instituted during the Nazi period that did not change in the East or West following the collapse of the Third Reich was the *Diplom* (equivalent to the M.A. in the American system) for physicists and mathematicians. It was more difficult to deal with measures that had been the product of political as well as scientific motivations, for instance, the separation in 1934 of *Habilitation* (the postdoctoral thesis) and *Dozentur* (university lectureship). This separation was a political move, guaranteeing the economic livelihood of some of those who had obtained *Habilitation* (and were politically acceptable). Another step, taken in 1939, was also a scientific necessity: after 1945 the so-called *Diätendozenten* (remunerated lecturers) were initially retained in both West and East—although this was not in accordance with the Western *Privatdozenten* model—while in the centrally controlled university system of East Germany, political criteria once again played a major role in selecting those cadres intended for a professorial career.[6] This issue was brought up in a faculty meeting at Humboldt University in Berlin on August 25, 1954, in a discussion about a new East German *Habilitation* regulation. The East Berlin mathematicians Karl Schröter and Heinrich Grell again politicized qualifications: "Herr Schröter agrees with Herr Grell that one should not support applications for *Habilitation* in cases where there is reason to believe the applicant may emigrate. Herr Grell noted the preference of the State Secretariat for applicants whose social backgrounds vouch for their reliability."[7]

Governments in the West and in the East propagated a "democratic" development for the universities, a move intended to make a clear break with the Nazi dictatorship, which had given the word "democracy" a bad name. In East Germany, "democracy" was increasingly understood as the "democratic centralism" called for by the SED. Those elements in the university system of Hitler's Germany that had been a clear expression of Nazi ideology were abolished in the Soviet Occupation Zone and later in the GDR, at least during the initial years. For instance, the role of the rector as the "Führer" of the university and the strong position of the lecturers *(Dozentenschaft)* were abolished. It was only in the late 1960s, through the Third University Reform, that similarly strict methods of administration were reintroduced. Signs of the retraction of democratic self-government became visible as early as 1952 in Rostock. In 1951 the mathematician Rudolff Kochendörffer, in his capacity as the dean of the faculty of mathematics and natural sciences, sent a memorandum to the newly established State Secretariat for University Education, and copies to SED chief Ulbricht and to all deans of faculties of mathematics and natural sciences. In it he pointed to the danger that innovations such as the planned ten-month course of studies and the inclusion of social science and Russian sections in professional qualifying examinations would pose to the level of professional training. He then went on to contend that "the election of the rector by the senate, and not, as hitherto, by the council of all professors of the university, constitutes a totally unjustifiable restriction of the democratic rights of the faculty."[8]

In 1945 there were no widely held democratic traditions among university professors either in the West or in the East, traditions in line with Western parliamentarianism or the Weimar Republic and transcending the principles of self-government within the universities.[9] The postwar pressure for *pro forma* political cleansing, and the preparation of vast numbers of "whitewash certificates" that were easily recognizable as pure favoritism, surely did not encourage scientists to examine critically their own role under National Socialism. A particularly outlandish example is the verdict (dated July 22, 1948) of the mathematician H. L. Schmid about his colleague Heinrich Grell: "He is the one German who probably suffered the greatest persecution and disadvantages from the past regime." This was said notwithstanding the suffering of émigrés and the fact that Grell joined the NSDAP in 1933 before he became embroiled in a private conflict with the party.[10]

Formal criteria, such as membership in the NSDAP, initially played a dominant role in the process of de nazification. Needless to say, membership also paved the way for the perpetuation of the traditional mindset among scientists, which favored the authoritarian state. It remains an open question whether democratic development in the Federal Republic did in fact substantially enhance the personal courage of its scientists, beyond individual examples, such as the authors of the 1957 Göttingen Manifesto of Atomic Physics. It must be asked whether and in what way scientists made use of the greater freedom that exists in a democracy.

Rebuilding the system of higher education in East and West had to be accomplished "from above," through the intervention, first, of the occupying authorities, and then of the respective political authorities. Given the earlier manipulation of the German people under National Socialism, the concept of "socialist education" in the age of "re-education" may not have sounded illegitimate at first.

Antifascism and the Legitimation of Political Power

Following the 1952 Resolution for the Construction of the Foundations of Socialism, the SED was the decisive political force in East Germany. Through the Party's science and education *(Wissenschaft)* department and the politically subordinated State Secretariat for University Education, the SED also substantially directed the development of science. Much of SED politics and policy, including its science policy, especially in the 1950s, must be seen against the backdrop of the GDR's self-conception as the "legitimate antifascist state." Because of their antifascist past and their experience of having been, unlike the great majority of the German people, "on the right side" during the Nazi era, many SED politicians believed they now had the right, if necessary, to rule against the will of the majority. This psychological fact cannot be underestimated.

We must now examine what this officially proclaimed antifascism accomplished with respect to science in East Germany, and where it fell short; where it was truly concerned with the history of National Socialist Germany, and where it became more a propagandistic vehicle for the prevailing Marxist-Leninist ideology.

Two functions of this antifascism undoubtedly contributed to the legitimation of the East German system, and their effect on science was particularly grave: the *pedagogical function* vis-à-vis students and scientists; and the *demarcation function* vis-à-vis capitalism, especially in opposition to the Federal Republic. The discussion that follows, however, will look not only at antifascism as an ideological instrument of power, but also at all the ways in which scientists reflected on and reproduced the Nazi dictatorship. Examples will be drawn from the history of mathematics in East Germany, primarily from a political-biographical perspective.

The Pragmatism of Antifascism

Given that the East German leadership made a clear break with Germany's political past, it is not surprising that science policy in East Germany was, from the outset, significantly shaped by a scientifically and ideologically veiled voluntarism and pragmatism. The East deliberately renounced the pool of experience the Federal Republic used to restore partially the conditions that had existed prior to the Nazis. The West Germans, by contrast, sometimes idealized federalism, which had in fact existed in the Weimar Republic in the cultural and scientific sphere.[11]

According to the narrow conception of history held by many SED functionaries, monopolistic capitalism was solely responsible for the rise of fascism, and the anti-Semitism of the Nazi regime was frequently regarded as merely a derivative of this phenomenon. This view allowed for a very pragmatic response in cases where scientists had been heavily involved in the Nazi system. Since the majority—not all—of the scientists who were severely compromised by the Nazi period had emigrated to the Western zone, SED politicians were relieved of the need to examine closely the role of science in the Nazi state. Even in East German historiography of later years this discussion was never very intense. On the one hand, some scholars of fascism had little use for scientists, who had rarely been among the clear opposition or ruling classes of the Nazi state. On the other hand, science in East Germany, which was closely linked with the SED's ideology of growth, was labeled "essentially humane and progressive," and this also played a role in the lack of attention paid to its past. Moreover, works on the Nazi period that were critical of science were always dangerously close to being sociocritical reflections on the conditions in East Germany, given the indisputable parallels in some of the political mechanisms.

The case of the number theorist Helmut Hasse, who was offered a chair at the university in East Berlin in 1949, is a particularly striking example of the unscrupulous way in which East German communist functionaries wooed an exceptionally competent scientist compromised by his conduct under National Socialism.[12] Banned by the British authorities from teaching in Göttingen, Hasse offered his services to the University of Marburg, located in the American zone. In a letter he wrote to the Marburg mathematician M. Krafft, dated February 20, 1948, Hasse told Krafft that he would prefer an appointment in Marburg, even though negotiations about a job in Berlin were going well.[13] The physicist and science functionary R. Rompe, of the German Central Administration of People's Education in the Soviet zone, wrote to the Soviet Military Administration, in Berlin-Karlshorst, regarding Hasse's appointment. The following is from the draft of April 10, 1947:

> As for his political attitude, we know that *at times in 1933 he expressed anti-Semitic views but later retracted them, and* had a serious disagreement with Professor Dr. Bieberbach, the representative of "German Mathematics," in the matter of a Jewish colleague. Professor Hasse was a marine officer in World War I, and was active in World War II in administrative services with the same rank. *There is no denying that he has a certain militaristic attitude. That is why we endorse Professor Hasse's readmission with the stipulation that his teaching be limited to the specific training in the senior semester until he has completely proven himself politically.*[14]

The italicized passages in the above draft were omitted from the letter sent on May 2, 1947. Instead, the letter contains the following statement:

> In view of the professor's outstanding qualifications in his field, we recommend his readmission to teaching and research, for we believe that Professor Hasse never took a genuinely pro-Nazi stance; in any case, today he is ready to become involved in training the new generation of scientists.[15]

When the Soviet Military Administration failed to respond positively, J. Naas, an administrator, followed up in a letter to P. Wandel, the president of the German Central Administration for People's Education: "Moreover, it can be assumed that the democratic education of the students will be furthered by Herr Hasse."[16] Still, the Soviet Military Administration, in a letter dated March 10, 1948, from the Central Administration to the academy, turned down the request because of Hasse's political past. However, the following note, written in pencil and dated March 31, 1948, was added:

> Mr. Naas requests that you consider this letter as not having been written. Wandel, Rompe, and Naas, in the presence of Hasse, agreed to make every effort to get Hasse a professorship as soon as possible. If necessary, Wandel himself will plead the case in Karlshorst.[17]

Hasse was eventually appointed professor at Berlin University on May 3, 1949.

In Hasse's case, the SED therefore made a practical separation of professional competence and political responsibility (even more so than the Soviet authorities who were in charge at the time). In much the same way, the SED was quick to talk of "humanistic bourgeois scholars" in other cases, and to construe from supposedly selfless and disinterested scientific research a "higher morality" possessed by scientists and operative also outside of science. This put the SED—which almost to the very end of the East German state regarded science as "essentially humane and progressive," even if it was, unfortunately, sometimes "abused" under capitalist and fascist conditions—firmly in line with the scientists' own ideology. The SED continued to repeat general themes of postwar justification for the alignment of science and ideology, and in so doing fostered the willingness of scientists to accommodate themselves to the new regime. As we shall see, however, this sharp separation of professional accomplishments from political activity applied only to a specific period and to the older generation of leading scientists in the 1950s.

Moreover, higher political concerns (in Hasse's case the need for exceptional professional competence) were invoked to justify why the regime dispensed with a critical examination of the past. For example, even West German mathematicians like G. Hamel and W. Blaschke, whose conduct in the Nazi period had been highly opportunistic and who, like Hasse, were no supporters of the political system in East Germany, received the highly endowed "National Prize," which was established in 1949. Incidentally, the choice of the name National Prize was, at the least, an act of political insensitivity, given the fact that Hitler had set up a prize of the same name in 1937 in deliberate opposition to the Nobel Prize. Here, too, higher political interests came into play: in the early 1950s, the prize was awarded as a means of maintaining the unity of the German nation and as an expression of the East German government's responsibility for that unity. After receiving the prize, the geometer Blaschke wrote, ironically, how amazed he was "that a communist state, of all things, wanted to turn me into a capitalist."[18]

Of course the interests of scientists, like those of F. Willers, who tried after 1948 to loosen the law of the Allied Control Council to allow the publication of articles on the fundamentals of aerodynamics in his *Journal for Applied Mathematics and Mechanics,* were not sufficient by themselves to provoke support for science at the level of the central government.[19] It was only after the founding of NATO and the Warsaw Pact in the mid-1950s that the need for applied research in military mathematics became apparent. The needs of the East German aircraft industry between 1954 and 1961 also turned the leadership's attention to improving science and research. A functionary of the SED's Science Department wrote in 1956:

> It is well known that many mathematicians during the capitalist-fascist period made their great knowledge available to Hitler's war machine. In the process they gave little or no thought to the abuse of science under the Hitler regime. During the Hitler period they acquired substantial expertise in fields of military importance, such as the aircraft industry, aerodynamics, and so on, which still exists today unused in the minds of the scientists. It is our task to prevail upon these scientists, through close, comradely, and constant contact, to make this expertise available to the workers' and peasants' state, Germany's future, and to transmit this hidden knowledge to students.[20]

A higher political interest also came into play when scientists came out strongly in favor of the new political system, especially since membership in the SED by mathematics professors remained very low until the late 1950s; only five out of thirty-six were SED members in 1958.[21] In such a case, the scientists promoted by the Party did not have to demonstrate exceptional professional competence. The most extreme example of which I am aware is that of Max Draeger, the author of "Mathematics and Race" (1941), an article published in the National Socialist professional journal *Deutsche Mathematik*, edited by L. Bieberbach and T. Vahlen. Draeger joined the SED after the war and subsequently became a professor in Potsdam and a member of both the Presidential Council of the Kulturbund and the Advisory Council for Mathematics at the State Secretariat for University Education. It was only a letter by the East Berlin logician Karl Schröter that forced Draeger to step down in 1957.[22] It is not clear from the files whether the East German authorities were informed about the full extent of Draeger's Nazi activities.

The Pedagogical Function of Antifascism

While frequently criticizing the old "submissive spirit" of scientists and their apolitical stance during the "capitalist-fascist period," the East German regime was breeding the same spirit under the opposite political banner. A key quote, which inadvertently describes this creation of a new submissive spirit, comes from a political evaluation of mathematics professors by the SED's Science Department on March 21, 1958: "[Most] are making an effort to meet their obligations as teachers at an East German university, though one cannot say whether their heart is in it. In part the well-known submissive spirit plays a role in this."[23] We can assume that the SED politicians who read this, and who had completely internalized the class warfare scheme, did not see how outrageous this passage was. Incidentally, this helps us understand that administrative socialism by necessity had to reproduce certain features of the Nazi system.

The old intelligentsia's political accommodation to the new situation was undoubtedly dominant in the 1950s, and it was also promoted by special privileges, some of which already incurred the disapproval of the younger generation of scientists. The State Secretariat for Higher Education reported to the Science Section in 1958 specifically with regard to mathematics professors: "Many have the impression that the scientists, especially the older scientists, are being bought with higher salaries . . . What is important is less money for scientists, more money for science."[24] Still, the reactions of leading scientists at the time to official policies and politics must not be seen *only* from the perspective of timid accommodation, or *only* from that of purely professional interests.

In this regard, as well, the prehistory of East Germany, the era of National Socialism, is important: after the collapse of Hitler regime's and their own traditional edifice of values, quite a few mathematicians—and other scientists, too—were willing to make a new beginning intellectually, to make a contribution to this endeavor through their scientific work and thereby simultaneously legitimate mathematics. It is clear that in the critical postwar period, individual scientists pondered the meaning of scientific thinking more intensely than ever before. The intellectual fascination of Marxist dialectics, often conveyed through Lenin's *Philosophical Booklets,* the novelty of these previously unknown ideas, gratitude toward the Soviet liberators, the obvious international importance of Soviet mathematics, and the undeniable priority of the practical questions of reconstruction—all these circumstances undoubtedly created an intellectual atmosphere that made many mathematicians receptive to a radical new discussion of the philosophical-ideological premises and consequences of mathematical thinking. For example, Schröter, though not a member of the Party, cofounded the Marxist journal *Deutsche Zeitschrift für Philosophie (German Journal of Philosophy)* in 1953.

This willingness to rethink politics in a fundamental way was surely limited to a handful of leading older scientists, and presumably waned somewhat as the years went by. That said, it was above all the relationship of Marxist ideology to science itself, radically different from that of vague Nazi ideology, that created possibilities for further coalitions between the Party and the "bourgeois intelligentsia." In connection with the so-called scientific *Weltanschauung* (world view) of the working class, the SED often acknowledged that the rationalism of mathematics and the natural sciences was a crucial prerequisite for mastering nature and society and solving the tasks of the future. Older mathematicians remembered well the disregard the Nazis had for basic sciences like mathematics, an attitude expressed most clearly in the expulsion of leading scientists after 1933 and even in the neglect of applied mathematics at the universities prior to the war.[25] They could recall offshoots of National Socialism's racial theory, like Ludwig Bieberbach's *German Mathematics,* which sought to intrude upon the conceptual sphere of mathematics itself, though it had no true cognitive consequences.[26] While *German Mathematics* provided a "justification" for the expulsion of Jewish mathematicians (and was, in that regard, historically significant), the ruling ideology of the GDR was, of course, more interested in *preventing* the growing exodus of leading scientists to West Germany.

As far as ideological interference in internal mathematical debates is concerned, only the last reverberations of the Stalinist "idealism debate" in mathematics reached East Germany. One example is the sharp attack by the Soviet geometer A. D. Alexandrov against the "bourgeois Hilbert" (the famous German mathematician) in the November/December 1952 issue of *Forum.* In contrast, the article by T. Gnedenko and L. Kaloujnine, published in 1954 under the title "The Struggle between Materialism and Idealism in Mathematics," was already more like a defense of D. Hilbert's axiomatics against attacks by incompetent, supposedly "Marxist philosophers."[27] They dutifully denounced the so-called idealistic ideological conclusions that B. Russell and the Vienna circle of positivist philosophy drew from mathematics. The repercussions of Stalinism for mathematics, still little studied, thus tend to be discernible more in the peripheral fields of logic, arithmetic technique, and cybernetics than in the more prominent disciplines.[28]

In the Nazi state, mathematicians, when it came to science policy, often felt the need to go along with accepted epistemological ideals. Some mathematicians would, for example, propagate the "implacability of mathematical thinking."[29] In East Germany, where mathematics was officially recognized for its intrinsic value, they did not have to do this. In some sense representatives of basic science in East Germany, unlike in Nazi Germany, even enjoyed a higher social prestige than engineers, as Chapter 8 shows. However, this recognition also had something to do with the new social prominence the basic sciences had attained worldwide. SED functionaries gave ideological justifications for some measures "favorable to science" and thus used them as political propaganda for the prevailing *Weltanschauung,* even though those measures were a necessary part of the worldwide social and scientific modernizing process.

Increased interest in mathematics on the part of Marxist ideology (as compared with Nazi ideology) also involved the potential danger of greater control and the possibility of incompetent interference on the part of the state. This tension between interest and control characterized much of the relation-

ship between the bourgeois intelligentsia and the Party, and here the historical experience of National Socialism once again played an important role.

Perhaps the greatest potential for ideological conflict lay in the question
whether and to what extent political engagement (on behalf of the new
system) had to be joined to professional work, or whether good professional
work in itself was a positive thing for the system. Of course, scientists preferred the latter view. Needless to say, older scientists could still remember the
Nazi slogans about "political science" when they heard talk about "progressive science" in East Germany. In the 1950s, it was not only SED functionaries but also younger up-and-coming scientists who at times made a connection between "progressive *Weltanschauung*" and advances in knowledge.
Older scientists, in particular, had to feel threatened by such a connection.
For instance, an evaluation by the Science Department, written with the
collaboration of young mathematicians and dated February 10, 1956, had
this to say about mathematical statistics in East Germany:

> There is an academy institute for statistics, and it is exceedingly weak
> politically. It is headed by Professor Lorenz. Statistical data on mice and
> other unimportant things are pursued at this institute. Lorenz is no
> theoretician; he resists theory and uses outdated methods.[30]

In this report, political conservatism was in some way connected and corre-
lated with professional conservatism, which is not to say that either criticism
in itself was unjustified. The mathematician Kochendörffer early on recognized the danger of linking such criteria. In his capacity as dean of Rostock
University, he called a "Directive for the Drafting of Statutes at Universities
and Colleges," passed by the Ministerial Council on August 28, 1952, merely
a basis for discussion and wrote: "There is no need to speak of 'progressive'
science, since the concept of progress (of human understanding) is part of the
concept of science."[31] A 1967 history of Rostock University, which dealt
harshly with Kochendörffer, who had left East Germany in the mid-1960s,
made the following remarks regarding the previous document: "Here the
concepts 'progressive science' and 'progress of human understanding' were
equated in an unacceptable way."[32]

As we have seen, the SED considered professional competence (an area
where Lorenz fell short) a sufficient substitute for political activism in its
dealings with the older intelligentsia in the 1950s. However, that was not so
for the generation of students in the 1950s, who were supposed to become the "party-loyal" professors of the 1960s and 1970s.[33] For instance, the
Science Department wrote the following complaints about a mathematician
in Leipzig:

Nor can it be said that the comrade professor . . . in Leipzig is working vigorously at improving the cadre of comrades. From now on we must ask such comrades what they think about this situation, and whether they do not have the obligation as comrades to train cadres loyal to the Party and the government and to keep them at the institute as assistants.[34]

The problem of the political engagement of the next generation of scientists was thus another, early area of conflict between the Party and the old intelligentsia. Memories of their own political past, of the political pressures during their student years and assistantships in the Nazi state, weighed heavily on scientists when it came to this issue. Now the older scientists had to come to terms above all with the new educational ideals that were bound up with Marxist ideology.

Under National Socialism priority from the outset was given to the political activities of students and assistants over their professional competence. In East Germany, by contrast, SED functionaries preserved the fiction that there were no real areas of conflict between professional and social engagement. For instance, it was often assumed that scientists who were politically active in support of the Party would be able, with appropriate efforts, to make up for professional shortcomings later on. Functionaries generally tended to ascribe such deficiencies to the fact that these scientists had less time to devote to their fields. Of course that conviction on the part of many SED science functionaries had a lot to do with the assumption that science was directed toward specific purposes, and with their need to control it. But it also reflected a strong underestimation of the importance of individual talent and the firm belief that scientific abilities could be taught, convictions that were often tied in with socialist educational ideals. This set the stage for conflicts, especially with mathematicians, who had first-hand experience of the unequal distribution of talent in their field, both in their day-to-day work and in the high drop-out rate in mathematics.

As early as 1949, Hasse, who had just been appointed to a professorship in Berlin, "lamented that exceptional talents cannot be admitted outside of any formal classification," and he called for a so-called "clause for talented people" *(Begabtenklausel)*.[35] Unlike in the Nazi state, however, this kind of conflict was negotiable, not only on the pragmatic level (which the Nazi state, too, was frequently compelled to recognize), but also on the purely ideological level. For example, the educational ideology of the functionaries was redirected into an expansion of mathematics classes at schools, which brought mathematics additional legitimation. In December 1962, the Politburo of the SED passed the "mathematics decree," which meant a considerable boost to the math curriculum at secondary schools.[36] The fact that the highest political authority in East Germany acknowledged the "esoteric science" of mathematics with such a decree made it clear to many mathematicians that their science held a place incomparably higher than it had during the Nazi years, and to some extent enjoyed greater recognition than it did in the Federal Republic. Moreover, since the problem of training the mathematical elite seemed to have been solved satisfactorily in East Germany through the "Mathematics Olympics" and several "special math classes," it is understandable why some mathematicians outwardly renounced earlier "elitist" reservations about the socialist educational ideal. We read this in a 1960 political evaluation of the logician Schröter (Berlin):

> With this work, Professor Dr. Schröter, together with the assistants under him, has furnished proof that in mathematics, too, any person of normal talent can produce satisfactory results, provided he receives intensive training. He has thereby dealt a decisive blow to the theory (which he himself once advocated) that drop-out rates of 30 to 40 percent are the rule in mathematics.[37]

The Demarcation Function of Antifascism

It is easy to understand why the mathematics decree, passed shortly after the erection of the Berlin Wall, placed strong emphasis on drawing a line of demarcation against West German mathematics. For instance, the decree said, with some propagandistic exaggeration: "Mathematics classes in our socialist secondary school are in no way comparable to the lessons in arithmetic and geometry at the bourgeois Volksschule, which today in West Germany is still the school for the great majority of children."[38] Another essential function of antifascism was thus to draw a line of demarcation against capitalism, especially against the Federal Republic. And it is clear that the desire to divert attention away from some obvious similarities between the Nazi system and the East German system played an important role in this.

SED science functionaries also made propagandistic use of weak points that left the Federal Republic vulnerable to criticism: the continued reinstatement of scientists with a dubious political past, and the prevailing reluctance among German historians in the 1950s to write a critical history of the socioeconomic and sociocultural causes of National Socialism.[39] Emigrés from the Nazi period, for example, the mathematician R. Courant, who then worked in the United States, at times spoke out against the uncritical attitude of their West German colleagues.[40]

The majority of mathematicians whose conduct under the Nazi regime had left them politically compromised had moved to the Western zone at the end of the war; after all, anticommunism had been a basic component of Nazi ideology. However, the number of politically compromised scientists was smaller in mathematics than, for example, in medicine or some humanities.

Of course there were also many antifascist mathematicians in West Germany, among them former victims of National Socialism. Paradoxically enough, this situation, together with the aspiration of German mathematicians for national unity, which was politically supported until the mid-1950s, even led West German mathematicians to support East Berlin's opposition to West Berlin.

Since the mathematical culture in East Berlin was, on the whole, stronger than its counterpart in West Berlin, and since it was able to link up well with the great traditions of German mathematics (the traditional university and academy were in the Eastern part), West German mathematicians had better professional contacts with East Berlin in the early 1950s than with West Berlin. Added to this was the fact that Ludwig Bieberbach, severely compromised through his collaboration with the Nazis, lived in West Berlin. Bieberbach had good friends in that city, and there was a recurrent rumor he might become a professor again or receive a teaching appointment. This placed an additional strain on the relationship between leading West German mathematicians such as the long-time president of the German Mathematicians Association (DMV), E. Kamke, who had been dismissed because his wife was Jewish, and West Berlin mathematicians such as A. Dinghas.[41]

As East German politics slowly abandoned the goal of German unity from the middle of the 1950s on, however, the charge of fascism was dragged out with greater frequency against West Germany. The undiminished exodus of qualified people played an important role when such accusations were raised in science.

Even a mathematician like Heinrich Grell, who at times criticized East German science policies, made such accusations against West Germany on his own. During a trip to China, he wrote to his wife in a letter from Nanking (dated December 24, 1954):

Colleagues with a Nazi past can be sure that people here are extremely well informed about their political "accomplishments." And today, as well, they keep a close watch over people's conduct, for example, the exodus from the GDR to West Germany, and examine its causes. There were a few questions I could not answer to easily, as I did not wish to make statements that would have damaged the reputation of my German colleagues as a whole.[42]

Moreover, Grell's letter seems to express hidden feelings of guilt about his own political past under National Socialism, which was not free of inconsistencies, and he tried to overcome these feelings through loyalty to the "antifascist state" in East Germany.

Frequently the comparison with National Socialist Germany was also made to provide democratic legitimation for the GDR, especially to rebuff the accusation of totalitarianism. This becomes clear in some statements that East German science functionaries made about the anticommunist witch hunt in the United States in the early 1950s. On May 6, 1954, the executive committee of the Berlin Academy of Sciences discussed the possible appointment of the American mathematician and Marxist historian of mathematics D. J. Struik to a professorship in Berlin, and approved a letter by J. Naas, which said:

> I also ask the executive committee to pass a motion expressing its astonishment and regret that an outstanding and loyal mathematician like Prof. Dr. Dirk J. Struik is encountering such troubles in the United States. They remind us in Germany quite plainly of the same occurrences during the Hitler period.[43]

After the erection of the Berlin Wall in August 1961, and with the establishment of an independent "Mathematical Society of the GDR" imminent, West German mathematicians at the last joint East-West meeting of the German Mathematics Association in September 1961 in Halle had to listen to an embarrassing speech by a representative of the East German State Secretariat.[44] In it, he emphatically underscored the parallels between the Federal Republic and Nazi Germany:

> But everybody—and that includes you, ladies and gentlemen, who as mathematicians represent a part of German science . . .—must become aware of his place, role, and responsibility, if he wants to justify himself before our people and the future generations of scientists . . . It is a fact that in the Western part of our homeland, the same sinister forces that led Germany and the world into two horrible wars once again occupy the positions of power . . . That is why our government acted on August 13, this time in anticipation before it is once again too late.[45]

On March 22, 1962, the State Secretariat wrote an internal evaluation in preparation for the June 1962 establishment of the Mathematical Society of the GDR. It is not surprising that in it leading West German mathematicians were labeled "dangerous enemies" or "fascists."[46]

Conclusion

As we have seen, it is only within the context of political and professional interests that we can understand why SED functionaries and the older generation of mathematicians resorted—for intellectual and propagandistic reasons—to the history of science in Nazi Germany. Similarly, the current discussion about the necessity, possibility, and limits of a "comparison" of science in National Socialist Germany and the GDR is fraught with and driven by particular interests.

Needless to say, there are some areas that suggest themselves for comparative studies, such as international scientific communication, the limits on and use of human and material resources, and certain mechanisms of accommodation. I shall mention only the fact that in neither system were established scientists forced to make such a strong declaration of political loyalty as joining the Party. By contrast, younger scientists undoubtedly furthered their careers by taking that step, if they had not already done so out of political conviction. For instance, the East Berlin mathematician Kurt Schröder joined the NSDAP in 1939 to advance his career, but in East Germany he attained eminent positions—including the rectorate of Humboldt University—without joining the SED.[47]

However, problems in historical methodology are caused, among other things, by the absence of a shared basis of reference and comparison for the two systems of science, the very different worldwide stage of development of science after 1945 as compared with the 1930s, as well as the historical derivation of one system (East Germany) from the other (Nazi Germany).[48] That is why we should first say something about the latter aspect of the relationship between the two systems of science. It is an aspect that addresses concrete historical effects, and it goes without saying that it does not allow for hasty conceptual generalizations about "the essence" of the difference between East German and Nazi science. At the least, such a discussion can help destroy certain historical myths still in circulation, for example, that "dictatorships are hostile to science" or that "scientific activity has an oppositional quality."[49]

Some contemporary German mathematicians, representatives of the one science in East Germany that came out of the political "turning point" of 1989 relatively intact, seem to feel that their basic "apolitical" attitude has been confirmed by that very turning point. And through superficial comparisons between the state of mathematics in National Socialist Germany and the GDR, they contribute—no doubt unintentionally for the most part—to the historical revisionism that is gaining ground in Germany today. This revisionism brings with it the danger of relativizing the crimes of National Socialism. A leading East German mathematician, R. Kühnau, in a 1992 article entitled "The Situation of Mathematics and Mathematicians in the Former GDR," wrote the following about the *Communications of the Mathematical Society of the GDR:*

> They contain many highly valuable review articles . . . which are not known outside the GDR, though they deserve to be. But there are also politically saturated articles, which are sometimes even more embarrassing than what one finds in [the journal] *Deutsche Mathematik:* resolutions of the SED party congresses, Honecker quotes, and similar canonized blather.[50]

It is not surprising that a mathematician prefers the professional articles to the political ones. But when Kühnau compares the journal *Deutsche Mathematik,* edited by the Nazi Bieberbach, positively to the *Communications,* one should take a look at the political articles the former ran. Among them were the likes of Draeger's "Mathematics and Race" and other related concoctions, next to which all resolutions of SED party congresses seem rather harmless. In this case it is not difficult to divine the reasons for the failed comparison and trivialization of Nazi ideology: the eminent mathematician and fanatical National Socialist Oswald Teichmüller had published some of his best work on quasiconformal mappings in *Deutsche Mathematik.* That happens to be the field of the mathematician just quoted, as well. It seems reasonable to conclude that Kühnau was trying to propagate the ideology of the "apolitical mathematician" who gives little thought to how his science is organized and how its results are used.

These "apolitical mathematicians" overlook another fact: if their predecessors in East Germany had not laid the groundwork of accommodation, not only would the pursuit of their science have been obstructed even more in some areas, but also the positive stimuli to the cultivation of mathematics that were unique to this system would have gone unused. For instance, without the political decision to create the massive Berlin Academy Institute for Mathematics in 1946, the West German Science Council would most likely not have given East German mathematics such a positive evaluation in 1991. Moreover, if mathematics cultivates this "apolitical attitude" in unified Germany, where the battle for resources and public attention often demands political engagement, it will only hurt itself.

It goes without saying that in these situations we always demand responsible political behavior, the same demand we made of German mathematicians in the Nazi state and in the GDR. If, as has been seen in this chapter, the shadow of National Socialism fell on East German science, it is to be hoped that historiography, at least, will escape the dangers of simplistic and misleading comparison.

—Translated by Thomas Dunlap

4

Espionage and Technology Transfer in the Quest for Scientific-Technical Prowess

Kristie Macrakis

Unlike conventional political and military intelligence gathering and analysis, scientific, technical, and economic espionage in East Bloc countries sought primarily to enhance the economic and technological level of the country rather than to protect national security. Throughout East German history, scientific and technological espionage was seen as intrinsic to invigorating or supporting the country's often faltering economy. Communist Party slogans usually glorified the "scientific-technical revolution" and promoted technology as a productive force; the GDR aspired to be a technocracy. In MfS directives to the operational units, orders were given to better the economy by increasing scientific and technological espionage. At the intelligence school, new recruits were told that the task of scientific-technical intelligence was to "strengthen the economy."[1] Ultimately, the GDR sought to pull itself out of backwardness and reach the same level as the West scientifically and technologically. In the end, however, the GDR was unable to catch up to Western technology, let alone surpass it. As a former intelligence officer graphically put it, the goal of the espionage enterprise was to "make gold out of shit."[2]

This chapter will explore how espionage was used in East Germany's quest for scientific-technical prowess. Since a large part of scientific-technical espionage consisted of the acquisition of knowledge from the West and its transfer to the East, much of the chapter will deal with the inner workings of this covert technological transfer, including its history, organization, methods, and personnel. By definition, technology transfer is the physical transfer of technology over borders. Usually it does not include the assimilation of technology into the country itself. Although in East Germany a "switchboard" existed to transfer the knowledge and equipment from the espionage agency to scientists and industry, the MfS was not responsible for its assimilation into the indigenous infrastructure. Occasionally, however, reports were written by counterintelligence on the use made of MfS-acquired material and its benefits for research and development.[3]

An examination of science and technology through the lens of espionage will increase our general knowledge of the way the scientific enterprise operated in the GDR. The espionage units for science became an integral part of the scientific enterprise there, carrying through, supporting, or stimulating socialist party directives, serving the needs of scientists or industry, and keeping science under surveillance. Sometimes the MfS made independent contributions to science as the initiator of projects, but more important, it served as a conduit of knowledge from the West to the East. Most revealing is the way in which other state organs such as the Politburo and the myriad of science ministries cooperated with the MfS.

This chapter is organized into four parts: the first sets forth an argument about the relationship between the MfS, especially the scientific-technical units, and the Party and state. It presents the national-political reasons for the need for covert technology transfer as part of a quest for scientific-technical prowess and economic strength. The second part provides a description and history of the mechanism for technology transfer, namely, the major acquisition organs—the Commercial Coordination Unit (KoKo) of the Ministry for Foreign Trade and especially the Sector for Science and Technology (SWT). A major section then presents the way in which the SWT's evaluation unit functioned as a technology transfer switchboard. The Vienna Residency is showcased as a hub for technology transfer between East and West. The third part outlines the role of the agents and case officers as mediums for information transfer. The final part uses espionage in computer technology as an example of the GDR's illusionary quest for technological prowess. This example also illustrates the emergence of a "state security regime" and the cooperation between the MfS and other institutions in the GDR.

East Germany was a pivotal player in the acquisition of scientific knowledge and technical know-how from the West and its transfer to the rest of the East Bloc, in particular the Soviet Union. Espionage opportunities were easily facilitated in West Germany, with whom East Germany shared a common language and culture. Unlike East Germany, West Germany by the 1960s and 1970s had achieved international standards in science and technology, and this knowledge and equipment had the potential to strengthen East Germany's crumbling scientific edifice, or so hoped the East German regime. Although East Germany operated from a strategically advantageous venue and had built-in advantages, such as its shared language and culture with West Germany, it nonetheless faced many barriers to its scientific-technical development after World War II and its subsequent entry into the Soviet Bloc.

Some knowledge could be legally exchanged between East and West, but most technology could not be legally imported or transferred to the East Bloc because of the embargo (the Coordinating Committee, or COCOM, lists) and industrial restrictions. Secrecy prevented internal company information and military science and technology from a free-flowing information exchange. As a result of these limitations in knowledge exchange, espionage agencies created units for scientific-technical intelligence to meet the country's needs in a growing global technological community.

How It Worked

Sword and Shield

Like many twentieth-century espionage agencies, the MfS was a child of the Cold War, and therefore its mission and character were shaped by political goals. Created in 1950, soon after the 1949 founding of the GDR, it characterized itself as the "shield and sword" of the Party.[4] As the metaphor implies, the MfS was designed to protect the SED in a defensive way, as the leading power center, and to play an offensive role; the image evokes analogies to the military. The MfS was militaristic and bureaucratic in its organization, with departments, subdepartments, and groups led by officers given military ranks (such as colonel and general). Even MfS lingo was similar to that used in the military (for example, *Aufklärung* refers to reconnaissance, intelligence operations, and so on).

During the forty years of the GDR's existence, the MfS's task was to safeguard the Party's goals, and this was done by executing the SED's resolutions *(Beschlüße)*. Countless orders and directives open with the statement that certain actions need to be taken in support of an SED resolution at a party meeting or congress. For example, as will be seen below, in 1956 when the scientific-technical evaluations unit was founded, the MfS was responding to the SED's mandate during the Third Party Conference to attain international standards in all areas of the economy; the SED ordered quick developments in science and technology. The MfS responded to the Party's mandate by sending out a directive to increase greatly the volume of scientific-technical information available to East German scientists and industry. Documents from the 1980s also display the stimulus-response relationship between the SED and the MfS in the areas of hi-tech transfer—especially the Microelectronics Program—and particularly in programs developed to respond to Ronald Reagan's Strategic Defense Initiative (SDI).

In addition to responding to and supporting SED resolutions, the MfS also provided the SED leadership with information on problem areas. In order to protect the interests of the Party, the MfS often provided this information to party leaders prior to meetings and before they made their resolutions. The MfS often recognized the importance of certain fields, such as computer technology, before the Party did.

The Politburo member and economy czar Günter Mittag played a pivotal role in cooperation between the scientific-technological enterprise and the MfS. He not only received the MfS's information, but also actively worked together with the Ministry for Trade and its head of Commercial Coordination, Alexander Schalck-Golodkowski. The various science ministries, including the Ministry for Science and Technology and the Ministry for Microelectronics, also worked together with Mittag. For example, during the 1980s the MfS (both the foreign department and counterintelligence) worked closely with Schalck, Mittag, and the Ministry for Electronics acquiring Western embargoed computers.

Occasionally the MfS proposed its own projects, which did not always meet with the approval of the Party. For example, the MfS's Chief Intelligence Administration (the *Hauptverwaltung Aufklärung,* or HVA), its foreign intelligence unit, and specifically the Sector for Science and Technology, had gathered information from the West on the plastic polyurethane with the intention of using the information to build a polyurethane factory. Markus Wolf, the legendary head of the MfS's foreign department (HVA), and his staff were enthused about the advent of plastics, "a wonder material," recalled Wolf in 1995, and were interested in introducing this chemical technology into the GDR. The SWT had acquired the blueprints for the formula from one of the successor firms to I. G. Farben, the chemical company. The party leadership, however, was not interested, and it took some persuasion and understanding of "human weaknesses" to convince them of its importance. Apparently, the MfS bought Walter Ulbricht, then the head of state, a plastic couch on the occasion of his birthday! This gift was enough to convince him of the worthiness of the project. It was not, however, part of the state plan, and so counterintelligence, which thought this project would sabotage the economy, was against pursuing it.[5]

Major Acquisition Organs: KoKo, SWT, and Department XVIII

In East Germany there were primarily three channels for acquiring Western technology . The semilegal, and sometimes legal, way to import Western goods was through the Ministry for Foreign Trade and its Commercial Coordination Unit (KoKo), founded in 1966 and which was also connected with

Günter Mittag's economics department at the Central Committee; Schalck was the head of KoKo. Although the Ministry for Foreign Trade was not officially an espionage unit, its leaders were either full-time or unofficial staff members. Schalck was the thirteenth highest-paid MfS employee (the MfS had 85,000 full-time staff members) at the time of the collapse of the GDR, with an annual salary of 54,700 GDR marks.[6]

KoKo laundered hard currency to be used by the GDR and the MfS and founded dummy front companies to do business with Western firms. KoKo reportedly gave the Sector for Science and Technology 400 million West German marks in cash between 1987 and 1990. West German intelligence (*Bundesnachrichtendienst*, or BND) believes that more than 90 percent of GDR microelectronics technology was acquired illegally by Schalck.[7]

The second most important acquisition organ was the Sector for Science and Technology, housed in the HVA, the foreign intelligence unit of the MfS. While KoKo concentrated on acquiring equipment, especially computers, the SWT during the 1950s, 1960s, and early 1970s was engaged in typical espionage work—the acquisition of blueprints, plans, scientific information, and so on—via agents scattered throughout the West. By the late 1970s, with the introduction of the Microelectronics Program, the SWT was also heavily involved in "importing" computer technology through its Department XIV, established in the late 1970s for importing embargoed goods. The SWT also established front companies in order to transfer technology to the GDR through "Western" companies.

The SWT also performed other functions, such as surveying science and technology in West Germany. Its analyses included information on institutions as well as their personnel. By definition, the task of scientific-technical intelligence included the "acquisition and analysis of the scientific-technical achievements of the enemy." There was a constant surveillance of science in the hopes that the results might be used to benefit the military or society in general. The SWT also collected information regarding Western nuclear weaponry and military technology, hoping one day to anticipate the "surprise moment."[8]

The MfS counterintelligence unit for the "protection of the economy" (Department XVIII) also occasionally worked in the "operational area" of West Germany with the HVA and other state organs to acquire information from the West. Although the counterintelligence unit had a working group for travel cadre and foreign cadre (Reisekader and Auslandskader) since 1964, an independent group for the operational area was not founded until 1979. It became an independent department (XVIII/14) in 1980. Counterintelligence had some liaison officers and advisers from the Soviet Union and worked together with other East Bloc counterintelligence organizations, known as "brother organs."[9]

Cooperation

Since the fall of the Berlin Wall, the public has learned a lot about Alexander Schalck-Golodkowski's KoKo, in part because Schalck "defected" on December 6, 1989 (before East Germany was absorbed by the West), and told his story to a variety of Western intelligence and police agencies, including the criminal police and the BND. His defection and the information he shared made headline news. Even before the Wall came down, West German Intelligence knew about many of his activities and those of West German companies such as Siemens and Leybold, which were involved in selling embargoed goods to the GDR.[10] Schalck's activities ranged from gun-running to trade in antiquities to illegal technology transfer. The German Parliament undertook a massive investigation of Schalck's activities, culminating in a recommendation and report in May 1994.[11] Schalck had a close working relationship with the HVA and its leaders, Markus Wolf and Werner Großmann. He met with them at the beginning of each year to determine the year's activities and was an influential figure at the HVA. The HVA and KoKo cooperated primarily in economic and scientific-technical espionage—that is, with the HVA's Sector for Science and Technology. The SWT received funding from KoKo, used some of its business contacts in the West for operational purposes, and used the unit to camouflage its import activities, primarily through dummy front companies like Interport and Intertechna.[12]

Department XVIII also worked with KoKo and the SWT in transferring Western technology to the East. Although the department's chief function was security and counterintelligence and it ended up acting as the political police, it also took part in scientific-technical espionage. Just as the HVA/SWT had agents in the West who passed on secret information or embargoed goods, so too did Department XVIII. A recent estimate puts the number of informants, or IMs, for that department at 2,140 in 1989; Alfred Kleine, the last director of the department, reported that it had 15 agents in the West, primarily involved in acquiring embargoed goods. In addition, Department XVIII passed on information to the HVA and gave it tips on possible embargo contacts.[13] Department XVIII/8 for electronics played a major role in breaking the embargo, especially after Western intelligence increased its vigilance in the late 1980s. The head of XVIII/8, Artur Wenzel, worked closely with Schalck. After the fall of the Wall, a former department head in XVIII/8 delivered 96 diskettes to the BND, including dossiers on embargo dealers. The traditional delivery routes for hi-tech items via neutral countries like Switzerland, Sweden, and Austria began to be replaced with routes through Thailand.[14]

In addition to the security measures of a typical counterintelligence department, such as, for example, opening "Operational Files" (Operative Vorgänge, or OVs) and Operational Person Control files to place those suspected of having contact with Western intelligence agencies under surveillance, Department XVIII also cultivated and recruited high-ranking officials as IMs. For example, Wenzel placed IMs in strategic jobs and deliberately directed the professional development of several people, including Gerhardt Ronneberger (IMS Saale and head of Trade Area 4) and Wolfram Zahn (IMS Rolf and acting director of the Combine Microelectronics). Department XVIII even had a so-called MfS firm, Günther Forgber.[15]

As will be shown in the case study on microelectronics, the acquisition organs also worked closely with the Ministry for Electrotechnology and Electronics (MEE), which often funded expensive embargoed microelectronic equipment. In fact, the decision about which acquisition channel to use was made by a coordination group at the ministry consisting of Karl Nendel, the secretary of state, Siegfried Stöckert, a KoKo official (who was also an IM for the MfS), Ronneberger, and Zahn. At least since 1986, the system worked in the following way: industrial representatives and scientists filled out applications "for the support of research and development tasks," which seemed to originate with the SWT, requesting certain goods or plans. These requests were then passed on to the appropriate ministry or representative; thus the applicant did not always know the goods were to be acquired through the MfS or KoKo. The request was classified as fulfilling a state need such as a state plan topic, a defense project, an economic necessity, or an operational offer from the MfS (that is, something to which one of the operational departments had access). An operational department was then assigned the task of procuring what was needed.[16]

For the acquisition of microelectronic components, a list was given to the Ministry for Electronics. Nendel then discussed the list with KoKo and SWT and gave the requests for acquisition to Trade Area 4 of KoKo (see Figure 4.1).[17] Some embargoed goods were more difficult to acquire than others, and the so-called Special Acquisition Organ was created to handle these cases.

At least since the 1980s, the acquisition of embargoed Western technology was a top priority in fulfilling the state-sponsored microelectronics program. This marked a shift in espionage style from agents stealing blueprints to suppliers delivering equipment. This change also became quite expensive; indeed, Schalck's agency not only contributed to technology transfer

Figure 4.1 Acquisition lines for imports of embargoed goods (1986–1990)

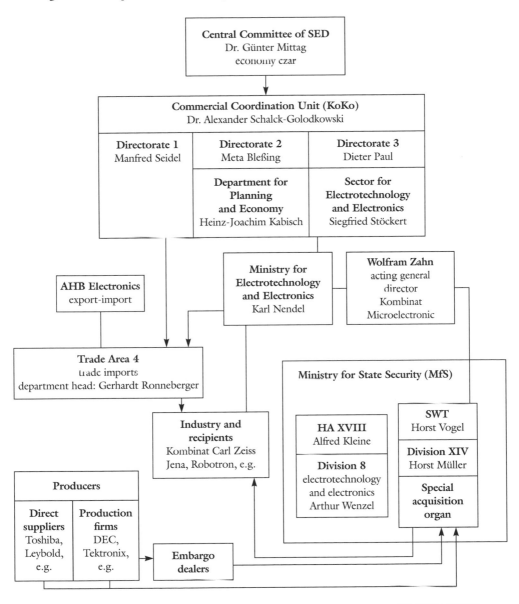

After "Beschlußempfehlung und Bericht des 1. Untersuchungsausschusses nach Artikel 44 des Grundgesetzes," *Deutscher Bundestag, 12. Wahlperiode,* Drucksache 12/7600, 1994, p. 254.

through trade contacts but also provided the state with a large amount of hard currency.

Some suppliers of embargoed goods—mostly businessmen—knew they were dealing with espionage personnel; others did not. Many deals were brokered through the numerous dummy front companies that were actually HVA or KoKo companies. Reportedly, both at home and abroad there were some 160 KoKo firms involved in Schalck's activities.[18] The goods were then funneled to the appropriate scientist or industry. Many of these companies were founded in the early 1960s, and West German Intelligence was told of their existence by defectors before the Wall came down.

Several of the important SWT firms included Interport, Industrial Representative, and Intertechna, GmbH (*Gesellschaft mit beschränkter Haftung,* a capitalist company that uses its own capital to cover its investments and is usually not found in communist countries!). They were founded on the initiative of the SWT in 1965 and 1969, but were bureaucratically placed under KoKo. They were located in East Berlin. Like many of the firms, the bureaucratic organization was complicated. For example, Intertechna was not legally owned by the MfS because it belonged evenly to VEB Kombinat Zentronik Sömmerda, VEB Kombinat Robotron, and the KoKo firm, Interver International Representative, GmbH. Horst Müller, the head of the SWT's Department XIV, managed the personnel issues and business activity, although an HVA informant, Herbert Brosch, was the director of the firm officially. In reality, this company was a pure import outfit used to acquire key technologies and microelectronics for the MfS and GDR industry. Its West German business partners included IBM Roece Inc., Vienna, and Siemens.[19]

By contrast, Interport was a pure HVA firm officially involved in the buying and selling of "old-timers," but its real activity was the acquisition of prototypes and technologies from modern agricultural equipment to microelectronics. Robotron was its chief beneficiary. Among other things, Interport took part in importing a hard disk for a Robotron combine branch office. Since Interport had no foreign trade rights, it was officially the KoKo firm Intrac.[20]

The Sector for Science and Technology: A Mechanism for Technology Transfer

HISTORICAL OVERVIEW The expansion of the Sector for Science and Technology (so renamed in 1971 from Department V) from a handful of staff members in the 1950s analyzing West German industry to a large (some 400 staff members by 1989) and respected scientific-technical espionage unit by the end of the GDR, reflects both the importance of science and technology for the GDR and the unbridled expansion of the MfS.

Heinrich Weiberg, who studied chemistry, is usually credited with having founded the sector when it was a department for scientific and technical intelligence in the Institute for Economic Research (*Institut für Wirtschaftswissenschaftliche Forschung,* or IWF), a cover name and organization for the Foreign Secret Service (*Aussennachrichtendienst,* or APN).[21] Weiberg, who was born on January 20, 1911, had served in the German *Wehrmacht* and was in Soviet captivity between 1940 and 1949. After a brief stint working as a chemist in a sulfuric acid factory (1949), he became the head of personnel in the Ministry for Heavy Industry (1949–1951) and finally the head of Department II in the HVA (1951–1954; he first entered the MfS on September 1, 1951). Between 1954 and 1959 he was deputy director of Department IV. On January 2, 1959, he became the head of Department V, the forerunner of the Sector for Science and Technology, located in the newly founded Department XV, as the HVA was called then.[22]

Although many of the details are still shrouded in mystery, it is generally agreed upon that the Soviet Union played a crucial role in the founding of this unit. The founding generation (including Weiberg) had spent time in the Soviet Union, and advisers were on hand from Karlshorst, the Soviet compound where the KGB had its offices. In fact, the Soviet Union had, and Russia still has, an acquisition organ—Directorate T—very similar in nature to the SWT.

The Soviet Directorate T was housed in the first directorate, which was responsible for foreign espionage. Its staff was highly educated and had been recruited from the best universities. At least 500 of its employees had a scientific or technical background. Directorate T worked together with Line X, which operated in embassies abroad—in legal residencies, trade missions, and consulates—and had as many as 20,000 spies who were involved with illegal high-technology transfer. Directorate T had and has subdepartments for the atomic industry, chemical industry, computers, airplanes, biology, bacteriology, weapons systems, and general questions of political-military strategy.[23]

In 1960 alone, Directorate T stole, bought, or smuggled 8,029 classified documents, blueprints, and schemas and 1,311 pieces of equipment from the West.[24] While East Germany's chief operational area was West Germany, the KGB's special target was the United States. The KGB was asked by the Central Committee to prepare a plan for working against the United States. The KGB submitted a report including plans to plant agents in American scientific-technical centers, universities, industrial corporations, missile-building companies, and electronics, aircraft, and chemistry installations.

The KGB planned to use other countries to pursue its espionage against the United States, including West Germany. Agents were instructed to find their way into scientific, industrial, and military institutions with direct connections to America or to American technology. It also planned to start "brokerage firms" in America, England, and France to obtain classified technical information and to buy state-of-the-art American equipment.[25]

In 1973 Directorate T apparently became one part of a system of *spetsinformatsiya* (special information) used to acquire militarily significant technology. According to information that a Soviet defector with the prophetic name "Farewell" supplied to French counterintelligence in the early 1980s, the acquisition efforts were directed by the Military-Industrial Commission (VPK), which coordinated defense research and development and production. The VPK set collecting tasks for five acquisition agencies: the KGB, the GRU (military intelligence), the State Committee for Science and Technology, a cover unit in the USSR Academy of Sciences, and the State Committee for External Economic Relations.[26]

Other East Bloc countries were a conduit for science and technology transfer to the Soviet Union. In 1980 alone, Eastern European intelligence agencies contributed 54.1 percent of the total information provided to the KGB's Directorate T. The average percentage over the years seems to have been 20.2 percent. Undoubtedly, East Germany contributed the largest share, owing to its operational advantages. Apparently 80 percent of the acquisition organs' material was open-source information and not secret. The KGB complained that its Berlin Residency tried to pass off American company reports as classified.[27]

Robert Rompe played a pivotal role in the relationship between the KGB and the MfS during the 1950s, especially in the area of applied physics. Rompe had already worked for the KGB in the Soviet Union and had forged a personal connection between the scientific community, especially the GDR Physical Society, and the MfS.[28] In fact, Rompe's industrial espionage activities probably date back to the Third Reich. While being interrogated by the Central Committee about his alleged connections to Noel Fields, Rompe admitted to being engaged in "illegal activity before and after his entry (1932) into the [German Communist] Party." During that time he had gathered "material on new work in the Berlin electrical industry. It was called BB *(Betriebs-Berichterstattung)*—company reporting . . . later I worked at Osram [a Berlin electrical company]." Another source says he began his work after joining the Party and gathered information from Osram in the area of electricity while socializing with a "circle of comrades who were working illegally."[29]

Rompe, who was born in Leningrad, was known as the GDR's "high priest of physics." He was part of the crème de la crème of East German science, a member of the circle of scientists who would attempt to rebuild East German science after the war. He was politically acceptable and was considered one of the "progressive scientists" by the central committee. He was also a member of the SED's Central Committee. He was useful and important to the MfS for both strategic questions and personnel issues. Strategically, he worked together with Weiberg, a close friend, on building up intelligence, and on the personnel level, he decided whom to cultivate and recruit. Because of his towering position in a country as small as the GDR, he exerted an enormous influence on training and placing students. According to the defector Werner Stiller, Rompe soon became one of the sector's most valuable agents and was personally handled by Colonel Willi Neumann, the deputy head of the sector.[30]

The most direct connection between the SWT and the KGB was the Soviet liaison officer who received information from the sector, often in the form of lists of the material collected ("accompanying material lists"). Much of the information was then sent to Moscow.

While the Soviet Union forged contacts in East Germany during the 1950s, the MfS started to build an informant network to infiltrate West German companies and began to establish methods of evaluation (see below). During the mid to late 1950s, espionage seemed to concentrate on industry, although atomic physics was also a strong focus.[31]

Stealing, buying, or smuggling scientific-technical embargoed goods had not yet become a major emphasis of GDR espionage. In the late 1950s Weiberg's Department V was created for scientific-technical intelligence in the HVA. In 1962 Paul Bilke's group for scientific-technical evaluation was incorporated into that department. Gradually, the operational areas 1, 2, and 3 emerged, covering physics and chemistry, computers and space, and military technology, respectively.

The MfS had already installed personnel for "cadre and security questions" at the Office for Technology and the Office for Atomic Research and Technology in February 1956 in order to "prevent enemy elements from penetrating important companies and to keep state secrets." Four months later, in June 1956, the MfS's own scientific-technical evaluation unit was created to work together with its counterpart in the Office for Technology.[32]

During the early 1960s there was a general expansion in tasks and resources for the science sector, and a special subdepartment (Referat) was created in Jena in 1962. Jena has a tradition of scientific excellence and was the home of Zeiss, a company world renowned for its excellence in optics. The Stasi recognized that Jena was an ideal location for the Referat because of the community of scientists affiliated with the university, institutes, and big companies.[33]

During the late 1960s Erich Mielke, the head of the whole Stasi, ordered increased coordination in the procurement of prototypes, models, and documentation in military technology, including conventional weapons, airplanes and missiles, military electronics, ABC weapons, defensive weapons, and military vehicles.[34] In 1969 thirty-four new positions were created in new subdepartments and the sector was reorganized to separate the collection of military technology from economic tasks. There was also a renewed emphasis on the collection of prototypes in military technology.[35] This change may have been a reaction to the Prague Spring uprising in 1968.

During a major reorganization in July 1971, institutional forms were solidified, and Department V became an independent Sector for Scientific-Technical Intelligence *(Aufklärung)* with three operational departments: XIII, XIV, and XV, as well as one devoted solely to evaluation: V. This change coincided with Erich Honecker's becoming head of state; given that Honecker had been responsible for overseeing state security in the Central Committee, it is not surprising that the security service expanded. Like Ulbricht, Honecker glorified science and technology and promoted the advancement of "key technologies" such as microelectronics. The expansion of the sector also reflected the same trend in the Soviet Union, where there was a renewed emphasis on the importance of science and technology. The 1970s saw a marked increase in scientific-technical espionage emanating from East Germany and the Soviet Union. Weiberg was now head of the sector and Wilhelm Neumann was his deputy. Horst Vogel, who became head of the entire sector in 1975, became head of Department XIII, Werner Witzel head of XIV, and Gerhard Franke head of Department XV. When Vogel became head of the whole sector, Gerhard Jauck succeeded Vogel as head of Department XIII.[36]

OPERATIONAL DEPARTMENTS   The three operational departments of the SWT housed different scientific, technical, and economic disciplines and had between 188 and 197 staff members (including operational officers, drivers, and secretaries) by the end of the GDR. The evaluation unit was proportionally larger and had some 113 officers, while the leaders of the SWT numbered some 74 people.[37] The sector was unique in that it had its own evaluation unit. Some material from the sector was also given to the HVA's evaluation department (Department VII) for analysis.

Department XIII was responsible for both basic and applied research in atomic physics, chemistry, medicine, biology, genetic technology, and agriculture (see Figure 4.2). Before the Wall fell, the West knew more about Department XIII than about any other department because of Stiller's defection. He provided details of the department's work in the area of atomic physics and technology, exposing key agents at the Karlsruhe Nuclear

Figure 4.2   Sector for Science and Technology: Operational departments

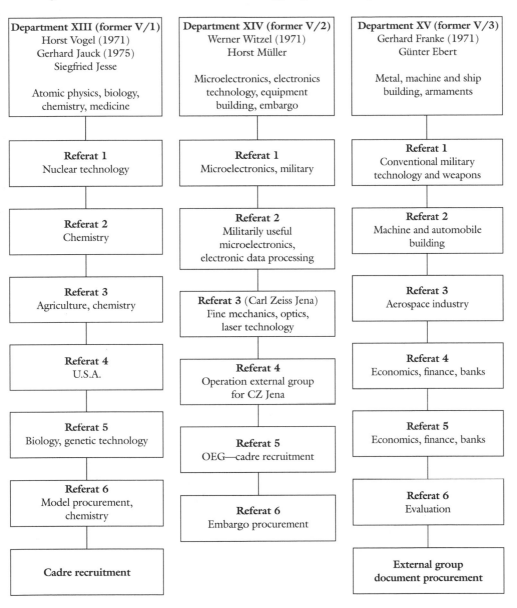

Department XIII (former V/1) Horst Vogel (1971) Gerhard Jauck (1975) Siegfried Jesse Atomic physics, biology, chemistry, medicine	**Department XIV (former V/2)** Werner Witzel (1971) Horst Müller Microelectronics, electronics technology, equipment building, embargo	**Department XV (former V/3)** Gerhard Franke (1971) Günter Ebert Metal, machine and ship building, armaments
Referat 1 Nuclear technology	**Referat 1** Microelectronics, military	**Referat 1** Conventional military technology and weapons
Referat 2 Chemistry	**Referat 2** Militarily useful microelectronics, electronic data processing	**Referat 2** Machine and automobile building
Referat 3 Agriculture, chemistry	**Referat 3** (Carl Zeiss Jena) Fine mechanics, optics, laser technology	**Referat 3** Aerospace industry
Referat 4 U.S.A.	**Referat 4** Operation external group for CZ Jena	**Referat 4** Economics, finance, banks
Referat 5 Biology, genetic technology	**Referat 5** OEG—cadre recruitment	**Referat 5** Economics, finance, banks
Referat 6 Model procurement, chemistry	**Referat 6** Embargo procurement	**Referat 6** Evaluation
Cadre recruitment		**External group document procurement**

Compiled from Cadre Card files, MfS structure, BStU; Rainer O. M. Engberding, *Spionageziel Wirtschaft: Technologie zum Nulltarif* (Düsseldorf, 1993), p. 86, and Peter Siebenmorgen, *"Staatssicherheit" der DDR: Der Westen im Fadenkreuz der Stasi* (Bonn: Bouvier, 1993), pp. 326–327.

Research Center and at IBM. He also showed how institutions like the Physical Society in East Germany and the Karlsruhe Center were penetrated by the MfS. The former was interlaced with MfS people and the latter was an "object" about which information was collected. In addition, the information collected from Karlsruhe was passed on directly to the Soviet Union. Not only did the Soviet Union want to know if West Germany was building nuclear weapons, but the knowledge was also useful to its nuclear power industry.

Department XIV worked primarily in the area of computers both for military and for civilian use. The department worked closely with the company Carl Zeiss Jena in computers, optics, and laser research.

Department XV was primarily responsible for military, space and aviation, and machine and vehicle-building technology, and therefore acquired information with the most practical applications. This department was also involved in economic espionage in banks and economic institutes. It becomes clear from agents caught before and after the Wall fell that Department XV had sources at Messerschmitt-Bölkow-Blohm (MBB), West Germany's most important armaments firm, and had acquired the complete set of blueprints for West Germany's tank *Leopard 2* and fighter plane *Tornado*. The military technology was passed on to the Soviet Union.

In addition to the operational departments, several working groups were formed. The largest and most important of these was Work Group 3 (headed by Erich Gaida), founded in 1978 to acquire military technology and equipment. The evaluation unit, to be discussed below, was now named Department V and continued to be headed by Paul Bilke until 1976, when Harry Hermann was appointed director.[38]

By the end of the GDR there were approximately three hundred officers in the SWT, up from some thirty-five officers in the 1950s.[39] By this time more of the staff had had university training in science or technology. During the early years, officers' backgrounds were mixed. Whereas Weiberg had a scientific background, Neumann had been a policeman. During the 1970s there was a drive to recruit officers with the *Diplom* or even a Ph.D. Before this drive, Referat 1 of Department XIII, for example, consisted of a book printer, someone from a guard regiment, a wood engineer, a long-time employee of internal security, a Marxist-Leninist graduate, and a mechanical engineer.[40]

The MfS usually recruited stalwart communists or, later, people from working-class or farming families. Throughout the GDR period, membership in both the SED and the Free German Youth (*Freie Deutsche Jugend,* or FDJ) was a prerequisite for employment in the MfS. This was all part of the GDR's "affirmative action" program.

Horst Vogel, who became the head of the sector in 1975 and greatly influenced its direction and development, did not initially have training in science or technology but later went on to receive a degree in engineering. Vogel, who was born in Theissen (Saxony) on May 11, 1931, came from a working-class family and first learned a trade as a locksmith; he attended the university between 1952 and 1955 without earning a degree. He joined the MfS on March 1, 1955 when he was twenty-four, and attended the HVA school during the academic year 1955–56. He returned to the university— the Technical University of Chemistry in Leuna-Merseburg—where he studied chemical economics from May 1, 1968, until August 31, 1970, when he received a *Diplom* in engineering. After returning from his stint in Merseburg, he was a subdepartment head in HVA/V, as the SWT was called then, and quickly rose in the ranks, becoming deputy director of the HVA in 1987. One former subordinate and defector has characterized him as an "espionage man in heart and soul" and an intimidating leader.[41]

In 1979 a crisis occurred in the sector when a first lieutenant from Department XIII, Werner Stiller, defected to the West. He brought with him secret documents and a vast knowledge of the history, structure, organization, and operational activity of his department, as well as a general knowledge of the sector and the foreign intelligence units. Until this point Western intelligence had known very little about the sector; indeed, they had known only that economic espionage had increased during the 1970s. Stiller's defection was a real coup for West German intelligence, and it was celebrated for many years after. Stiller's betrayal of his own agents not only led to their arrests, but also provided the West Germans with clues to pursue other agents planted in West German companies and research institutes. Within months seventeen agents had been arrested. The Stiller defection also called renewed attention to the extent to which East Germany had penetrated West German governmental agencies and institutions for science and technology.

To throw even more salt in the Stasi's wounds, Stiller, together with the BND, published his memoirs in 1986. In this autobiographical book, he describes how he recruited and ran agents. The book also provides a detailed account of the personalities in the sector, its history, its tasks, and its successes. The operational details are fascinating. For the first time, the names of his former staff members were exposed along with their activities. Most of the material on the science sector is true, but information concerning the length of Stiller's cooperation with the BND is inaccurate. In the book Stiller claims that he was a double agent for at least seven years when, in fact, he cooperated with the BND for a short eight months in order to defect to the West.[42] As a result of this disinformation, the Stasi spent a lot of time after his defection tracing back any and all leads for years. All his former internal agents were interrogated, and the MfS collected 295 pages of newspaper articles from the

West on the Stiller coup. To characterize the extent of his evil deeds, Stiller's code name in the counterintelligence files was "Jackal."[43]

By the early 1980s, another hurdle had been placed in front of the MfS as the CIA began "Operation Exodus" in an attempt to reduce the flow of illegal imports to East Germany, other East Bloc countries, and the Soviet Union. As one intelligence officer put it, "It became increasingly difficult to bypass the embargo." As a result of this attempt to stop embargo smuggling, the SWT set up a new group headed by Horst Vogel in 1987 to increase efforts at overcoming this obstacle.[44] This came at a time when the GDR launched its ambitious microelectronics program. Nevertheless, the SWT continued to be quite successful at establishing new acquisition lines.

A Technology Transfer Switchboard

One of the most important departments for passing on scientific and technical information from the operational units to scientists and industry was the evaluation unit (Department V; see Figure 4.3). It was here that the collection of material was coordinated, evaluated, and the sources made anonymous. During the mid-1950s, the evaluation unit emerged as a technology transfer switchboard between industry and scientists, on the one hand, and the MfS, on the other. It seemed to make up the core of the outfit, playing a more important role than the nascent operational departments. At that time, the evaluation unit used general operational departments of the HVA instead of relying on its own departments, as it did later. By the end of the 1950s the SWT had about thirty-five members, of whom about twenty were affiliated with the operational departments.[45]

The evaluation unit's organization and responsibilities remained fairly uniform throughout the GDR's history, and were consistent with Mielke's goals when he ordered the creation of a working group for scientific-technical evaluation in 1956. The impetus behind the group was the mandate during the Third Party Conference that the GDR attain international levels in all areas of the economy; the SED thought quick developments in science and technology would help to achieve this goal. The Party's slogan was "Modernization, Mechanization, Automation." The five-year plan was seen as the beginning of a new industrial revolution brought about by the use of nuclear energy, heavy industry, and the further development of technical progress.[46] The Ministry for State Security responded to the Party's goals by sending out a directive to increase greatly the volume of scientific-technical information and documents available to East German scientists and industry and to support the SED's goals. Paul Bilke was appointed the director of the group.[47]

Unlike traditional Western espionage, the evaluation of material in the

Figure 4.3 Sector for Science and Technology: Department V

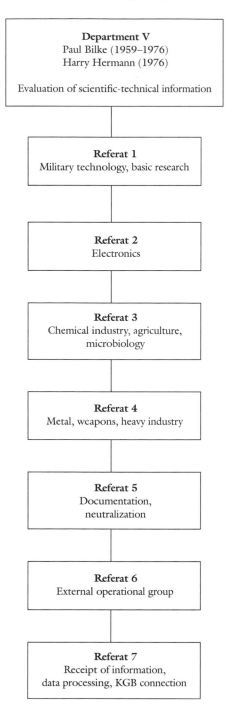

Department V
Paul Bilke (1959–1976)
Harry Hermann (1976)

Evaluation of scientific-technical information

Referat 1
Military technology, basic research

Referat 2
Electronics

Referat 3
Chemical industry, agriculture,
microbiology

Referat 4
Metal, weapons, heavy industry

Referat 5
Documentation,
neutralization

Referat 6
External operational group

Referat 7
Receipt of information,
data processing, KGB connection

Compiled from Cadre Card files, MfS structure, BStU; Rainer O. M. Engberding, *Spionageziel Wirtschaft: Technologie zum Nulltarif* (Düsseldorf, 1993), p. 86, and Peter Siebenmorgen *"Staatssicherheit" der DDR: Der Westen im Fadenkreuz der Stasi* (Bonn: Bouvier, 1993), pp. 326–327.

GDR was centrally organized, and the needs of industry were funneled through the MfS's evaluation unit. All scientific-technical material that reached the MfS and its district branches was sent to the working group—later an independent department—for evaluation. The group's first priority was to collect all material—including prototypes, blueprints, scientific papers, and so on—available in the area of science and technology, "neutralize" (the MfS's word) so that the source would not be known, and pass the information on to industry or research institutes. Every precaution was taken to secure the protection of the sources; "strict source protection" was often written in block letters at the top of documents. The group was sensitive to the economic focus of industry, scientific developments in the GDR, and the army's need for equipment.[48]

The staff members of the evaluation unit, who were all expected to be knowledgeable in the area of engineering and technology, had constant contact with the various science, technology, and industry ministries. This close relationship was supposed to help industry use the material effectively; industry in turn was encouraged to give the MfS "concrete tips" on further material to collect. Informants were recruited from the ranks of the "technical intelligentsia" in order to help evaluate the material.[49]

Evaluators were also recruited from the ranks of scientists and engineers, who were not told that they were evaluating stolen or covertly acquired material. The material was usually passed on from the ministries, which served as mediators between the MfS and the scientists and industry. For example, the MfS had personnel at the Ministry for Science and Technology who passed the information on to scientists.[50]

Each staff member was required to keep up with the literature, attend lectures in his or her field, and keep abreast of scientists' research plans. In this way the staff members would continually expand their knowledge and have an overview of the state of technological developments in the GDR. This then helped the evaluation unit create assignments for the informants or agents (IMs), who passed the material on to the operational units through their case officers. Theoretically, the evaluation group at first had an overview of all the operational unit's possibilities in procuring scientific and technological material, though later on there was much greater compartmentalization.[51]

After receiving the material, the staff members of the evaluation unit had to give the operational department an estimation of its value. Each document received a grade ranging from I, the highest, to V, the lowest. In the mid-1970s an evaluation of "I" meant the document or instrument had a minimum value of 150,000 West German marks. If the GDR were to develop the research or buy the product itself, it would have to pay that amount in hard currency. At the regional office in Leipzig, the SWT drew in 32 percent of all information for that office, and half of the information received marks of I or II.[52]

Since it was relatively easy for the SWT to quantify the worth of the information and material collected, it was considered one of the most successful units in the HVA. Reportedly, for example, the SWT strengthened the economy by an order of 150 million hard currency marks, with an investment of only 2.5 million marks. During the early 1980s, an average of 3,000 pieces of information had been acquired by the SWT with an input of 1.2 million DMs and 400,000 East German marks, while the economic usefulness for the GDR had been about 300 million East German marks.[53] The GDR did indeed save on research and development costs, though the main motivation for gaining information was to attain "world stature" in science and technology. In the end, in most fields, the GDR was just struggling to keep up.

Vienna as a Hub of Technology Transfer

During the 1950s, Vienna, Austria, which in 1955 signed a special treaty of neutrality, emerged as a crossroads between East and West and became a center of international intrigue and espionage (see Figure 4.4). It also became a conduit for the flow of Western technology to the Eastern Bloc.[54]

In order to bypass the embargo, the SWT began to use neutral countries as intermediaries for accessing high technology. Austria was an especially successful place for this activity in the 1960s, 1970s, and 1980s for East Germany, the Soviet Union, and other East Bloc countries. The GDR often imported through Austria, Switzerland, and Sweden scientific equipment that had its origins in countries that forbade the export of embargoed goods to East Bloc countries. For example, in the late 1960s, the GDR was interested in importing equipment to help its polyester industry. East German industry wished to have a whole facility delivered from Uhde-Hoechst in West Germany. The Stasi assumed they would have to get this material via Austria, because it "presumably was on the embargo list."[55] Austria was used by the SWT both as a neutral country to export materials into the GDR and as a place to establish an illegal residency.

Establishing both legal and illegal residencies is a major activity of most espionage agencies. The legal residencies are usually at embassies, whereas the illegal residencies can appear in various sectors. In the case of the "Vienna Residency," hi-tech companies and firms were used as front companies for activities directed by the SWT.

One of the most important residencies for the SWT, and for technology transfer, was in Vienna. Activities at the residency began in 1971, when an SWT officer, Horst Müller, recruited Rudi Wein, an Austrian businessman

Figure 4.4   The microelectronics connection: Vienna

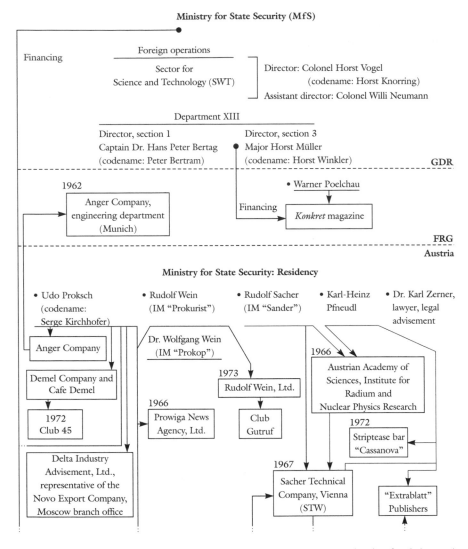

Ministry for State Security (MfS)

Financing

Foreign operations

Sector for
Science and Technology (SWT)

Director: Colonel Horst Vogel
       (codename: Horst Knorring)
Assistant director: Colonel Willi Neumann

Department XIII

Director, section 1
Captain Dr. Hans Peter Bertag
(codename: Peter Bertram)

Director, section 3
Major Horst Müller
(codename: Horst Winkler)

GDR

1962

Anger Company,
engineering department
(Munich)

Financing

• Warner Poelchau

Konkret magazine

FRG
Austria

Ministry for State Security: Residency

• Udo Proksch
(codename:
Serge Kirchhofer)

• Rudolf Wein
(IM "Prokurist")

• Rudolf Sacher
(IM "Sander")

• Karl-Heinz
Pfneudl

• Dr. Karl Zerner,
lawyer, legal
advisement

Anger Company

Dr. Wolfgang Wein
(IM "Prokop")

Demel Company and
Cafe Demel

1973

Rudolf Wein, Ltd.

1966

Austrian Academy of
Sciences, Institute for
Radium and
Nuclear Physics Research

1972
Club 45

1966

Prowiga News
Agency, Ltd.

Club
Gutruf

1972

Striptease bar
"Cassanova"

Delta Industry
Advisement, Ltd.,
representative of the
Novo Export Company,
Moscow branch office

1967

Sacher Technical
Company, Vienna
(STW)

"Extrablatt"
Publishers

(continued on facing page)

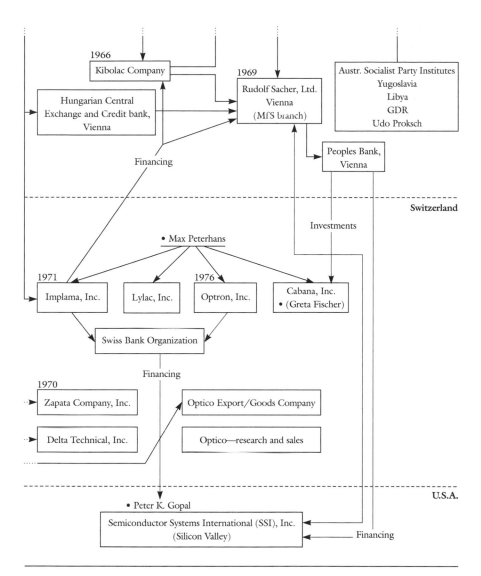

1966
Kibolac Company

Hungarian Central
Exchange and Credit bank,
Vienna

1969
Rudolf Sacher, Ltd.
Vienna
(MfS branch)

Austr. Socialist Party Institutes
Yugoslavia
Libya
GDR
Udo Proksch

Financing

Peoples Bank,
Vienna

-------- Switzerland

Investments

• Max Peterhans

1971
Implama, Inc.

Lylac, Inc.

1976
Optron, Inc.

Cabana, Inc.
• (Greta Fischer)

Swiss Bank Organization

Financing

1970
Zapata Company, Inc.

Optico Export/Goods Company

Delta Technical, Inc.

Optico—research and sales

-------- U.S.A.

• Peter K. Gopal
Semiconductor Systems International (SSI), Inc.
(Silicon Valley)

Financing

Courtesy of Bernhard Priesemuth

sympathetic to communism. Gradually, as more and more agents were added, Vienna became an intelligence colony of the GDR. Front firms were established with strong connections to the GDR; many of them banked in Switzerland. The majority of the information passed on to these front firms originated in Silicon Valley in America.[56]

Rudi Wein was a good friend of Udo Proksch, who has been portrayed as a shady character and has been implicated in the sinking of the ship *Lucona* in the Indian Ocean. Together, Wein and Proksch set up a high technology firm called Kibolac. By the end of the 1960s Kibolac had successful business and trade contracts with East Germany. Many other firms involved in illegal technology transfer were also set up, including Rudolf Sacher, Inc., and Lylac.[57]

At the end of 1968, even before the Vienna Residency was established, Rudi Wein, Rudolf Sacher, and Karl-Heinz Pneudl, a physicist co-opted into working with the two men, went to East Germany to meet with their handlers. They were greeted by Horst Winkler, who introduced himself as an officer of the Ministry for Science and Technology, a common cover for SWT personnel. Winkler was in reality Horst Müller, who at that time worked for the SWT Department XIII. He received them warmly, arranging their hotel accommodations and visas and setting up numerous business appointments. At the time, they did not know they were dealing with intelligence personnel.[58]

The contract with Winkler and Müller led to a series of successes for Sacher, Wein, and Pneudl. In 1968 they received a patent from the East German patent office for an electronically controlled heating system modeled after a West German prototype. For this work the Kibolac firm received a fee of 1.2 million shillings.[59]

By 1971, a formal contract was signed in East Berlin by Rudi Wein in the presence of his lawyer, Karl Zerner, and the heads of the SWT, Horst Vogel (cover name Horst Knorring) and Willi Neumann (cover name Willi Peter).[60]

In 1972 Müller took a posting at the East German Embassy in Berne, Switzerland, under the cover of first diplomatic secretary. Dr. Peter Bertag, the head of the American unit in Department XIII, replaced Müller as operational officer for the Vienna Residency. When Stiller defected in January 1979, he blew the cover of the Vienna Residency. Stiller was not involved in running the residency, but he learned about it by reading the files of Peter Bertag, his officemate, upside down across his desk.[61]

A Conduit to the Soviet Union: Carl Zeiss Jena

Well before the birth of the GDR, Carl Zeiss Jena enjoyed a long tradition of excellence in Germany and was known around the world for its outstanding optical equipment. During GDR times much of its tradition was preserved, and it soon became one of the country's most successful companies. As early as the Occupation the Soviet Union recognized the importance of CZ Jena and began exploiting its scientists and their work for its own purposes. As became common in the relationship between the Soviet Union and the GDR, many agreements were signed for scientific-technical cooperation. For example, in 1981 the GDR signed an agreement to take part in military ship building with the Soviet Union. Carl Zeiss Jena took a leading role in this project, and the cooperative agreement led to the Politburo Resolution of May 1983 to build CZ Jena into a center of excellence in research, development, production, and export profile, while establishing basic technology in microoptic electronics. Chief areas the GDR aspired to focus on were laser technology, optoelectronics, and infrared technology. Although the Politburo Resolution sounds general and just briefly mentions "special production," in fact most of the work involved a secret military technology project code-named "016" to manufacture an infrared seeker head for ship missiles (Project 152).[62]

Soon after the announcement of Ronald Reagan's Strategic Defense Initiative (SDI) in 1983, the Soviet Union and the GDR launched a highly secret "anti-SDI" program, code-named "Heide" in 1985. There is no public mention of this program in SED resolutions, and I have only discovered its existence in the MfS files. In 1984, the MfS sent out two orders relating to military technology: one involved the HVA and the acquisition of military technology, and the other involved internal security (primarily Department XVIII) to protect the security of research, development, and production of modern, strategically important weapons systems. The latter was code-named "Precision," and both were part of the "anti-SDI" program.[63]

Although the GDR's production of military instruments was small compared with that of other Warsaw Pact countries, there was an increase in R & D programs in the 1980s. CZ Jena took a leading role in the area of optoelectronics, optics, fine mechanics, and microelectronics; the emphasis for military technology was on instruments involved with navigation, position finders, and seeking instruments, and the most important programs included the multispectral camera, the laser distance measurer, and lithography instruments for microelectronics. By 1988, the infrared seeker head program for Soviet missiles was halted because the Soviet Union did not need them anymore, and CZ Jena planned to concentrate more intensely on microelectronics.[64]

The SWT's Department XIV worked closely with Carl Zeiss Jena in computers, optics, and laser research. Jena used the SWT mostly to acquire technology related to microelectronics and military technology with an emphasis on lasers. In an interview with the author, Markus Wolf called Jena "a child of the HVA." Some paper evidence from the foreign department survives to substantiate this claim, and enough other evidence, including the cooperation between Department XVIII and Jena, and the fact that a branch office was set up in 1962, exists to make it believable. For example, the MfS strongly supported Carl Zeiss Jena after the Politburo Resolution in 1983 to build Jena into a center of excellence. The foreign department provided Carl Zeiss Jena with embargoed microelectronic components, computers, and laser technology. It also supported the secret military technology Project 016.[65]

Between 1977 and 1989, CZ Jena received goods and blueprints totaling more than 8 million DMs from the SWT. Judging from the orders put into the SWT, CZ Jena could have received more than 14 million DM's worth of goods if all the acquisition requests had been realized. Between July 1977 and January 1990, Carl Zeiss Jena placed 164 orders for blueprints or equipment.[66] CZ Jena's applications tended toward microelectronic components and laser-related items or equipment needed for Project 016. For example, many requests were submitted by Wolfgang Biermann, general director of the combine, for prototypes of a 1-megabit chip, and by 1989, a 4-megabit chip prototype. Occasionally, CZ Jena (along with other firms, academy institutes, and university departments) even specified which company had the equipment. For example, Carl Zeiss Jena applied (indirectly) to the MfS for a laser system from the American firm Line Light Laser and computer components from Hewlett Packard in the United States.[67]

The importance of CZ Jena to the state is further underlined by another Politburo Resolution for R & D at CZ Jena in 1985 and 1986 and a visit Erich Honecker made there in May 1986. At Jena, Honecker was interested in discussing and visiting areas involved in military technology, microelectronics, and light-wave conductor technology. On the occasion of Honecker's visit, the MfS prepared a list of items it had provided to CZ Jena as part of its "operational support" of the combine in the area of military technology and microelectronics in a jointly written report by SWT V (the evaluation department) and XVIII/8. The MfS had provided the most help to the project on intelligent seeker heads for ocean guided missiles. Scientists at the combine estimated that research information from the United States covered approximately 95 percent of its own basic research, and therefore they could develop their own systems much more quickly. The MfS listed about seven items it had procured for the scientists, including an infrared window for the seeker head (called a DOM), a heat detector, and a miniature CO_2 wave conductor laser for a laser radar. In the area of microelectronics the MfS had provided the plans and technological know-how for the 64-kilobit chip.[68]

A subdepartment (Referat 3-headed by Reichmuth), or "branch office," of Department XIV was stationed at the Combine Carl Zeiss Jena in order to service the firm directly with information and material. Not only that, but CZ Jena also had its own evaluation group stationed at the branch office. Zeiss Jena was probably the most successful and internationally known company in East Germany. The MfS's early recognition of the operational potential of Jena was fully realized by the 1980s. Wolfgang Biermann, head of Zeiss Jena and a computer specialist, had close ties to the SWT and to the Politburo and profited immensely from covertly acquired computer materials. Biermann was instrumental in working on Erich Honecker's dream project—the development of the 1-megabit chip. (The HVA had, in fact, acquired the chip from the West.)

Of Agents, Case Officers, and Safe Houses

Most of the romance and adventure associated with espionage relates to the activities of the agents in the field. While James Bondean adventures in divided Germany were not the stuff of everyday routine, the context led to East Bloc–style adventures. More common than chases in British sports cars equipped with hi-tech gadgetry or dangerous diving adventures in the Caribbean were slow train rides across the border, meeting intelligence officers in Eastern European capitals, and depositing material in train station lockers in East Berlin.

Although the term "agent" is commonly used in the West to refer to people who were run by case officers at headquarters to collect material at home or abroad, the term is noticeably absent from East German MfS files and the language of intelligence officers. Rather, the term used is *inoffizieller Mitarbeiter* (informal staff member), or IM for short. There were domestic and foreign IMs who acted as both informants and agents. Running IMs was considered to be the chief method for effective espionage, together with answering the political-operational question "Who is who."

At the HVA school students were told that "the most important weapons are IMs in the operational area" (chiefly West Germany, but all NATO countries, the United States, and China).[69] To answer the question "Who is who," information was used from informant reports and compiled in a dossier. These dossiers could in turn be used to blackmail people, among other things.

Just as the term "agent" was not the stuff of bureaucratic jargon in the Ministry for State Security, so too is the term "case officer" absent from the files (at least until the 1980s). People who worked in the operational divisions of the SWT were termed "operational staff" (*operative Mitarbeiter*) or "officials" *(Sachbearbeiter)*. Operational work included running agents, recruiting in the "operational area," that is, West Germany, setting up recruiters in the FRG and the GDR, recruiting on the basis of politics and ideology, and recruiting under foreign flags. It also included work with prospective-IMs (possible agents), infiltration abroad, cooperation with sources, migrations from East to West Germany, residency work, directions for IMs should they be confronted by enemy police, and instructions on how to make contacts. Other work included the analysis of objects (for example, the Karlsruhe Center for Nuclear Physics or Rheinmetal), group analysis of people, analysis of a place or country, observation, and investigations.[70]

The HVA and SWT had thousands of agents in West Germany who were either West German nationals who agreed to become "West IMs" or East Germans who migrated to West Germany under the orders of the Stasi. West German prosecutors were investigating two hundred West German agents in the wake of unification. These agents often worked for West German firms, research centers, or universities and provided material to the SWT. Thus they were agents of technology transfer from the West to the East.

The 1950s was a time of heavy agent recruitment, especially at the universities. Students at East German universities were recruited on the basis of their sympathy for the regime, and then pursued their studies at West German universities. Later many of them became important in their fields and provided the MfS with useful intelligence information.

One of these celebrated "agents for the future" recruited in the 1950s was the "atom spy" Harold Gottfried (codename "Gärtner," or "Garden"), who was caught in 1968. Gottfried always maintained, and still maintains, that he spied out of conviction for communism; he received only fifty marks a month for his work. When asked in an interview if he would have told East Berlin had he found out that West Germany had a nuclear bomb, he responded with an emphatic yes.[71]

Gottfried was eventually stationed in the Karlsruhe Nuclear Research Center, where he worked between 1963 and 1968 as an engineer in the group for technical security for reactors and was a member of the company council. He forwarded all material that passed through his hands to the MfS through dead-letter drops in a bunker in Karlsruhe or through a courier; sometimes he even stashed the material behind mirrors in the bathroom of interzone trains. Especially noteworthy was the material he collected from German and foreign sources on the development and building of the fast breeder reactor. When he was discovered, West German police found a minox camera, film, a radio for secret transmissions, and tape recorders, among other incriminating spy paraphernalia, in his bathroom. Gottfried sent hundreds of microfilms to East Berlin.[72] The MfS continued to place other agents in Karlsruhe after

Gottfried's cover was blown by a double agent named Alois and the CIA. Penetrating the Nuclear Research Center in Karlsruhe was a top priority for the MfS.

During the 1950s many agents spied out of conviction. Some were later motivated by a desire to bring the two Germanies together; thus they refused to recognize the political division and associated policies such as the embargo. By the 1970s and 1980s, however, more and more cases emerge where the agent spied for the money. For example, Peter Köhler earned approximately 1.1 million West German marks for his espionage activity in computer technology between 1978 and 1989. Köhler, a West German who worked at Texas Instruments, delivered information on the development and manufacture of semiconductor technology to case officers from the SWT's Department XIV. He may have been East Germany's best-paid agent.[73]

The case of Gerlinde, Dieter, and Kerstin Feuerstein is of particular interest. Gerlinde Feuerstein worked for the MfS beginning at least as early as 1960 in the area of aviation. Around 1973 she brought her son Dieter into the business. In 1984 Dieter began to work at MBB as a system engineer in airplane construction and was supervised by the case officer Kurt Thiemann (Department XV). Dieter's ideologically motivated wife Kerstin was also brought in around 1976–1977. Dieter had access to secret NATO material between 1985 and 1989, including documents on the tank *Jäger 90*. He also delivered material relating to the fighter plane *MRCA-Tornado*. Together Dieter and his wife filmed hundreds of pages of secret material and passed it on to East Berlin hidden on interzone trains. After 1985 they met with an instructor, the East IM Uwe Albrecht, five times a year in Italy, Austria, and Eschborn. In addition, they met regularly with the case officer around New Year's in East Berlin. After Dieter delivered his material to Thiemann at the end of 1989, he expressed the wish to continue his work with the KGB. Instead, he was caught in 1990.[74]

Before the fall of the Wall and the dissolution of the MfS, most of the West's knowledge about the Sector for Science and Technology had been pieced together from Stiller defection material and the interrogation and sentencing of agents who had been caught. Given that most of the SWT operational files disappeared soon after the Wall collapsed, these agents continue to be a good source of information about the SWT. Exposed agents can provide information on where the GDR had placed agents, what kind of scientific-technical material was collected, the methods of meeting, the extent of the cooperation, the recruitment methods of agents, and the MfS's relation to the Soviet Union. Western intelligence also benefited from the apparent "defection" of an SWT officer in 1990 (the case was not made public because it would have caused a great deal of unrest among other officers if they had known of a traitor in their ranks). Other officers have given tips to the Federal Criminal Office or have been cross-examined as witnesses at agent trials. A card file including the cover names and superior officers of all Western agents seems to have fallen into the hands of the CIA, and West German officials have had to go to Langley to take notes on them for their legal investigations.

West German intelligence has identified at least 274 agents run by the SWT in West Germany since the fall of the Wall in areas ranging from computer technology, chemistry, atomic physics, the air and space industry, military technology, machine building and tank development, and even banking and finance.[75]

After the Stiller defection in 1979, the West had a good deal of knowledge about Department XIII, Referat 1, for atomic physics. In addition, given that more than half of all captures by Western agencies since the mid-1960s had been in the area of computers, the West knew of this "hi-tech" espionage and had instituted stronger counterintelligence operations.[76] After the fall of the Berlin Wall, legal investigations were made into all SWT officers and more agents were caught, providing the West with a more complete understanding of Departments XIV and XV as well.

The Quest for Technical Prowess in Computer Technology

Espionage in the area of computers is probably the most well known aspect of Soviet and East Bloc hi-tech espionage. By the late 1970s and early 1980s stories about agents caught and machines smuggled littered the pages of the *Wall Street Journal,* the *New York Times,* and other leading newspapers. Several books on the topic came out in the mid-1980s. From the early 1960s until the late 1980s, half the agents caught in scientific-technical espionage were working in the area of computers. The number of agents in this area reflects the extent of the effort. As the Soviet electronics industry failed to take off in the 1960s, it increasingly turned to smuggling computers, especially IBM models, to copy and manufacture in the Soviet Union. By the late 1970s and 1980s, breaking the embargo in the area of computers was one of the highest priorities for acquisition agencies.[77]

In his 1986 book *High-Tech Espionage,* Jay Tuck examined half a dozen cases of large international rings of professional smugglers who shipped electronics equipment to the Soviet Union. The rings spanned continents, reaching from Silicon Valley and Hong Kong to South Africa and West Germany. Several

West Germans were involved in these rings, including "Megabucks Müller" and Werner Bruchhausen. While Tuck mostly examined cases of illegal technology transfer to the Soviet Union through seemingly innocuous business deals, the role of other East Bloc countries in this transfer, especially East Germany, becomes increasingly clear.[78]

The Soviet Union had already begun to copy the IBM System 360 with the assistance of East Germany's Robotron Computer Center in Dresden in 1968. A top-secret team working in Moscow took apart stolen IBM machines and reverse-engineered them to create their own model. The "Ryad-1" (Ryad is Russian for "series") was available on the market in 1972. The Russian model used the IBM operating system DOS as well as IBM peripherals and software. When these models were delivered to East Bloc countries they came with a Russian translation of the IBM service manual.[79]

East Germany not only served as a conduit for technology transfer from West Germany to the Soviet Union, but also integrated high-tech espionage into its own computer industry. The most well known place for computer technology was Robotron in Dresden, although Carl Zeiss Jena and the Dresden Institute for Computers were comparatively more important. During the 1960s East Germany was acquiring IBM prototypes for Robotron, and by 1970 they had acquired, taken apart, and reverse-engineered at least a dozen computers. By 1973 Robotron machines were being mass produced at the rate of 80–100 a year.[80]

During the pre-1989 era, West German intelligence had become increasingly aware that the Soviet Union and East Germany had targeted IBM in Stuttgart as a source of high technology. In 1970 Gerhard Prager, an IBM employee, had been caught copying magnetic tapes and passing them on to the MfS. In 1979 Gerhard Arnold had been betrayed by Werner Stiller. Stiller has always maintained that Arnold's material was the most important of any of his agents', even though it mostly consisted of internal IBM company plans, designs, equipment, data sheets, and software. His material apparently helped modernize the East German army, and Stiller considered him the father of computing in the GDR.[81]

IBM-Stuttgart, however, was far from the sole source for the East German computer industry. By the late 1970s, Intel and Texas Instruments microchips had already been integrated into East German computers on a large scale. This is not surprising, for the MfS had had major sources at Texas Instruments since at least 1978. Since the Wall has come down additional sources have been discovered at IBM in Essen, as well as at Siemens and Digital Equipment Corporation in America.[82]

The enormous success of hi-tech espionage was followed with alarm by the American intelligence community, and "Operation Exodus" was launched in

1981 to stem the flow of illegal COCOM items to the Soviet Union and the East Bloc. The program was initiated soon after the Reagan administration began to investigate the extent of Western high-technology transfer to the Soviet Union.

Western intelligence's increased vigilance coincided with Erich Honecker's illusionary drive to achieve world-power status in technology through computers and the production of the GDR's own 1-megabit chip. In 1981 Honecker announced a ten-point program at the tenth Party Congress to "draw the bulk of [the GDR's] microelectronic needs from domestic production by 1985."[83] The wish to attain world-power status in computers reached a feverish pitch in 1986 on the occasion of the Eleventh Party Congress, at which microelectronics figured prominently.

Although Ulbricht had shown an interest in computers during the 1960s, in the 1970s the party leadership's interest in the new technology increased dramatically. A cursory glance at the Central Committee's minutes reveals a steady flow of resolutions during the late 1970s, and by the 1980s, there was a new resolution each year.

In 1964 the Central Committee passed a resolution to develop a program for electronics during the period 1964–1970. During the same year the State Commission for Electronics was founded and embodied the state-controlled electronics industry. By 1968 the data-processing machines "R 300" (presumably the IBM copies) were being introduced in the most important companies and institutions in East Germany.[84]

During the late 1970s, the GDR came to the realization that it was far behind in computers and developed intensified programs to move forward. The MfS was instrumental in advising government leaders on how to proceed. In April 1977 the MfS prepared information on computers in preparation for the Sixth Meeting of the Central Committee. Copies of this information were distributed to Otfried Steger (Ministry for Electronics), Schalck, Department XVIII, and Markus Wolf. The MfS thought that the Ministry for Electronics' suggestion of buying fully assembled factory equipment for the production of modern electronic hardware was an exceedingly important concept for the 1980s, since "every area of life is increasingly influenced by the state and application of microelectronics." The MfS advised the Central Committee to turn to Japanese corporations (and maybe "nonsocialist" countries in Western Europe) for cooperation (American firms were out of the question owing to strict embargo regulations). They recommended simply buying the licenses and complete factory set-ups. Although the GDR had signed numerous bilateral agreements with the Soviet Union, the USSR did not prove to be a cooperative partner. Soviet "development and production potential" was shared only in a limited way with socialist countries because of "military-strategic considerations." This program was to be carried out in strict secrecy because of the potential for sabotage by the "enemy." The program was to be directed by the Ministry for Electronics with the support of the Trade Ministry (and Schalck) and the relevant departments of the Ministry for Science and Technology. GDR leaders were also aware that they should aim for their "own production" in this area to avoid becoming dependent on nonsocialist countries.[85]

Although bilateral treaties on computers had been signed in 1982 between the GDR and the Soviet Union for work on a 1-megabit chip, the Soviet Union had again proved uncooperative, forcing the GDR to move ahead on its own. Honecker saw the 1-megabit chip as a symbol of world-power status for the GDR, a goal that later proved to be illusionary. The MfS was aware that the goal was unrealistic and considered the Politburo's motive to be a thirst for power.

One-Megabit Illusions and the Toshiba Connection

During the mid-1980s the GDR's efforts to develop an indigenous microelectronics industry reached a climax: the goal was for the country to produce its own chips by 1989. A large-scale program for microelectronics was developed using KoKo and the MfS as part of a system to acquire computer hardware and software. This program required illegally importing goods on the embargo list at prices 30 percent to 80 percent higher than normal. Money for this program was plentiful and Western currency was made available by Schalck and by revenue from GDR firms. Around the time of the Eleventh Party Congress in April 1986, 1.3 billion West German marks was earmarked for microelectronics over five years.[86]

By the 1980s Japan and the United States had emerged as leaders in the world of computers, and even highly industrialized nations like West Germany had trouble keeping up. In the West the computer industry had grown from individual company initiative, and the companies generated their own capital and innovative strategies. In contrast to that of Western countries, East Germany's computer effort was centrally orchestrated and financed by the state. It is no wonder that the GDR economy was so strained by the time communism collapsed! The country had invested fourteen billion marks in microelectronics by 1989.[87] And by then it had not even come close to mass producing a 1-megabit chip. Honecker saw the 1-megabit chip as a symbol of technological prowess and an expression of the GDR's abilities.[88] When East Germany collapsed, GDR leaders recognized that their efforts had been futile and that they had dug a billion-mark microelectronics grave.[89]

The program for microelectronics operated within what I earlier called the "state security regime" and required close cooperation among the Politburo, KoKo, the Ministry for Microelectronics, the Ministry for State Security, and several key scientists. Schalck worked closely with the head of the Ministry for Electronics, Karl Nendel. Nendel provided KoKo and the MfS with information and wish lists from scientists and industry. In turn, Schalck provided hard currency. The Microelectronics Program required an especially high number of imports on the embargo list at a time when Western vigilance had increased. For example, of the materials necessary to produce the 256-kilobit chip, 60 percent were on the embargo list.[90] Gerhardt Ronneberger, the acting director of the Foreign Trade Company Electronics, was the head of its imports division and was to direct the planned import of equipment and know-how for the 256-kilobit chips from nonsocialist countries (Nicht Sozialistische Wirtschaft, or NSW).

During this period of intense concentration on microelectronics, scientists and politicians exaggerated the GDR's potential to become a competitive nation in computers. For example, Biermann, the general director of Carl Zeiss Jena and a member of the Central Committee, wrote to Honecker in Neues Deutschland that Jena had achieved the preconditions necessary to produce a 1-megabit chip. As a result, ministry leaders and the MfS began to assess the GDR's technological level, and ways to better it, in computers. As part of this assessment, reports were written on the preconditions necessary to produce a 1-megabit chip, and the question was asked whether it was possible to do it in the GDR. Although the MfS and the Central Committee had heard from informants that the answer was no, GDR leaders persisted in pursuing this national political goal. MfS informants reported that there were political reasons for saying that the 1-megabit chip could be produced.[91]

The relationship between the GDR and the Soviet Union in the area of computers was analyzed and proposals were made to deepen "cooperation," though it was clear that the Soviet Union had profited (as it had in other areas of science and technology) more than had the GDR. The MfS reported that cooperation between the Soviet Union and the GDR on the 1-megabit chip had failed because of the Soviet Union's "unwillingness" to work with GDR researchers and "insufficient conditions" in the GDR. As a matter of fact, Zeiss Jena had built the Soviet Union an electronic ray exposure machine (Elektronenstrahl-Belichtungsanlage), ZBA 20, that was useful for producing chips, yet the company did not keep one for itself. Reports later questioned the reasons for this. As it turned out, the ZBA 20 was only part of the machinery needed to produce the chip, and other elements were missing in the GDR.[92]

In December 1985 Schalck proposed a project to Nendel that involved importing 256-kilobit bit chips and equipment from nonsocialist countries into the GDR. Schalck ordered the Foreign Trade Company Electronics to import the know-how and production facilities for a complete factory to produce twenty to thirty million 256-kilobit bit chips yearly without the GDR's own development work. This was the riskiest computer project the GDR had ever undertaken, and there was no guarantee of success. The equipment sought was protected by the strictest embargo laws. In order for the GDR to realize its ambitious goal of producing its own chips and eventually semiconductors, Schalck proposed that the GDR work with one of four countries and their firms: Japan (Toshiba), Korea (Samsung), Taiwan, or China. Because the GDR had successfully worked with Toshiba before in the area of embargoed computers, Schalck favored that company. He found it flexible and cunning in devising strategies to bypass the embargo. Toshiba's motive for helping the GDR, according to Schalck, was the hope of receiving more big business contracts.[93]

By March 1986 Toshiba agreed to cooperate with the GDR provided that no official contract was signed. The plan was to realize the project covertly and in two steps: Toshiba would first deliver a complete original template including a description of how it worked and parameters for a 64-kilobit chip. In the second step Toshiba would work on a special sketch of a 256-kilobit chip that would differ from Toshiba's well-known one. This step would occur a year later. Toshiba suggested that the GDR use its own production facilities (whether imported or self-made) for this, but was ready to provide assistance. Representatives from the company insisted on visiting Jena's and Dresden's computer institutes when no GDR citizens were present. This was arranged through Nendel and Ronneberger. Ronneberger had had close business ties to Japan for many years, and later recalled that "the door was always open at Toshiba."[94] Toshiba and another company had received 7.8 million dollars for their help with the 256 kilobit chip project.[95]

As the project was coming to completion in 1987, Toshiba faced close scrutiny and criticism because of its widely publicized sale of milling technology to the Soviet Union, allowing the installation of super-quiet submarine propellers. The U.S. Senate voted to ban Toshiba imports, and members of Congress smashed a Toshiba boom box on the Capitol lawn. The Japanese government responded by barring Toshiba's machine-tool company from doing business with the Warsaw Pact for one year.[96]

Toshiba's semiconductor project with the GDR was not well known at the time, and I have found only one short newspaper article and a short East German ADN news clip from the GDR's Washington correspondent mentioning it. Republican congressmen alleged that Toshiba also made illegal sales to East Germany in 1986, and attempted to export an assembly line to produce semiconductors. By March 13, 1988, this allegation could not be proven.[97]

Meanwhile, representatives from East Germany (from AHB Transinter) had several conversations with the general managing director and manager of Toshiba and the assistant general manager of Mitsui in Japan in November 1987. Toshiba representatives told of their knowledge of investigations by the U.S. Congress and their belief that the CIA had sources in East Germany. Toshiba warned the East Germans not to use the Toshiba masks for 64-kilobit chip production in Erfurt.[98]

Because the spotlight was on them, Toshiba/Mitsui made a request to Ronneberger in February 1988 that all material relating to 64-kilobit and 256-kilobit technology be destroyed immediately. The 64-kilobit templates were also to disappear. In addition, they asked for assurances that institutes in the GDR would not produce 64-kilobit and 256-kilobit chips using Toshiba technology. On February 9 representatives of Toshiba and Mitsui met with Ronneberger and several scientists in Erfurt at the microelectronics combine and allegedly destroyed all the material at a nearby garbage dump; the template was mechanically and later chemically destroyed.[99]

Ronneberger, however, did not tell the Toshiba representatives that they had only destroyed copies. Toshiba technology was still in the hands of Dr. H. [a scientist] and for a cheap price too: the Japanese returned the 7.8 million dollars after the material was allegedly destroyed.[100]

By the mid-1980s the GDR's attempt to build its own chip industry had failed miserably in Dresden. It simply did not have the technical know-how and equipment to do it. By the time the GDR had produced the 16- and 32-kilobit chips they were outdated, and capitalist countries were already onto the new generation. By 1989 the GDR was able to produce 90,000 256-kilobit chips, whereas Austria, a small country, had produced 50 million.[101] By contrast, Japan was already mass producing one million megabit chips per month by April 1986.[102]

Erich Honecker had been especially eager to move the GDR into a position of technological prowess by latching onto computers. He was delighted when Biermann wrote to him in January 1986 announcing that Jena would be able to produce a 1-megabit chip by 1989. He called it the "nicest New Year's present" that he had received.[103]

By the fall of 1988 Biermann wrote to Honecker again (this time personally and not through Neues Deutschland) announcing that he was presenting him with the first functioning model of the 1-megabit chip one year earlier than promised in 1986. In this way Biermann thought the GDR had joined the "international elite" in science and technology.[104] Several weeks later Honecker met with Gorbachev in Moscow to discuss questions of cooperation in science and technology. There he presented Gorbachev with the first "sample" of the 1-megabit chip developed in Jena. Although there was discussion of cooperating on large-scale production of the chip, by the time the

Wall fell, the GDR had not produced any 1-megabit chips.[105] It turned out the sample was just a dummy acquired from the West.[106]

Scientists Who Benefited

By the mid-1980s, then, close cooperation existed among the MfS, the Central Committee (Mittag), the Foreign Trade Ministry and KoKo (Schalck), the various science-technology ministries, and scientists and their institutes or combines, especially in the area of embargo breaking.

Several scientists profited from, and worked closely with, the various ministries and the Central Committee. As we saw, Wolfgang Biermann took an active part in influencing policy as a Central Committee member and the general director of Zeiss Jena. In addition, during the same time period Volker Kempe profited from the support of the MfS.

Kempe was the director of the Academy Institute for Cybernetics and Information Processing between 1977 and 1990. A child of the GDR, he was born in Berlin in 1939 and received his masters at the Worker's and Farmer's Faculty of the University of Halle. He also studied in Moscow.[107] After the collapse of communism he turned up in Vienna and was named manager of the year.

Kempe is just one of presumably many scientists who benefited from the support of the MfS. In 1986 he provided the MfS—Department XVIII, SWT, and KoKo—with lists of journals and equipment he needed for his work. The MfS passed these lists on to the various acquisition organs, which were to "import" the material for Kempe. In exchange for this support, Kempe was to work on solving software problems, including neutralizing a software package called "Medusa," which had been acquired from Vienna, and adapting it to GDR conditions. This was just the beginning of the establishment of a central institute for software.[108]

Supporting scientists like Kempe was part of the GDR's massive Microelectronics Program. Kempe had apparently written to Erich Honecker with his suggestions for this program. As a response, Günter Mittag arranged a meeting with Kempe on January 30, 1986, several months before the April 11th Party Congress in which microelectronics figured prominently, and around the same time as the Toshiba and Biermann story unfolded. At this meeting, Mittag enthusiastically supported Kempe's suggestions, which included setting up a central institute for software development. Mittag saw political and ideological importance in Kempe's ideas, which he considered a basis for the scientific-technical revolution. After this meeting Kempe received the support of the MfS (although he may have been receiving support even earlier).

Kempe's institute was just one of many in the GDR that received embargoed 32-kilobit Vax 750 and 780 computers for CAD/CAM (Computer Assisted Design/Computer Assisted Manufacture) use. The institute acquired 275 16-kilobit personal computers from IBM, and a few systems were imported legally from Siemens. Forty-seven of the embargoed computers were acquired from the strictly protected American firm DEC. These embargoed goods were acquired through SWT's Department XIV—(Its "Area for Equipment Imports of the Combine Microelectronics and a Special Import Area of the AHB Electronics"), but the orders were given by the state Leader's Group for Key Technologies.[109] Horst Müller, the head of Department XIV, was awarded the "Banner of Work" for his achievements in acquiring DEC VAX computers.

In addition to the secrecy surrounding methods of illegal acquisition, all personnel who used the computers, especially CAD/CAM computers, or were in some way involved with computers, were sworn to secrecy. Secrecy began to be slowly lifted in some areas, such as the legally imported Siemens computers. Nevertheless, many scientists who worked outside of secret military projects and who would have been normal civilian scientists in a Western country were called "bearers of secrets" (Geheimnissträger).[110]

Conclusion: Implications of Operational Technology Transfer for Science and Technology

This chapter examined how espionage functioned and was used in the GDR's quest for scientific-technical prowess and economic strength. It argued that the MfS worked hand in glove with the SED and other institutions such as the science ministries in order to transfer science and technology from the West to the East Bloc and the Soviet Union. Aside from an example of how a scientist benefited from the support of the MfS, it deliberately did not address the separate and important issue of the influence of this operational technology transfer on science and technology—that is, the diffusion of the technology tranfer into the East German science system.[111]

On paper it would seem that the espionage system ought to have been an effective way to strengthen East German science and technology, given the barriers the country faced as a result of embargo restrictions and isolation from the West. There is no doubt that the "technology transfer switchboard" (SWT's evaluation unit) was well organized and had the potential to support scientists. As one intelligence officer reminded me in an interview, however: "The SWT was not a magic shop." In actuality, things did not work so smoothly. Although a scientist could order some plans or equipment from the MfS, this system—although it was centrally organized—did not work like a mail-order house. It might happen that the MfS had a source at a company that delivered material, and this information was passed on whether it was on a shopping list or not. There are cases of expensive material or equipment being delivered to the East that could never be copied or reproduced.

This brings us to another important issue in copying prototypes from the West, or using blueprints or plans in industry. It is not unusual for a small country to copy and then improve upon foreign technology. The Japanese are famous for their success in this regard. Examples abound from history of successful technology transfer from one country to another. Espionage-led technology transfer even goes back to the eighteenth century, when France undertook large-scale industrial espionage against Britain while importing its textile technology. The rivalry between Britain and France reminds one in some ways of that between East and West Germany—thought clearly France had more success in becoming a leading modern industrialized state.[112]

The successful improvement on acquired technology clearly varies from country to country. East Germans themselves questioned the efficacy of copying Western technology. For example, according to an informant, a scientist from Jena complained that scientists in the GDR were "sleeping through" international developments, and this had implications for their own science: "Why are we only "copiers" and never "creators?" he complained.[113] As was plain to the East Germans themselves, copying can stunt creativity and lead to laziness. Indeed, the drive to create something new using one's own initiative is lessened with the knowledge that there is no need to do the work if one can acquire it. Finally, this way of operating does not stimulate a country to become an innovator. In the short term, companies can save on research and development costs, but in the long run, acquiring knowledge may have adverse effects on productivity for the reasons outlined by the Jena scientist.

Unfortunately, there exists neither material that would aid a historian in evaluating the extent of the technology transfer in a variety of fields, nor an overview of which companies profited from plans acquired from the West. Judging from workbooks left behind by the SWT from the late 1980s, companies such as Robotron and Carl Zeiss Jena benefited the most from illegally imported equipment because they were privileged and worked on projects supported by the state, including the Microelectronics Program and military technology (including lasers) used to resist Reagan's Strategic Defense Initiative Program.[114]

Let us take the example of the computer industry. Simply acquiring and copying an IBM computer is not enough to develop a computer industry or even to work with computers. It often took months or even years for scientists in the GDR to acquire parts illegally, and without those parts they could not make proper use of the machine (even diskettes were put on the acquisition lists!).[115] By the time a machine or part arrived, it was often considered obsolete in the fast-moving computer industry.

The example of the huge investment in acquiring Western microelectronic components and computers in the 1970s and 1980s raises other questions. Both KoKo and the SWT's Department XIV were responsible for "importing" embargoed Western goods. The GDR was handicapped by these restrictions but nevertheless had the unrealistic goal of achieving world-class stature in computers. While other countries, like West Germany, could legally buy computers and equipment for manufacturing chips (Ion-Implanters, for example) this option was closed to the GDR, and it tried to circumvent the restriction. Put simply, without the modern computer equipment available to most industrial nations, the GDR could not hope to reach international levels in technology. Striking, however, in the whole espionage outfit, is the willing and wide-scale cooperation of West German nationals and West German and Japanese companies with communist countries.

Both KoKo and the Sector for Science and Technology were thus effective and needed acquisition agencies for scientists, industry, the MfS, and the Politburo. Some of the covert activity they undertook would have been perfectly legal in the West: importing computers and parts (embargoed) and passing on open-source material to scientists (through an espionage agency). The GDR also suffered, however, because of its own security regulations. After the Wall was built, the MfS became a way to keep East Germany hooked up to the international community of scientists. Not only did the many reports by agents and the travel cadre keep the MfS informed of possible recruits, but Western books, and other published and unpublished material, were often brought to East Germany through espionage channels (though this was not a high-priority activity and was not rewarded by the evaluation system).

Although much of the knowledge and equipment from the West was transferred to East Germany, a very large portion also went to the Soviet Union. For example, the Soviet Union could benefit from espionage in nuclear physics and technology more than East Germany could. Jena seemed to act as a steady conduit of knowledge to the Soviet Union, especially in military technology. The case of the combine Carl Zeiss Jena is of particular interest with regard to the cooperative project with the Soviet Union to work on a guided missile for ships. Clearly, acquiring military technology benefited the East Bloc's balance of power.[116]

The acquisition agencies operated on many levels, from responding to and implementing SED policies, to working together with other institutions, to recruiting and running agents, to acquiring information. Daily operational activities led to the emergence of a spy culture that often lost sight of the initial goals. This becomes clear when one considers how much time a case officer spent recruiting and running agents, as well as compiling dossiers in order to answer the question "Who is who?"

Although the SWT and KoKo had always worked closely with the science ministries, by the 1980s the level of cooperation among the MfS, the Politburo, and the science ministries in acquiring computers is striking. It reveals the extent to which East German society had become a state security regime, and the way in which they had to cooperate to aid East Germany's quest for technological prowess in computers. This quest was sparked by the Politburo's dream of achieving the West's level in science and technology. The MfS knew that this illusionary goal was not possible, but supported the Politburo's wishes nevertheless.

II

Institutions

5

The Foundations of Diversity: Communist Higher Education Policies in Eastern Europe, 1945–1955

John Connelly

The mature German Democratic Republic occupied a special position within Eastern Europe: in no other place did Marxist-Leninist ideology so thoroughly penetrate society.[1] This fact was particularly evident at universities, which were unique in East Central Europe for their failure to lead, or even participate in, general societal mandates for change. In Poland, Czechoslovakia, and Hungary members of universities, particularly students, took prominent roles in the movements of the years 1956, 1967–1969, 1980–1981, and 1989. East German students were conspicuous by their absence from the events of June 1953 and September–October 1989, however.[2] In East Germany, the Party controlled higher education more carefully than anywhere else.[3] Given that East European societies were supposedly constructed according to identical blueprints, the peculiarity of the East German case demands explanation.[4] Why did East Germany take a separate path? How did the East German regime succeed so well in creating an intelligentsia beholden to the needs of a socialist state?

One might argue that East Germany's special position in Eastern Europe derived from its special political logic. East Germany was the only state in the Bloc that had to legitimate itself primarily as a socialist country, and in direct, unceasing contrast to a neighboring capitalist state. Nevertheless, the SED had consciously to mold the elites who would give life to this broader political logic. If the Party succeeded it was because it understood the challenges involved in producing supportive elites. These challenges were essentially twofold: after establishing Soviet-style structures to govern the elite-producing institutions, that is, universities, communists had to begin carefully selecting the new elite from underprivileged social groups; and they had to weaken, then replace, the old professoriate, especially in the humanities, social sciences, and law faculties. The former set of measures would permit the con-

125

struction of a loyal elite, for the social climbers would owe their professional success to the Party. The latter would enable the ideological re-forming of this new elite.

It paid for East European regimes to recognize these challenges early, because the Stalinist period was a time of opportunity that would not be repeated. From 1947 to 1955 communists possessed a full range of devices—including political terror—to realize their goals.[5] They could intervene more deeply into societal processes than ever before or since. If they did not manage to break up existing structures in this period, they would not receive another chance.

In this chapter I will argue that the SED benefited not only from recognition of the challenges in transforming higher education, but also from early attention to these challenges. It was thus able to use most effectively the common structures of control imported from the Soviet Union in order to establish communist hegemony in society. I will first look in detail at those common structures, paying particular attention to the substance and methods of transfer of Soviet models of higher education to East Central Europe. Then I will focus on the ways in which Poland, Czechoslovakia, and East Germany responded to the challenges of transforming their systems of higher education, concentrating first on policies aimed at socially recasting the student body, then on changes within the professoriate. Did the East German leadership enjoy structural advantages that complemented its early activities on the higher education "front"? Did its fraternal parties suffer from certain disadvantages in this formative phase of building socialism?

The Soviet Model: Planned Education

By 1953, the structures of rule in Polish, East German, and Czech higher education appeared identical. These were strictly hierarchical, top-down, and two-tiered: Party-state. In each case a Central Committee functionary was in charge of issues of science, culture, and education. Each oversaw a department of the Central Committee apparatus that coordinated party education policy among ministries, lower-level party organizations, and institutions of higher education.

The three systems of higher education were re-formed in compliance with Soviet models of organization. This is apparent in the establishment of separate ministries for university-level institutions and in the fitting of these institutions into the party-state command hierarchy. Soviet models were also visible in special courses of Marxist-Leninist indoctrination, and in the attempt to permeate all academic subjects with "scientific socialism." Curricula, even for subjects like classical philology or mineralogy, were simply translated from the Russian. Perhaps most important, the Soviet model was reflected in student admissions policies geared toward increasing the numbers of students from "worker-peasant" backgrounds.[6]

All of these policies were adopted in order to facilitate Soviet-style planning. In each society, the plan for higher education was directly subordinated to the goals of the overall state plan. No area of university operations was immune to planning: in 1949 the chief SED culture functionary Anton Ackermann even compared fulfilling plans for students to fulfilling plans for potatoes—with the exception, as one SED wit noted, that students "don't grow as fast."[7] Once students had been admitted to universities, the Party attempted to control every moment of their lives. What had once been relatively autonomous university bodies, like academic senates and faculty councils, now occupied lower levels in an administrative hierarchy subordinated to the plan. The rector was transformed from *primus inter pares* to a sort of manager.

In each East European society the plan was formulated in numbers and stressed heavy industrial growth. Thus higher educational resources were shifted from humanistic and legal education to technical subjects. A parallel trend was the breaking up of academic institutions along narrowly defined disciplinary lines, and making them more amenable to control from a central administration. With compartmentalization, institutions of higher education increasingly lost the function of imparting liberal education, and increasingly became institutions for specialized training.

As early as 1949–1950 resources were shifting to technical fields throughout East Central Europe. In 1950 the Communist Party of Czechoslovakia (KSČ) closed the law faculty in Brno as well as the pharmaceutical branch of the medical faculty in Prague. That same year engineering schools were opened in Plzeň and Ostrava, and a chemical school was established in Pardubice.[8] In careful replication of Soviet models, these new schools were subordinated to the ministry responsible for their branch of the economy. This was supposed to aid planning: ministries would know the requirements for their sectors and could requisition students accordingly.

Multiplying numbers of narrowly educated technical specialists reflected Stalinist conceptions of the new society: it would be built by engineers and not by bourgeois remnants like pharmacists and lawyers. The numbers of Czech students studying technical subjects rose precipitously: of 9,080 students entering university-level institutions in 1949–50, 2,983 (32.8 percent) embarked upon technical studies; the plan for 1950–51 called for 4,120 (50 percent) of the 8,260 entering students to take up technical studies.[9] Large faculties were split up; for example, in 1951 the philosophical faculty in Prague was divided into philosophical-historical and philological faculties, and its natural sciences faculty was divided into mathematical-physical and biological faculties the following year.[10]

SED General Secretary Walter Ulbricht took an interest in higher education that was unusual for an East European leader. He repeatedly interceded in favor of worker-peasant studies. In 1950, he personally directed the creation of a communications school in Dresden.[11] The tendency to weaken existing university education in law and social sciences had been in evidence since 1946–1947 in East Germany, when the SED and the Soviet Military Administration in Germany (SMAG) set up competing Marxist "social science faculties" in Rostock, Jena, and Leipzig, as well as the Walter Ulbricht Administration Academy in Forst-Zinna. East German higher education also witnessed a promotion of the technical and natural sciences at the expense of the humanities. The plan for 1953 allotted the Technical College in Dresden almost one-third of the total state budget for higher education. It received more than twice as much as Humboldt University, and as much as Leipzig, Jena, and Rostock combined.[12] From 1951 to 1955 the number of students in the technical sciences had risen 463 percent, while the numbers of those studying philosophy, languages, or the arts had risen only 112 percent.[13]

Polish higher education witnessed similar developments. Because of Poland's unmatched wartime destruction, the shift to technical subjects was even more pronounced there than in other East Bloc countries. In 1937–1938 Poland had 5 universities with 30,000 students; ten years later the number had grown to 7 universities with 52,000 students. But the number of technical and engineering colleges grew even more quickly, doubling from three to six, and two to four, respectively. Faculties in the technical sciences increased from 16 to 40 and the number of students from 7,888 to 21,000.[14] In 1951 at every Polish university save Poznań; humanities faculties were broken up into philological and philosophical-historical faculties. Similarly, mathematical-natural science faculties were split up into several components. After 1951 the law faculty in Toruń could no longer accept students. New schools were set up to train legal experts and weaken university-taught law: the T. Duracz Law School was opened in April 1950 and the Central School of Foreign Service in September of that year.[15] In 1953 the Ministry of Higher Education planned to break the Warsaw Polytechnic into six separate schools: the "main schools" for architecture, civil engineering, chemical engineering, electronics, machine-building, and communications.

Soviet-style planning not only changed higher educational landscapes across Eastern Europe, but also dramatically altered the structures of students' and scholars' lives. Planners hoped to transform conceptions of time, and control the most mundane activities. Standardized curricula had been debated in East Germany almost since the war's end, and in certain ideologically central faculties guidelines for curricula had been introduced as early as 1947, usually under the supervision of Soviet education officers.[16] But it was not until 1951 that such curricula took effect. Highly specialized Soviet-style training replaced a somewhat anarchic Central European regime in which students had been able to choose the lectures they attended. The SED needed not well-rounded scholars but "specialists who functioned according to the plan's discipline." By 1953 students attended 28–32 obligatory hours of lecture weekly, and had to work 12–14 hours daily to keep up. Free German Youth (FDJ) groups were charged with making sure students followed their study plans even after classes.[17] In 1951, the German seven-month academic year was lengthened to ten months.[18] Technically, students were still left with two months of vacation, including winter break, yet increasingly the FDJ began to take control of this time as well.[19]

Students in the Czech Lands and Poland were similarly subjected to the rigors of planning. For the first time ever, students' attendance at lectures was made obligatory, and these lectures were reorganized according to centrally mandated curricula. Students received crushing work loads. Attempting to end "liberalistic autonomy" in higher education, communist functionaries forced students to spend as many hours in the classroom as workers did on the factory floors.[20]

In the wake of plan fulfillment, the quality of teaching declined. Teachers, like their students, had insufficient time to prepare lessons. The number of students per full professor at East German universities increased dramatically, from 54.5 in 1948 to 74.5 in 1952. With the greater need for teachers, qualifications decreased. The percentage of university teachers without a doctorate rose from 18 percent in 1948 to 28 percent in 1952; in that same period the total number of teachers increased from 1,236 to 1,743 (41 percent). At Leipzig the number of non-academics on the teaching staff rose steadily from 9 percent in the winter semester of 1946–47, to 29 percent in the fall semester of 1951–52.[21] The qualifications of teaching staffs dropped precipitously in philosophy and history, where chairs were created for "professors" who had not even acquired doctoral degrees.[22]

In the Czech Lands the picture looked much the same: in 1937 there had been 17.9 students to every college-level teacher in Czechoslovakia. By the fall of 1952 the ratio was 37.4 to 1, and by 1956 it reached 48 to 1.[23] The professional qualifications of teachers declined; as a result of heavy purging of faculties in 1948, something once almost unknown became rather common: people without habilitation were appointed professors.[24] The decline in qualifications was perhaps least severe in Poland, because the Party had not dared challenge the position of the old professoriate, even in ideologically sensitive fields like law and history.

The transfusion of Soviet experience proceeded smoothly in each of these societies because state and party apparatuses were full of people with first-hand knowledge of the Soviet Union. Yet East Germany held a slight advantage, for the number of party-state bureaucrats with knowledge of the Soviet Union was greatest there, and the GDR was the only place in communist-ruled Europe where Soviet advisers ever directed the day-to-day operations of education.[25] The first head of the German Education Administration (DVV), Paul Wandel, had taught in the Comintern school in Moscow, and had been a member of the Soviet Communist Party. His deputy for higher education, the physicist and communist Robert Rompe, was born in Russia and knew that country intimately.[26] Graduates of Soviet antifascist camps were found at all levels of state and party administration.

Officially, Soviet occupation officers strove to do no more than implement the Potsdam Accords' "four d's": democratization, demilitarization, de-nazification, and decartelization. They carefully avoided describing their program as "socialist." Yet their very understanding of "democracy" and "de-nazification" tended to favor preparation for the socialist construction of the 1950s. "Democratization" of universities, for instance, meant increasing enrollments of worker and peasant students and adding the "scientific world view" of the working class to curricula. "De-nazification" meant strengthening the party that was by definition antifascist: namely, the SED.[27] Gradually, the SMAG transferred more and more responsibilities for "democratization" to SED-controlled education ministries in the five East German states.

In the Polish and Czech apparatuses, direct experience with the Soviet system was usually limited to top functionaries. The ministers themselves, Stanisław Skrzeszewski and Zdeněk Nejedlý, had lived in the Soviet Union during the war, and so had the top men in the party apparatus: Edward Ochab, Jakub Berman, and Gustav Bareš. But middle-ranking functionaries did not know the Soviet Union first-hand, having been too young for war-time political activity. Younger functionaries could fill gaps in knowledge through frequent fact-finding trips to the Soviet Union, however; and they could learn from frequent "delegations" of Soviet academicians that visited the countries of East Central Europe.[28]

The Challenge of Making the New Student

The Polish, Czech, and East German Communist Parties thus modeled their higher educational systems on that of the Soviet Union. But the Soviets did not closely supervise this process. In Poland and Czechoslovakia there were no "university officers" in embassies or military missions who would check on the day-to-day running of university-level institutions. Rather, the Soviets trusted Berman and Bareš to ensure the proper transfer of their system. Frequent delegations from Moscow visited Eastern Europe, yet the Soviet visitors would occasionally advise against undue haste in copying their system. Miloslav Valouch, in charge of higher education in the Czechoslovak Ministry of Education, reported to his Central Committee counterpart, Luděk Holubec, in mid-1952 that Soviet visitors "told us of a number of cases where tactful behavior and patient cadre work in the end won over reactionary professors in Soviet institutions." And Valouch complained that very often in the Czech discourse opinions close to those of the Soviets were "criticized as harmful."[29] In practice, therefore, East European communists had to interpret the meaning of Soviet experience for themselves. Even in the shadow of Eastern Europe's most violent show trials, Czech higher education functionaries were debating what Soviet experience would mean for them.

The difficulties of interpreting Soviet experience were most apparent in student admissions policies, which were arguably the central component of Soviet-style higher education policy. On a superficial level, the three parties' policies indeed seemed exact replicas of the Soviet model. All three created faculties that prepared "workers and peasants" for college, similar to the Soviet *rabfaky*.[30] All three favored children of workers and peasants in regular university admissions. But when it came to determining the precise percentages of workers to admit in any given year, or how exactly to set up worker-peasant faculties, or how long to "school" worker-peasant students, the three parties were on their own. Soviet history could not be a sure guide, because it had not followed a straight line. For example, Soviet admissions policy had strongly stressed class in the 1920s, but in the 1930s its emphasis shifted to performance. By the 1950s the majority of Soviet students were themselves not of working-class origin.[31] Native communists could not ask the Soviets how many workers to accept in a given year, how to "recruit" them, how to stack admissions committees, or how to formulate entrance examinations. Yet any one of these issues could prove decisive in determining the success of worker-peasant studies and in the transmission of "Soviet experience."

The challenges confronting East German, Czech, and Polish communists in their efforts socially to transform student bodies were similar. Throughout East Central Europe, "worker and peasant children" had been notoriously underrepresented in universities, as well as in high schools, which provided preparation for universities.[32] University establishments were in each place conservative, and viewed experiments in admissions policies with suspicion. But in several important respects the SED possessed advantages in its ability to come to terms with these problems.

East Germany was the only place in East Central Europe where from the outset a communist-dominated education administration could influence decisions to admit individual students. From the very beginning students in East Germany could be chosen according to explicitly political criteria; indeed, the SMAG could refuse any student who did not fulfill their "de-nazification and democratization requirements." At first, admissions policies had been geared toward strict exclusion of people tainted by the past, such as former NSDAP members or *Wehrmacht* officers. Those with antifascist credentials were favored. But by 1946 the designation "antifascist" was expanded to include postwar political engagement.[33] Students responded by joining one of the three remaining "antifascist, democratic" parties, especially the SED, which after 1948 was referred to as the only "consistently antifascist democratic" party.[34]

In order to increase rapidly the number of worker-peasant students, communists and Social Democrats began as early as 1945 to set up high school equivalency courses for people of modest social background. In the beginning such courses lasted one year, and their character varied by region; they were better funded and more numerous in places of traditional communist and Social Democratic strength, like Saxony. The first large cohorts of worker-peasant students commenced university studies in the winter of 1946–47. In 1949 the SED fully standardized these courses, renaming them "worker-peasant faculties" (ABF). They lasted three years and were placed directly at institutions of higher education, so that worker-peasant students could immediately take advantage of university resources.

Poland's communists were numerically weaker than their comrades in East Germany or Czechoslovakia, having about one-third the per capita party membership in 1947. In addition, the Polish population was far more hostile to communism than either the East German or the Czech population. Thus the Polish communists felt a need to move slowly and carefully in the implementation of socialist policies. At universities, they decided early to pursue a strategy of compromise, leaving professors in place, even in the social sciences and law faculties, while concentrating on student admissions.[35] The old professoriate was meant to help train a new elite drawn from the worker and peasant strata. Failure to purge the professoriate would make fulfillment of this task unlikely, however. The largely anticommunist professoriate remained intact and culturally dominant at universities, and very few students entered the Party. By the late 1950s, more than 60 percent of Poland's students were of worker-peasant origin, but less than 10 percent belonged to the Party. The old professoriate was free to inculcate the new students with its values and standards—even within the strictures of the new curricula. In East Germany, by contrast, as early as 1947 up to one-third of the students in Leipzig, Jena, and Halle were party members.

Czech communists did not become involved in higher education policy in the early postwar period. Higher education was not one of their priorities, as is indicated by their voluntary surrender of the education ministry after their May 1946 election victory, in return for the ministry of domestic trade. Yet even while it held the education ministry, the Communist Party did nothing to advance the cause of worker-peasant higher education. In complete contrast to their counterparts in Saxony, Communist party cells in Czechoslovakia did not promote study among workers, nor did the KSČ Central Committee do anything to initiate activity below, in the way its Berlin counterpart had encouraged hesitant communists in Mecklenburg to foster worker-peasant studies. Czech communists were strangely silent on the need for social transformation of student bodies.

These early differences in emphasis persisted through the Stalinist period. Neither the Polish nor the Czech Communist Parties gave worker education the same priority that the SED did. The Czech Communist Party instituted a form of the ABF after the 1948 seizure of power, but never committed the institutional resources to it that the East German Party did. Whereas more than one-third of the entering classes of East German universities came from the ABF in the early 1950s, less than 10 percent came from the Czech equivalents (called DKs) at their high point.[36] Admission requirements were never made stringent enough to guarantee majorities of worker students. Worker students could not make use of university resources, because the courses were set up in small towns far from the universities. By the mid-1950s the Czech leadership had abandoned special measures for enrolling workers at universities, and therefore never supplanted the influence of "bourgeois" elements.

Perhaps the most important reason for this failure to use universities for creating a new elite was the KSČ's unmatched hostility toward and suspicion of universities. Rather than make workers students, it preferred to make students workers. In 1949, more than one-quarter of Czech university students were expelled and sent to work in production. This was a squandering of academic potential without parallel in the region.

The Polish Party likewise failed to devote resources to worker education comparable to those mobilized by the SED. Not until 1951 were worker courses attached to universities. Worker-peasant students were not as carefully chosen as in East Germany. The SED leadership had mandated that Party members and trade unionists recruit factory workers for study at the places of production; the Polish United Workers' Party (PZPR) leadership did not similarly mobilize lower-level organizations, and therefore could not fulfill its "plan" for worker-peasant students. Poland's communists discontinued worker courses in the mid-1950s, because they thought that reformed high schools would ensure a steady influx of worker-peasant students. They thus failed to utilize worker-preparation courses to give worker-peasant students carefully ideological instruction, and to ensure that they would feel they owed their social advancement directly to the Party.

The result of these varying worker-peasant courses was different student bodies. Though Czech, East German, and Polish societies all had about roughly equal proportions of worker-peasants, the numbers of worker-peasant students differed markedly by the mid-1950s, which marked the height of social revolution in East European higher education: from about 60 percent in East Germany to 55 percent in Poland, and 40 percent in the Czech Lands.[37] Only the SED had shown the will to make full use of higher education to create a new elite.

The Challenge of Making the New Professor

An equally important element of personnel policies for which Soviet advisers could provide little guidance was faculty formation: purging, hiring, and training professors. The East German Party carried out the greatest purges and had the strongest role in selecting new professors; it therefore succeeded in transforming not only the structures but also the content of higher education. The key to East German success was the early start supported by the Soviets, who insisted that universities be fully de-nazified.

The principle behind de-nazification of professors in the Soviet zone was simple: anyone who had belonged to the NSDAP could not remain on a university faculty. In early February 1946 the SMAG released a command through the DVV requiring universities to "remove all Nazis immediately from teaching."[38] This meant more than a two-thirds reduction of teachers active in the 1944–45 winter semester. The number of full professors teaching in East Germany fell from 615 to 279.[39] In Berlin, Leipzig, and Rostock, approximately 85, 76.5, and 55 percent of professors and docents were dismissed, respectively.[40] Professor P. V. Zolotukhin, the director of the SMAG education department in Karlshorst, said that it was better to start small but properly at the universities.[41]

The education administration in the Soviet zone gradually adopted a more conciliatory approach toward former Nazis; the demands of training a new generation of scholars left it no other choice. Protocols of academic senate meetings from the late 1940s suggest greater leniency toward medical professors than toward professors in the humanities and social sciences. The number of former Nazis among GDR university teachers in the mid-1950s bears out this conclusion. Of the full professors of medicine teaching in the GDR in 1954, 45.8 percent had been NSDAP members. The next highest figures were recorded in technical (41.9 percent) and agricultural, forestry, and veterinary sciences (41.2 percent). The ideologically vital disciplines of economics, law, and Marxism-Leninism, philosophy, and pedagogy counted 16.7 percent, 11.01 percent, and 11.8 percent NSDAP members, respectively.[42] These figures tell us little about the degree to which the personal, let alone cultural, continuity of East German universities was maintained, however. For we do not know how many of the former Nazis integrated into East German teaching staffs had formerly been university teachers. Because they feared that their compromised pasts might be revealed, former Nazis figured among the SED's "most submissive servants."[43]

Regardless of limited continuities in personnel, especially in the natural science, technical, and medical faculties, the rigorous de-nazification of East German universities broke their continuity as institutions of cultural reproduction. The Soviet refusal in late 1945 to recognize universities' efforts at self–de-nazification spelled the end of these universities' autonomy. They lost the ability to determine whom they would hire, and therefore to determine academic standards. The Soviet Military Administration had to approve not only the appointment of every professor, but also the admission of every student in its zone.

In the Czech Lands and Poland, by contrast, professors and students reopened universities without outside prompting. Czech institutes of higher education had been closed by decree in November 1939 after large-scale student demonstrations against Nazi occupation. The early days of May 1945 witnessed what the Czechs called a "national revolution," namely, a popular uprising to reassert Czech rule. Student insurgents took control of universities, and by early June classes had resumed. In Poland university professors began reopening their universities after the Red Army liberated Polish university towns. Because of their leading roles in underground conspiracies during the war, Polish professors held great moral capital. The first university to reopen was the Catholic University, in Lublin, in August 1944; it was followed by Jagiellonian University, in Cracow, in early 1945. Though the Communist Party played a decisive role within the central educational administration from the beginning, it could not strongly influence the reconstitution of higher education because it possessed very few cadres on the "higher education front."

In neither Czechoslovakia nor Poland did a decisive purge of professorships take place in the immediate postwar years. In both places universities operated according to prewar regulations, which guaranteed academic autonomy. In both places communist academicians were in exceedingly short supply, and could not be moved into vacant positions. The period 1945–1948 was one of broad concerted efforts to reinvigorate academic structures that had been inoperative for six long years, and to begin replacing the experts lost to the war. This proved an almost apolitical task.

A policy of compromise toward the old professoriate persisted in Poland through the Stalinist period. The PZPR was numerically smaller than its neighboring parties, and preferred not to alienate the Polish intelligentsia through large-scale purging. At worst, communist officials transferred Polish professors to other universities—for example, from Cracow to Wrocław—or kept them from teaching. But they retained their professorial positions, and could continue examining students. Some even conducted seminars in their apartments. They had experience in conspiratorial education from the occupation years, and so did their students. The academic milieu was controlled largely by the conservative old professoriate, and very few professors joined the Party; in 1954, only 7.5 percent of Poland's professors belonged to the Communist Party.[44] Those who joined the Party tended to feel primary loyalty to the university community, because they had belonged to it longer.

In East Germany, the opposite tended to be the case: professors entered university posts from the outside, and felt stronger loyalty to the SED. For example, the rectors in Warsaw, Cracow, and Wrocław, Wasilkowski, Marchlewski, and Kulczyński, had belonged to their respective higher educational milieus for decades, and felt themselves primarily representatives of their universities. At the same time, the rectors in Leipzig (G. Mayer), Greifswald (H. Beyer), Halle (L. Stern), Rostock (E. Schlesinger), and Jena (O. Schwarz) were recent imports into their academic environments. When the professors of Rostock University protested the restructuring of higher education in 1952, they found their rector standing against them. Several years later the medical students of Greifswald, protesting against the conversion of their faculty into a military-medical faculty, likewise discovered that their rector opposed them.[45]

Of the three countries under consideration, only Poland experienced a de-Stalinization in 1956, and thereafter the few "reactionary" professors who had been forbidden to hold lectures returned to teaching. The old professoriate maintained its hold on the culture of universities. Professors could pass on their values and beliefs to students, even while lecturing according to Soviet-style curricula.

Czech professors had a radically different experience of Stalinism. In Czechoslovakia the Communist Party was relatively stronger than in Poland, and in the weeks following the February 1948 seizure of power dozens of professors were dismissed from their positions, particularly in the philosophical and law faculties. The committees that carried out these purges were controlled by students. For a time, the term "studentocracy" was used to depict the new university regime. It signaled the KSČ's unmatched disdain for the culture of higher education. In 1968, some of the dismissed professors were still young enough to take up teaching again. After the crushing of the Prague Spring, however, they, along with many of their erstwhile purgers, would fall victim to "normalization," that is, to a new round of repression.

Conclusion: Stalinist Diversity

In higher education East European societies maintained their separate identities, even in the Stalinist period. In East Germany, the old professoriate was thoroughly purged, and a new type of worker-peasant student began to predominate at universities. In Poland, worker-peasant students likewise made significant inroads into higher education, yet because the PZPR had not purged the old professoriate, the new students continued to be inculcated with traditional scholarly standards and cultural values. Czech communists achieved an almost opposite outcome: they purged the old professoriate, but did not achieve a significant change in the sociocultural make-up of the student body. The students of 1968 were largely drawn from the same sociocultural backgrounds as those of 1948.

These failures meant that neither the Polish nor the Czech communists would create loyal elites. Far from being ideologically reformed, Polish worker-peasant students learned to share their professors' reservations toward communism. If they themselves had brought such reservations to the universities to begin with, so much the worse for the PZPR. Czech students continued to be drawn from the classes that had invested most heavily in Czech democracy, and felt themselves to be keepers of Czech student traditions. One tradition was to demand the maintenance of democratic norms. The students who took to Prague's streets in the fall of 1989 were commemorating those who had taken to those streets fifty years before to protest an earlier totalitarian regime.

The reasons for the Czech and Polish communists' failures are complex. In the Polish case one can point to the weak Party and widespread anti-Soviet sentiments, and to self-confident professoriates, who had been strengthened by opposition to both Piłsudski and the German occupier. In the Czech case, one can likewise point to the wartime experience as a strong factor in shaping postwar outcomes. Yet in Czechoslovakia, in contrast to the Polish case, the war weakened the professoriate vis-à-vis the students, because professors had failed to mobilize opposition to the Nazis. Furthermore, one can point to traditions of strong vocational training, which lessened Czech workers' desire to embark upon higher education, and to the unmatched anti-intellectualism of the more popular Czech Communist Party.

Yet clearly each of these explanations begs further questions, and is at best an interim conclusion. The reasons for the SED's successes are more apparent, however. The war had delegitimized university professoriates and permitted the clearest break in continuity witnessed in any academic community in the emerging Soviet Bloc. The responsibility for effecting this break was not left entirely in the hands of native communists: in East Germany, unlike anywhere else in East Central Europe, Soviet education officers played a direct role in removing the old professoriate. And for several years after de-nazification, they insisted that extremely restrictive policies govern the reentry of former Nazis into professorial communities. As early as 1945, East German communists were in the business of creating their own elite—both in and through higher education.

Another component of Soviet-sponsored higher education policy ensured the gradual creation of a student body loyal to the new regime. As early as 1945 German leftist organizations were combing working-class communities for potential university students. The following year incipient worker-peasant faculties already accounted for one in three students at Saxony's university in Leipzig. The party apparatus continued its careful selection of worker-peasant students, who as a rule had little formal education and few ideological or cultural commitments, then bound these people tightly to the Party's mission through a combination of material rewards—such as generous stipends and rapid career advancement—as well as obligations to various organizations. East German worker-peasant faculties gave young people more complete technical and ideological training than faculties in any other East Bloc country.

Not only did the SED make an elite; it remade these people. Considerations of ideology aside, worker students were more likely to be loyal to the Party than were middle-class students because they owed the Party more.[46] Middle-class students even tended to consider higher education their right. After the departure of Soviet advisers in 1949, German communists continued building upon these foundations. By the 1960s a distinctly East German professoriate—drawn heavily from the working class and tightly bound to the SED—began to dominate East German higher education. During the 1960s the number of professors of working-class backgrounds advanced from 20 to 40 percent, and the number with SED memberships went from 35 to 60 percent.[47]

The SED gradually made higher education an effective lever for political control by exhausting its potential to reward and punish. It is safe to say that no other communist regime in East Central Europe so closely tied political conduct to educational opportunity. Both regime and population recognized education as the most valuable good in a socialist society.[48] It was reserved for those willing to conform to the regime's expectations: active membership in youth organizations, entry into the trade union and other political organizations, "voluntary" extra service in the army, in factory militias, and the like. Fear of endangering children's access to higher education even helped regulate adults' behavior, because children were made accountable for their parents' political mistakes.

The explanations for the peculiarities of East German higher education are thus complex. No doubt the unique political logic of the East German state left the SED no other choice than to define itself as "socialist." Yet as recent work on the East German economy has shown, there was more than one path to socialism in East Germany.[49] The Czech example shows that even a strong communist organization could fail to recognize the potentials within higher education for creating a supportive elite. An elite supportive of socialism did not emerge automatically in the GDR; it was painstakingly created over decades—essentially without break from 1945.

6

From German Academy of Sciences to Socialist Research Academy

Peter Nötzoldt

After World War II, the German Academy of Sciences in Berlin (GAS) held a special place in the world of German science.[1] Although it partly remained an all-German institution (until 1969) because of its membership, it simultaneously developed into a "socialist research academy" and the largest and most important research center in the GDR (with 131 staff members in 1946, 12,923 in 1967, and 24,000 in 1989). This chapter describes the most important stages of the transformation of a Learned Academy into a large-scale research institution up to the academy reform in 1968. It also delineates the ever-increasing influence of the Socialist Unity Party both as an external policy shaper and within the academy as more and more members joined the Party. It shows how the academy was groomed from the beginning to become East Germany's most important center for science as part of a policy that recognized the role science could play in reconstruction and in the economy. During the first twenty years of the GDR, there was a dramatic increase in the number of natural scientists (from 53 percent of all scientists in 1951 to 79 percent in 1969) and a reduction in the number of social scientists and humanists (from 47 percent in 1951 to 21 percent in 1969); there was also a threefold increase in research institutes.

The Foundation Year for the German Academy of Sciences

Most scientists in Berlin had left the city before the end of the war. The Prussian Academy of Sciences (PAS) had lost 75 percent of its active members by the time it resumed its activities. Of the academy's two classes, originally equal in size, the class for mathematics and natural sciences had become weaker, because it had been most affected when the Kaiser Wilhelm Society (KWS) moved its institutes out of Berlin. Thirteen KWS scientists had belonged to this class (constituting 35 percent of total membership), including the Nobel laureates Max Planck, Max von Laue, Werner Heisenberg, Peter Debye, Adolf Butenandt, and Otto Hahn. With the exception of Hans Stille, none of these thirteen scientists was involved in rebuilding the PAS. These numbers reveal the magnitude of the stream of scientists that flowed from Berlin toward the West; they also illustrate the break that occurred in the established structure of disciplines and institutions. Those scientists who remained in Berlin were forced to take on a range of different responsibilities; at the same time they were afforded opportunities to advance their careers and develop their fields and scientific institutions.[2]

The members of the PAS who were in Berlin seem to have recognized these opportunities very quickly. This is not surprising, given that they had lived through the many initiatives the PAS had taken after 1929 to move out of the marginal position it had increasingly found itself in after the establishment of the KWS.[3] A mere eight weeks after the end of the war, a new academy leadership was in place, with the classical philologist Johannes Stroux as president.[4] New bylaws had been drawn up, and the leadership had submitted an application to the Soviet Military Administration (SMA) to resume its activities. Lectures began again and something like self–de-nazification set in: at the end of July 1945, the academy dismissed eight members, and another seven followed later, equaling 22 percent of the total membership.[5] Moreover, the academy immediately staked its claim for a role in shaping the future landscape of scientific research. As early as August 1945, the academy tried to take over Kaiser Wilhelm Institutes and other research institutes; in the fall it drew up the first plans for new institutes.[6]

The PAS justified these actions by pointing to the organization of science in other countries (such as the Soviet Union and France) and invoking historical claims.[7] The small group of remaining scientists placed their hopes in the "state," which promised to allow the academy to evolve into the "highest scientific institution" in the "nation." They thought the time for this was propitious, believing as they did that the "abundant resources of the Kaiser Wilhelm Society and the Emergency Association of German Science have dried up and will not flow again in the near future."[8]

But on October 27, 1945, the "state" in postwar Berlin, the Allied Military Command, set the budget of the PAS at zero. Previous support from the Berlin City Council was cut off. Moreover, the PAS was not yet placed under the German Central Administration for People's Education, an action that had been hinted at since September 1945; in any case, such a measure would have been inconsistent with the four powers' responsibility for the city.[9] More important, the PAS had no lobby in the Central Administration. Because of the political importance of people's education, this department was in the hands of the German Communist Party (*Kommunistische Partei Deutschlands,* or KPD) and the Socialist Unity Party from the beginning.[10] The head of the section for people's education in the city administration was Otto

Winzer, the future foreign minister of the GDR, who had returned from Moscow with the Ulbricht group. The president of the Central Administration was Paul Wandel, a close collaborator of Wilhelm Pieck during the Moscow exile and later the GDR's first minister of people's education.

Leaders from the German Communist Party did not return from exile in Moscow with a conception of how to develop German science in the postwar period.[11] In Berlin, only four communist scientists were available in the beginning: Robert Havemann, the physical chemist; Josef Naas, the mathematician and director of the German Academy of Sciences from 1946 to 1953; Robert Rompe, the physicist and later institute director and member of the academy; and Alfred Wende, the chemist and later institute director of the academy. They pinned their hopes on a quick revival of what was left of the KWS in Berlin. These institutes were then available for the task of rebuilding. The Soviet Military Administration and the SED thought this was where the need for action was greatest, since nearly all of Berlin's Kaiser Wilhelm Institutes were located in the American sector and lacked leadership. At first it seemed rather easy to put Havemann, sentenced to death by the Nazi regime, into a key position by appointing him "interim head of the KWS."[12] The questions that guided Havemann were, "What opportunities do we open up by exerting our own, considerable influence on the society?" "To what extent can the Kaiser Wilhelm Society be useful to our enemies?"[13]

The society's central administration had moved to Göttingen and was able to use Max Planck's authority to thwart the plan of the German Administration for People's Education to gain control of the KWS. Interest in the academy rose only at the beginning of 1946, once it became clear that this plan had failed and that there was a growing need to consolidate scientific institutions, some of which had no leadership. The academy had also already received unexpected support in November 1945 from the Soviet Academy of Sciences. Viktor Kulebakin, a major general and member of the Soviet Academy, had been sent to Germany "to establish scientific and working contacts with German scientific and scholarly institutions, in particular with the Prussian Academy of Sciences." Kulebakin, experienced in dealing with the "Soviet power," recommended that the PAS drop "Prussian" from its name and petition the head of the SMA "to place the academy under Soviet military administration . . . until a legitimate, parliamentary central government for all of Germany is established."[14] Thus pressured into making a decision, Marshal Zhukov of the SMA considered it "advisable to place the academy under the authority of the German Central Administration for People's Education."[15] In the summer of 1946 the former PAS reopened as the German Academy of Sciences in Berlin. The new academy was given the right to admit members from all over Germany and to run its own research institutes.[16] The biggest departure from the German scientific tradition was the introduction of the research institute as the central body of the academy. This involved the incorporation of old institutes (like the Kaiser Wilhelm Institutes) into the academy as well as the establishment of new institutes. By contrast, academies in West Germany, like those in Göttingen, Leipzig, Munich, and Heidelberg, continued functioning as learned societies.

This change met the needs of the academy members in Berlin and satisfied the Soviet Military Administration. The former wished to strengthen their position in the future German scientific landscape, whereas the latter wanted to create a research center under its influence in order to have access to research in Germany.

Rapid Growth and Reorganization in the Occupation Period, 1946–1949

Although it was funded relatively generously and had independent control of its resources, the GAS did not develop into a magnet for all of German science in the late 1940s, let alone into the "top scientific institution of Germany," as the SMA wished.[17] By the end of 1948, only six members had returned and twelve had been newly elected. Simultaneously, the ranks of members who had remained in Berlin were thinned by five deaths and the departure of seven scientists who accepted appointments elsewhere. The work of building up the research potential of the GAS focused on the takeover of existing institutions and the resumption of interrupted work. The GAS incorporated research institutions of the Reich, the universities, and the like, all of which were in need of some umbrella organization. It further incorporated research installations that the Soviet occupying authorities had set up but no longer wished to use—for example, institutes in physics (solid-state research, plasma physics, and optics) and the former Kaiser Wilhelm Institute for Brain Research in Berlin-Buch, which was earlier under the jurisdiction of the SMA's Health Department and was groomed to become a center for biomedical research. The chief goal of the Soviet Military Administration was "to create an organ in the area of science that [was] strong enough to play a role in science in unified Germany."[18]

The German Central Administration for People's Education began to expect tangible results from the GAS, which was continually growing—personnel rose from 131 in 1946 to 768 in 1948, and 33 scientific institutes had been established since the academy's reopening.[19] It called upon the GAS to make headway in "contributing to the reorganization of scientific life; . . .

joining scientific work with the practical needs of reconstruction; . . . contributing . . . to the recovery of Germany unity; . . . [and] resuming and expanding international contacts."[20]

The German Central Administration believed that the tasks of reconstruction and reorganization of scientific life could be achieved only "if the plenum of the academy has the right composition" in terms of disciplines and fields. The executive committee agreed, in spite of concerns that expansion, "given the current personnel shortage," was possible only by making "more modest requirements" of those chosen for election into the academy.[21] The academy, however, was in no hurry to implement the suggested changes. The sixteen newly elected members in 1949 did not change the composition in any substantial way. The sole SED representative who held a position of power in the GAS, Director Josef Naas, believed that the only way to achieve the desired "reorganization" (through the introduction of new classes in agriculture, technology, medicine, the science of society [*Gesellschaftswissenschaften*], and civil engineering) was a "directive" from outside, even if such a step was almost certain to initiate a "very critical phase of development" for the academy.[22]

The 1949 Cultural Decree of the German Economic Commission was formulated to achieve this goal of reorganizing the academy. The old members were to be won over for continued collaboration with extensive privileges provided by the academy. At the same time, an effort was made to alter the academy's composition by electing new members (member positions were doubled to 120) and expanding the number of classes (from the two original classes to two classes in the humanities and four in the natural sciences). The academy was to be "reshaped into the top scientific center of Germany." It was given a new main building, extensive institute buildings, and additional investment capital of several million East German marks. Academy stipends for members were doubled, additional salaries for directors were provided, and money for national prizes, low-interest loans, and so on was made available.[23]

In the summer of 1949, twenty-seven additional members were elected to the academy, twenty-two of whom were scientists. In addition to their "professional qualification," they also met the second criterion for membership, being "suited for fulfilling their duties to the state," at least formally.[24] Though this laid the groundwork for a future reorientation of the academy, little changed at first, as the academy took a wait-and-see approach. During the period of Soviet occupation, it had remained an elite Learned Society. Moreover, it became clear that the Learned Society was not able to handle the added responsibility for non-university research that it had assumed from the KWG and the Emergency Association. The thin layer of top scientists, which was barely replenished, as few scientists returned to Berlin, was not sufficient to meet these new obligations. The effects of the Cultural Decree reverberated only over the long term.

The Role of the Academy in Building Socialism, 1950–1952

Although the Soviet Occupying Powers had been interested in the academy and had, in fact, been responsible for its rebirth, the SED leadership had paid little attention to research outside of the universities during the Soviet occupation period. This changed dramatically after the establishment of the GDR. The top echelons of the SED fell into two increasingly polarized camps. In one camp were those who pleaded for a careful approach in dealing with top scientists, and who tended to see science as part of the national culture and something that embraced all of Germany. In the other camp were those who believed that in the face of a tense economic situation, science should be primarily a means to an end. Wilhelm Pieck and Otto Grotewohl, supported by W. Semjonov, the Soviet ambassador, represented the first camp, Walter Ulbricht and the economic planners the second. From 1950 on, efforts to draw academy research into economic planning or to separate some parts of research from the academy intensified.[25] The disagreements between the academy leadership and the Central Office for Research and Technology at the State Planning Commission became so heated that Minister President Grotewohl had to intervene in 1951.

Academy research was not able to escape the planning mechanism entirely, however. The new academy president, Walter Friedrich (elected in 1951), was able to prevent even more serious interference by shrewdly invoking the example of the Soviet Academy, where "the directives of the ministries can be influenced by the veto of the academy,"[26] and by emphasizing the academy's "task in achieving German unity."[27] Nevertheless, the academy had already lost research potential in civil engineering and agriculture when specialized academies were set up in these fields.[28] It is not possible to determine exactly to what extent the Soviet organization of research, which included specialized academies, influenced this development, but clearly the division of applied areas of research into separate academies and the emphasis on basic research at the German Academy derived from the Soviet model. Indeed, there is no doubt that the Soviet model influenced the formation of the academy in East Germany. For example, both academy presidents Stroux and Friedrich learned about the structure of the academy in the Soviet Union, and the SED leadership made a study trip to Moscow at the end of 1952 in order to incorporate information into the new academy institutes.

Whatever the Soviet influence had been at the beginning, once the GDR government had come into power, it presented itself as a generous supporter of science (see also Chapter 7). The government declared the 250-year cele-bration of the academy's founding in 1950 a national holiday and intended to use it as a way to express the supportive position of the GDR toward science and to demonstrate the "close relationship between German science and the people."[29] The SED leadership declared its intentions at the celebration: the academy was to become the country's "central and highest scientific associa-tion." Instead, the celebration led to the exodus of more West German mem-bers (one of the reasons given for their leaving was the politicization of GDR universities and the impending politicization of the academy); it was boy-cotted by all West German scientific institutions.[30]

As part of the increasing control and intervention of politicians, the gov-ernment wished to use the academy when making decisions about science. It would have liked to see the establishment of a committee made up of "progressive" (party members) scientists in addition to academy members. Instead, by February 1951 an "Organization for Research" was created and Academy Director Josef Naas, one of the few SED members in a leadership position, was named director in a governmental resolution.[31] Another step that pushed the academy under state control was its direct placement beneath the Minister's Council of the GDR; this in essence made it a subministry of the council and not an independent actor. In addition, the election of the physicist Walter Friedrich as Stroux's successor in April 1951 reflected the increasing emphasis of the academy on the natural sciences.[32]

During the early 1950s, there were other signs that the SED was influ-encing the development of the academy. At a Central Committee meeting in 1951 there was a discussion about "paying more attention" to "science-of-society" subjects *(Gesellschaftswissenschaft)*. The Central Committee de-manded that new institutes in history, philosophy, German language and literature, and law and economics be founded.[33]

A dramatic change in the profile of the academy occurred with the "Procla-mation of the Building of Socialism" at a party Congress in the summer of 1952. The academy was asked "what kind of contribution it would make to the building of socialism in the GDR, especially as it had remained behind other institutions (universities and other scientific institutes) in its ideological development."[34]

By the fall of 1952, the Politburo had organized a Central Committee Commission on Checking Over and Re-Organizing the Academy, which had only one academy member on it (Wolfgang Steinitz, a philologist and eth-nographer elected in 1951 and a German Communist Party member since 1927). This was just part of the "ideological fight" that began that fall in the academy.[35] The goal of the commission was to prepare a Politburo Resolution on the academy's task of building socialism and the election of new members.[36] The Politburo Resolution endorsed the continuation of an alliance with the "bourgeois scientists" and allowed the all-German character of the academy to continue. The academy was instructed to "orient" all of its further scientific activity toward "building the basis for socialism in the GDR," if it "wanted to retain it place in the science" landscape of the GDR.[37]

Ulbricht's policy gained the upper hand following the Proclamation on the Building of Socialism in the GDR in the summer of 1952. To be sure, some members of the academy, engaged in the required exercise of thinking about the academy's contribution to building socialism, came up with very odd suggestions, such as examining "whether Marx and Engels could be posthumously elected regular members of the academy."[38] However, the decisive point was that the academy leadership and the plenum of members accepted the plan to reorient the academy toward becoming a predominantly scientific institution. Members deliberated about the "contribution of the academy to the building of socialism" at the end of January 1953 at a special meeting that had been repeatedly postponed. Ulbricht and scientists who were not members of the academy also spoke at the meeting. Their statements were all very sweeping and gave little hint of the new course that had already been discussed on November 26, 1952, between the party leadership (Wilhelm Pieck, Walter Ulbricht, Paul Wandel, Fred Oelssner, Werner Lange, Kurt Hager, and Bruno Leuschner) and top scientists (Walter Friedrich, Josef Naas, Hans Ertel, Erich Thilo, Karl Lohmann, Hans Stubbe, Wolfgang Steinitz, and Erich Correns), and was later implemented.[39] Several factors were responsible for the change in course. There was the usual loyal behavior of scientists toward the state and the realization that the original parity between the natural sciences and the humanities no longer corresponded to the actual development of science. More important was probably the pressure from those scientists whose institutes had become affiliated with the academy but who had not gained access to the Learned Society. Established out of practical considerations, these institutes had less of an aversion to practical research than other academy institutes and subsisted in part on funds from the Central Office for Research and Technology. In 1951, this outside funding amounted to 2,750,000 East German marks and accounted for up to half of all research funds at some institutes.[40] The alliance of utility between the SED leadership, where Ulbricht in particular was placing his hope in the ability of science to solve economic problems, and scientists, who nourished this hope and promoted their disciplines with memoranda and the like, initiated a new stage in the academy's development in 1952–1953.

Integrating Research into SED Socialism, 1953–1957

The members of the plenum predicted "a profound transformation in the life of the academy" when they resolved "that the academy should become firmly involved in building socialism and choose its tasks in such a way that some of the main challenges confronting the GDR can be accomplished by science, and especially by the academy."[41] This transformation occurred on the personnel level, which we shall examine first, but much faster still on the structural level.

The political requirement that continued to be sufficient for membership in the academy was "loyal conduct" toward the state as defined by the categories the SED bureaucracy used to classify academy members: progressive-sympathizing; loyal-sympathizing; completely loyal; very critical but loyal; and completely apolitical—loyal.[42]

Following the election of new members in 1953, the position of the SED among the members seemed considerably improved. Previously six academy members had held SED membership cards, and four of them were either barely active or "completely inactive." Ten were added in 1953, and the group of "active" SED members rose to seven.[43] We must not exaggerate their initial impact, however, for they made up just under 22 percent of those members who were actually involved in the life of the academy, and that ratio did not change at first. Moreover, most of these scientists with an SED membership card proved immune to coercion. While they all took advantage of career opportunities, they did not let the Party discipline them, and some even fought among themselves. Some were considered opponents of Ulbricht and mouthpieces of the bourgeois scientists. More likely they had a much greater interest in becoming part of the bourgeois academy elite than in replacing it, since they were convinced that this process would be, at the least, a difficult and protracted one.

The Central Committee apparatus judged that its chances of exerting influence were even less among the academy's scientific collaborators. Although as many as 17 percent of the academy's 1,050 scientists were already members of the SED in 1956, their distribution was highly uneven, with a few strongholds and many weak positions. Moreover, the scientific authority of the institute directors, who for the most part had no party affiliation, continued to prevail over attempts at interference from weak party cells.[44] Even the Central Committee party organizer appointed specifically at the academy in 1953 was not able to change much in this regard.[45] The party structure at the academy remained poorly developed.

Still, the SED was able to strengthen its influence in two areas. Its share of votes in the executive committee and on other decision-making levels grew, as did its influence in the powerful administrative apparatus, where it had been occupying the directorate, the most important position, since 1946. The replacement of Director Josef Naas (1946–1953) in the spring of 1953—after he fell into disfavor with scientists and functionaries alike because of his high-handedness—did not change this situation in any way.[46] His successor was Hans Wittbrodt, the very person who in 1950–1951, as the representative of the Central Office for Research and Technology, had fought vigorously for a change in the academy's direction and focus. However, neither Naas and Wittbrodt nor the party apparatus were able to bring about by themselves a truly profound change in the life of the academy. It was the specialists returning from the Soviet Union, headed by the new academy president Max Volmer and Peter Adolf Thiessen, who in early 1957 questioned the academy's organization of research.[47] They called for more efficient research structures in the natural sciences and technology: "Industry must finally and quickly receive what it needs from research. Research must finally and quickly receive what it needs from industry."[48] Allies were readily available in the academy's faculties. For some time now the ground had been prepared in terms of personnel and material conditions. Seventy-four percent of new members elected between 1950 and 1957 were scientists. In 1957, 66 percent of the 1,044 scientists were working in thirty scientific institutes and using 68 percent of the total budget. The humanities had to share the remaining 32 percent of the budget with the expensive Learned Society, a huge administration, several publishing houses, and the like.[49] However, this apparent imbalance between the natural sciences and the humanities was quite in keeping with international developments; in West Germany, too, the natural sciences were favored over the humanities though not under the roof of a single institution.

In 1957 influential scientists felt secure enough to suggest ending the "detour via the academy" in building their research institutions. They proposed setting up a "Leibniz-Society for the Promotion of Science and Technology." This society was to have a "structure similar to the former Kaiser Wilhelm Society or the Max Planck Society in the Federal Republic," and was to exist independently alongside a traditional science academy.[50] They found ready support in the party apparatus of the Central Committee of the SED: the proposed structures promised more say for the Party, and functionaries were hoping this would allow them to neutralize the influence they feared West German academy members might exert on research activities at the top scientific institution of the GDR.[51] The final outcome of the institutional discussions was determined by SED functionaries and a few scientists. That fact is underscored by charges of "clique formation" inside the academy and within the Central Committee apparatus, by protests against personnel decisions that the apparatus made without consulting the academy, and by the expressed concern of members of the executive committee that the plan was a "first step toward dissolving the academy in its traditional makeup."[52] What was eventually implemented with the founding of the Research Association of Scientific, Technical, and Medical Institutes of the German Academy of Sciences in Berlin in May 1957 was a hybrid solution, in that a virtually independent research association with thirty-nine institutes was created under the roof of the academy.[53] Simultaneously, the Council of Ministers proposed founding a Research Council of the GDR, of which the physical chemist Peter Adolf Thiessen was named chairman in the fall of 1957. In this way, the SED leadership signaled the end of the special position of the academy in science policy, particularly since the Central Committee's *Wissenschaft* department had suggested placing the Research Council above the academy.[54]

The Humanities at the Academy, 1958–1962

After the establishment of the Research Association in 1957, the question arose of what to do with the rest of the academy, that is, the Learned Society and research in the humanities. The party apparatus of the Central Committee wanted to "straighten things out" in the humanities. Setting up another research association seemed the logical thing to do, but this was judged not feasible for a number of reasons: "a lack of socialist cadres for a managing committee"; little "adherence to the party line and political-ideological clarity" among scholars in the humanities, which had not been so necessary among scientists, given the content of their work; and finally, party functionaries' uncertainty about whether "scholars would go along in the first place."[55] It took more than five years for these obstacles to be removed. The Party took a pragmatic approach to the Learned Society and the academy plenum and classes—they were to be preserved as an "all-German Learned Society," for in scholarly terms "there is more to be gained from an all-German society" than from an exclusively East German society, and one could also "make use of this for the cause of peace and progress."[56]

The election of a new academy president in 1958 offered an opportunity to install the man who seemed best suited for accomplishing these tasks. SED leaders accordingly had two requirements for the candidate: he must be a "member of the humanities" and a "resident of East Berlin."[57] This was the first time a member of the SED was sent into the race: he was Werner Hartke, the rector of Humboldt University who only very recently (in 1955) had been voted a member of the academy in the fields of classical philology, ancient history, and archeology.[58] His election was pushed through with great effort in a turbulent round of voting. West German and East German members who were present on the day of the election were especially incensed about the requirements tailored to Hartke.[59] Nevertheless, this election was a decisive success for the Central Committee party apparatus. The new president of the academy, who was intimately familiar with the operating mechanisms of the Central Committee from his university days, sought the support of the SED leadership and promoted the expansion of the party organization at the academy. The reasons for this close cooperation lay in the dominance of the rapidly developing Research Association, which doubled institutes and scientists until 1961 and tripled investments during the same time. There were also quarrels with the Research Association in the natural sciences over lines of authority, difficulties in setting up the Institute Association in the humanities, and the effects of the building of the Wall in 1961. There were pressing problems, such as finding employment opportunities for staff members in West Berlin, or pushing through the expulsion of the "renegade" member Ernst Bloch. But there was also a question concerning a pillar of the academy's very existence: did East Germany still need this all-German scientific and scholarly institution? The "dual track"[60] of the academy's work, created by its all-German character and its function within the GDR, was ended in the summer of 1962 with the declaration, "The German Academy of Sciences is to be refashioned into a socialist academy."[61]

The building of the Wall had another consequence that boosted the courage of the political leadership to undertake a radical restructuring of the academy. Previously, the labor market for scientists had included all of Germany, and every year between 1.5 and 3 percent of scientists had taken advantage of this by "fleeing the republic." Between 1955 and August 1961, this amounted to a loss of 12–15 percent of the scientific potential of the academy. Once the Wall was up, this threat ceased to exist.[62]

The chief characteristic of the change in the academy during the 1950s was its transformation into a research institution with a predominantly scientific and technological orientation, in which basic research continued to be dominant. The de facto separation of this research potential from the academy meant a return to the kind of organization of research that had proved itself in Germany after the establishment of the Kaiser Wilhelm Society, and had reasserted itself after the war in West Germany.

In terms of its composition and values, the academy remained largely rooted in the educated bourgeoisie. Two developments in the 1950s made it inevitable that the traditional leadership committees would lose power: one was the rise of forces in the SED leadership who believed that it was more important to replace this elite of scientists quickly than to pursue an alliance with them; the other was the fact that the coming generation of scientists believed that change offered them better career opportunities.

The special all-German character of the academy was preserved. With few exceptions, members living in the Western parts of Germany retained their membership.[63] In the West there was less "concern about the dangers of infection,"[64] and the festivities in honor of Max Planck's one hundredth birthday in 1958 amounted to the first large "celebration of the entire nation," in the words of Max von Laue.[65] For East German scientists the special status of the academy seems to have opened many doors to the international community. Even when reformers were planning to intertwine academy research with the East German economy and were concerned that Western members might theoretically have a say in how this would be done, they chose a solution that preserved the all-German character of the academy.

At the beginning of 1962 conceptions for further transforming the academy were very radical, and some in the Central Committee even proposed dissolving it. The issue revolved around the way in which the academy had developed into a socialist-style academy, that is, a research organization with remnants of a Learned Society. One of the reasons for keeping the academy, however, was its all-German character, which had still been preserved to some extent in 1962. But another, more compelling reason to expand the academy, despite the SED's complaints about its form, was that it offered a way centrally to steer and control an important part of the GDR's research potential. By the late 1950s most academy leaders were members of the SED, although scientists' membership had stagnated at 21 percent. The most important success for the SED came in 1960, when a party structure was established within the academy; no major decision at the academy could be made without the approval of the academy's party leadership.[66]

At the beginning of the second decade of the GDR, faith in the ability of science to solve problems became increasingly more pronounced within the SED leadership. This faith became the hope that the nation could hold its own in the competition between rival states, especially between the FRG and the GDR. In 1961, for the first time voices inside the academy spoke of science as a "material productive force [and] primary instrument for educating people into a socialist consciousness."[67] That same year, the first Central Committee meeting after the building of the Berlin Wall addressed the future directions of research policy.[68] In early 1963, the Sixth Party Congress of the SED approved a new party program, which for the first time devoted an entire section to science. Science was now regarded as a "direct productive force" alongside labor, soil, and capital, and it was to take on a crucial role in the implementation of the New Economic System of Planning and Managing the People's Economy.[69] This development continued and led to the Third University Reform, the Academy Reform, and the Reform of Industrial Research between 1967 and 1969 (see Chapter 2). As a result of this trend and the changes brought about by the building of the Wall, the development of the academy during the second decade of the GDR was strongly shaped by forces outside of science.

The Building of a Socialist Academy of Sciences in the GDR, 1963–1968

When one compares SED resolutions from 1952 with those written ten years later, the following changes are apparent: the unity of German science no longer existed, and the mere thought of it was perceived as dangerous. The all-German function of the academy diminished dramatically.

Another major change was in the celebrated "policy of alliance" with "bourgeois scientists"; by 1963 only a very few top scientists benefited from this policy. The SED thought the time had come to transfer influence and power to the "new intelligentsia." Behind the slogans "Cadre questions," "Cadre reserve," and "Cadre development" lay a chief goal: to increase the number of scientists who were members of the SED.[70]

Ten years after the first Politburo Resolution concerning the academy in 1952, the SED formulated its task in a much more precise way and determined that in the future the academy would be shaped more by outside forces than by those from within. The goal to "transform the academy into a socialist academy" meant "taking part totally in all areas of its activities in the realization of socialism in the German Democratic Republic and . . . embracing the development of the socialist world systems through close cooperation with socialist countries."[71]

The documents generated for new academy bylaws demonstrate the extent to which the SED had anchored itself in the administration of the academy at the beginning of the 1960s. A discussion about the formulation of new bylaws in 1962 reveals the goals that the planners had in mind for this "socialist academy." The discussion was carried on over a period of one year, not, as previously, by an academic commission, but between the Central Committee apparatus and a few members of the party leadership in the academy.[72] We can identify four main ideas in the discussion: first, the elimination of remnants of the academy's all-German character; second, the increased influence of the state in setting the goals of research projects (it had little success in the past); third, the creation of leadership systems that were in tune with the development of society in terms of their structure and personnel; and fourth, the cutting back of traditional ties of research to the West through increased cooperation with the East. Concretely, this last meant finding ways to restrict drastically the voting rights of Western members of the academy without risking spectacular resignations, at least by the most prominent members. The link between the Learned Society and the research institutions in the humanities was to be severed, much as it had been in the natural sciences. The goal was research solely at the behest of society—it had to "make a continually growing, direct contribution to scientific-technological progress and development of the national socialist culture." Daily production targets, developmental tasks of industry, and basic research in the broad sense seemed of equal importance, and non-academic committees decided which area would receive more funding in each particular instance. The principle of "democratic centralism" was to be fully implemented. The main ideas were change the structure of leadership, reduce the number of decision makers, switch to individual responsibility, and install younger "socialist cadres" in positions of responsibility.[73]

In the statutes of May 1963, much of this seems reduced to what was "doable." For instance, basic research was given more importance than in the past and the total exclusion of West German members was avoided. The plan was for independent institute societies to exist in the Socialist Academy alongside the Learned Society; though these institutes were formally part of the academy, they were run by state or party committees. The Research Association was working "at the behest of the Research Council,"[74] which at this time functioned something like a Ministry for Research and Technology of the GDR.[75] As for the humanities, "the main orientation will come from the Party."[76] The members of the Research Council were not elected but appointed by the government, and they in turn appointed the executive committee of the Research Council. Although it was intended that the Research Council coordinate its activities with the executive committee of the academy, the plenum of members ceased to have any influence. Though the list of "bourgeois scientists" in the Research Council and its executive committee was long, it was easier to forge utilitarian alliances and create SED majorities in these committees than in the plenum of the academy. The second society of institutes, the Association of Academy Institutes and Establishments in the Social Sciences, was to "coordinate and supervise" research in the humanities in the GDR at the behest of the SED.[77]

At the outset these plans ran into opposition among the academics concerned. In the opinion of the Central Committee apparatus, such academics were "either overbearing or attached to their academy classes," which were at that time the problem child of the SED. The "proper composition" of the executive committee with the necessary SED majority was already causing headaches.[78] All these problems proved solvable, however, most likely because it was possible to create scattered SED bastions in the humanities with the help of the next generation of scholars. For example, in the class for philosophy, political science, jurisprudence, and economics, 73 percent of the regular members who resided in East Germany were already members of the SED in 1963. By contrast, the percentage of party members in the class for languages, literature, and art was only 36 percent. For mathematics, physics, and technology the figure was 12 percent; for chemistry, geology, and biology, 24 percent; and for medicine, 14 percent. The influence of the SED in the research institutes must be seen in a more differentiated way. Many so-called bourgeois scientists were still heads of the institutes. Among staff members at the research institutes the percentage of SED members ranged from lows of 4 percent (ancient studies) and 6 percent (cultivated plant research) to 20 percent (chemistry, physical chemistry, and geology) and 78 percent (economics) in 1962.[79] Despite the nominal dominance of the SED in the institutes, the reorganization of the academy in 1963 introduced the essential elements of the Academy Reform in 1968.

Given the current state of scholarship on the subject, there is no satisfactory answer as to why the Learned Society and the Institute Associations *(Institutsgemeinschaften)* were not separated. There were advocates for this measure among scientists and the SED leadership. It was felt, however, that "this step, logically the right one, can probably not be implemented at this time."[80] It is conceivable that the majority of scientists were reluctant to do this because association with the elite Learned Society enhanced their prestige and brought more openness for basic research. Another factor we need to consider is the balance of power in the SED party apparatus. Separation would have led to power shifts between the science section and the economic section of the Central Committee. Moreover, the external function of the academy underwent a fundamental change. While its all-German character seemed unnecessary, cooperation with the scientific institutions of the socialist countries, which had hardly materialized to that point, seemed a compelling necessity.[81] Compatible structures, especially those in alignment with the organization of research in the Soviet Union, could only be helpful for such cooperation.[82]

Five years after the introduction of the new bylaws in 1963, another reform of the academy—the so-called Academy Reform of 1968—adjusted the institution once again to developments within larger society. The reform was preceded by a drafting of the tasks for science at the Seventh Party Congress of the SED in 1967, and by an internal academy evaluation which concluded that "insufficient progress can be recorded in the effectiveness of our work, especially with regard to the most important problems."[83] The task of "fashioning the system of socialism in a comprehensive way" called for "the conscious use of the scientific-technological revolution." For the academy and its institute this meant "overcoming those principles of building, management, and organization that date, in a sense, from the manufacturing period of science."[84] In 1968, the new academy president, Hermann Klare, declared that the goal was a research academy in which "all scientific capacities" would once again be gathered together under "unified leadership."[85] The "special strength" of the academy was now seen to lie in "the intertwining of social science and scientific research that existed within it."[86] As an integrative element, this explicitly included the Learned Society. This was a new society, however, in which new majorities had been created between 1963 and 1967 with the election of thirty-seven regular members from very different scientific institutions in East Germany. Here the GDR's policy of demarcation from West Germany was beginning to take effect. The structures of research were changed once again, decision-making and authority were centralized, and the plenum and the classes were defined as merely advisory organs to a central academy leadership and the leadership of the SED.

The goal of research, too, was defined more narrowly in line with the SED's science policy: for the natural sciences it was "investigative and basic research to secure a long-term head-start in areas that determine the structure of the East German economy," for the humanities it was scholarly support in "directly guiding and shaping the development of society."[87] Unlike five years earlier, research and development undertaken for specific purposes were to be the exception. At the beginning of the second decade of the GDR, the academy had been close to being dissolved. At the end of the decade, structures emerged that would turn it into a compact, tightly organized "research collective." What made this possible was the generational change in all positions of leadership and the abolition of its all-German character. Positions of leadership were now predominantly staffed by scientists who were willing, out of inner conviction, to create a strong bond between science and the interests of the SED state. Outside interference or consideration for the feelings of West German members dropped by the wayside. In addition, the SED had gained a foothold in the academy through stable organizational structures. In 1967, 18 percent of all scientists (compared with 14 percent in 1960) and 54 percent of all scholars in the humanities (compared with 39 percent in 1962) were members of the SED. Among regular members of the academy from East Germany, party membership had risen from 27 percent in 1963 to 32 percent.[88] The tightly run party leadership of the academy not only reserved the right of final approval for all personnel decisions—from director of the academy to director of the archives—but also assigned projects to the president of the academy, who had no party affiliation.[89]

Opposition would have offered little hope for success: more than 80 percent of the seats in the most important working committee of the president, the executive committee of the academy, were held by the SED, and for the first time the post of first secretary of the academy party leadership was a sufficient qualification for membership in the academy's management.[90]

Important groundwork had thus been laid for the creation of the Research Academy of the GDR, that is, a national academy whose research goals were defined primarily by the economy and the development of society—a "socialist research academy" whose voting members came exclusively from East Germany. Thus the Academy Reform at the end of the second decade of the GDR can be seen as the culmination of a very dynamic process that preceded it, rather than the beginning of a new one.

—Translated by Thomas Dunlap

7

The Unity of Science vs. the Division of Germany: The Leopoldina

Kristie Macrakis

Throughout East Germany's history, the Leopoldina (the German Academy of Nature Researchers), Germany's oldest academy, characterized itself as an all-German institution adhering to the values of the unity of science in a time of national division.[1] It promoted the all-German clip theory—the idea that it functioned as a clip or clamp holding the two German scientific communities together—as a way to distance itself from the notion of a national, regional, communist science. It did not recognize the division of Germany; to the academy there was no GDR or FRG science but unified, value-free science unfettered by the boundaries of nations.

The Socialist Unity Party and the Ministry for State Security, by contrast, characterized the Leopoldina as a reactionary center for enemy activity that resisted the measures of the Party and the government. The Ministry for Universities and Technical Schools (*Ministerium für Hoch-und Fachschulwesen,* or MHF), which was responsible for overseeing the immediate affairs of the Leopoldina, never did succeed in exerting its control even though it financially supported the academy. Indeed, the Leopoldina was never transformed into a "socialist institution" (the SED's term) as were the Academy of Sciences in Berlin and the universities. It was an anomaly in the scientific landscape of East Germany.[2]

After the crumbling of the Wall, the Leopoldina was touted as the only place in the scientific landscape that "remained free of communist influence."[3] Historians should look upon such claims with suspicion. Our first job is to verify the truth of this statement, and if true, to examine the reasons for the Leopoldina's escape from communist influence. The best way to approach this question is to examine the intentions of the Communist Party and other state organs and to look at the way the Leopoldina reacted to those intentions and plans.

This chapter will analyze the relationship between the Leopoldina and its credo and East Germany's power triangle in scientific and educational matters: the SED, the MfS, and the MHF. This analysis will uncover how and why the Leopoldina managed to survive in a country that considered abolishing it. What function did the Leopoldina serve for the Communist Party in the GDR? Did it actually succeed in promoting and maintaining scientific contact between East and West while remaining an "all-German institution?" What does this story tell us about the nature of the ruling power structures in "real-existing socialism" and the possibility of "resistance?" Finally, how successful were the "state organs" as instruments of totalitarian control, and how useful is the concept of totalitarianism to describe East German socialism?

Clearly, an institution with the proclaimed political agenda of keeping science united in a time of division was seen in a different light by the Party and the state as their political goals and attitudes toward German unity changed. The fact that German unity was still a possibility in the 1950s allowed the academy to be refounded. Although the Party toyed with abolishing the Leopoldina as the lines of political division solidified with the building of the Wall, the academy soon fit quite well into an emerging internationalist ethos in the 1960s and early 1970s.

Character and Brief History

The Leopoldina's chief functions were organizing biyearly conferences, publishing scientific journals, and sponsoring monthly lectures, primarily for members. The Leopoldina was an international learned society, maintained no institutes, and had between 600 and 1,000 members from all over the world over the years 1952 to 1989. The members were the core of the academy (the Leopoldina boasted that Charles Darwin was a member). During the GDR period, names of members familiar to Westerners included Carl Friedrich von Weizsäcker, Alexander Oparin, Manfred Eigen, Adolf Butenandt, and James B. Conant, among others. The academy's headquarters had been primarily in Halle (since 1878), in southeastern Germany, and sometimes in Schweinfurt, West Germany (before 1878), where it had been founded. Until the library grew too big to move, the headquarters changed to follow the president. The vice-president was always West German, and most of its members lived in the Western part of Germany.

Founded in 1652 by scientists and medical doctors, the Leopoldina is said to be the oldest academy in the world. From the start it characterized its mission as transcending national boundaries. During the 300-year celebration in 1952, the period in which it was founded was also described as a time of territorial "splintering" in Germany. It saw itself as a unifying element during times of political division.

During the occupation period in the Eastern zone of Germany, all societies were shut down, including the Leopoldina. Its members met unofficially and attempts were made to interest the Soviet Occupation Authorities and the state government of Saxon-Anhalt in making the academy official once again. The *Works of Science* by Goethe began to be edited under the auspices of the academy in 1947, and lectures began to be held in 1948.[4] It was not officially reopened, however, until 1952. The long-time academy president and eminent physiologist Emil Abderhalden had died in 1950, and so Otto Schlüter, the vice-president, presided over the celebration.

At that celebration, the German Democratic Republic officially recognized the Leopoldina and offered to support its scientific endeavors. The academy was reopened at a time when German unity was still a possibility and the Communist Party was willing to do anything to advance and support science and technology. In fact, Otto Grotewohl, the GDR's minister president, wrote a letter of congratulations and best wishes to Schlüter and the members of the academy, stressing the importance of scientists at a time of "struggle" for "German unity."[5] Stalin's note of March 1952 proposing a scheme for reunification underscored the possibility of German unity.

The main activity of the Leopoldina—the yearly meetings—featured scientific topics that usually crossed traditional disciplinary boundaries. The president's interests seemed also to influence the topics selected for those meetings. As we can see from the chart listing the meeting topics in the postwar period, Kurt Mothes's presidency featured many themes from the biological sciences, whereas Heinz Bethge's reign emphasized the physical sciences:

Topics of Yearly Meetings
Mothes:

1957: The Virus Problem
1959: The Time Problem
1961: Energy (not held)
1963: Nerve Physiology
1965: Radiation
1967: Biological Models (Delbrück)
1969: Structure and Function
1971: Computer Science
1973: Evolution

Bethge:

1975: Systems and Their Limits
1977: Process Kinetics

1980: Space and Time
1983: Dynamic Structures—Unstable Processes
1985: Singularities
1987: The Elemental
1989: Anomalies[6]

In 1954, Kurt Mothes (1900–1982), the eminent biochemist, became the president of the academy and shaped its style and character for the whole GDR period. His specialty was the biochemistry of plants. He had been a professor of botany at the University of Königsberg from 1934 to 1945 and was a Russian prisoner of war until 1949; he remained president until 1974. He was politically conservative and had been an NSDAP member during the Third Reich.

Mothes's successor, Heinz Bethge (born 1919), is a physicist whose presidency ran from November 15, 1974, until June 30, 1990. Although different in character, Bethge was also fairly outspoken vis-à-vis the government. Born in Magdeburg, he received his first degree from the University of Halle in 1949, and a Ph.D from the same university in 1954. Unlike Mothes, then, he was educated in the GDR and did not carry the weight of a Nazi or bourgeois past. He spent most of his career at the University of Halle until he became director of the highly respected Institute for Electron Microscopy of the Academy of Sciences in Halle. His research focused on crystals. It does not appear that he was an SED member.[7]

The Power Triangle

The Socialist Unity Party stood at the top of what I call the East German power triangle in science, culture, and education, at least in the relations between the Leopoldina and "state organs" from the 1950s through the 1970s. In this case the Party did play a leading role and the other state organs were there either to carry through the Party's goals or to protect its interests. The Ministry for Universities fulfilled the goals of the SED and was responsible for overseeing and controlling (Betreuung) the Leopoldina through frequent direct contact.

The Ministry for State Security, often referred to as the "sword and shield" of the Party, had placed the members of the presidium of the Leopoldina under surveillance and operational control because of suspected "resistance against the measures of the state" and possibly breaking several GDR laws. Since the fall of the Wall the thesis has arisen that the MfS was not merely the sword and shield of the Party, but also a state within a state, that is, its own power center. While this may be the case for the later years of the GDR, the story of the Leopoldina supports the sword and shield interpretation of the Party at least until 1969, when the operational file was closed.

The SED and the Leopoldina

> I can say with absolute clarity that the SED is very interested in sponsoring our scientists.
>
> —KURT HAGER, SED'S CHIEF IDEOLOGUE, TO KURT MOTHES[8]

In 1958 Walter Ulbricht, the head of the Communist Party, made a legendary visit to Halle to talk to the intelligentsia. At this famous meeting he made it publicly clear where the Party stood on questions of science and the intelligentsia.

Ulbricht's visit to Halle had an impact because he was the head of state. He was, however, restating what had already been formulated by the Party as part of the New Course five years earlier in 1953. The New Course, or the "building of socialism," was a concept from communist theory that allowed for a transition period after the revolutionary overthrow of capitalism until the end of the full socialist system, a phase that leads to the establishment of the fundamentals of communism. During the early 1950s, the GDR saw the New Course as consisting of all-German science in a unified peaceful Germany, under East German rule.[9]

During the 1950s the Party promised the intelligentsia material security and support for their work. Simultaneously, it was ready to be "tolerant" toward bourgeois intellectuals in the face of initial resistance. In the 1950s and 1960s there existed the concept of the new and the old intelligentsia. The new intelligentsia was to consist of people from workers' and farmers' families. This was to be the cadre *(Kader)*, the new technical socialist intelligentsia. At an important conference on universities and science under the New Course in 1953, party leaders announced that they "wanted not a course *against* the partyless [*parteilose*] bourgeois intelligentsia, but a course to win over the partyless and bourgeois intelligentsia."[10] This was accomplished, in part, by offering material benefits but also through political discussions or conversations *(Aussprache)*, a method of persuasion. (I recall a scene in the film *Kinder, Kader, Kommandeure*—Children, Cadre, and Commanders—a movie using propaganda trailers, in which Ulbricht, with his pointed goatee, is swinging on a child's swing. The camera zooms in on him and he says, patiently and paternalistically, "And now time for an *Aussprache*.")

This method of persuasion was used during the 1950s by the Central Committee with many partyless scientists. For example, the Central Committee member Kurt Hager, the head of the *Wissenschaft* department, conducted these "discussions" with scientists like Gustav Hertz (the GDR's only Nobel prize–winning scientist), Hans Stubbe, and Kurt Mothes. Over the years, Hager had numerous discussions with Mothes, who never wavered from his all-German ideology.

Mothes was considered one of the "old, bourgeois intelligentsia" and was thus tolerated by the Party despite his diverging views. But this was not the only reason the Party never ostracized Mothes. During the mid-1950s, Walter Ulbricht had decided that a "modus vivienda" had to be found to cooperate with Mothes, or the "most prestigious scientific society" would move to Schweinfurt, West Germany. The government described its discussions *(Aussprache)* with Mothes over the years as "cautious."[11] Not only was the Leopoldina an all-German academy, but it had the option of moving its operations to the West, something the GDR feared and wanted to avoid.

This early emphasis on cadre and personnel is also reflected in the Party's interest in recruiting "progressive" scientists as members of the Leopoldina. A member of the Central Committee's Department of Science marked up a Leopoldina membership list with colored pencils to indicate if a new member came from a socialist country (red pencil) or from a "capitalist foreign country" (blue pencil). The membership list showed that in 1959 only three new members came from the GDR, five from the Soviet Union, twelve from "capitalist foreign countries," nineteen from West Germany, and thirteen from neutral countries.[12]

Hager, who was the party leader responsible for the Leopoldina, described his relations with the academy as "only friendly and good, which does not exclude differences of opinion in one or another thing." Hager, all government officials, and most scientists had the highest opinion of the Leopoldina. Hager thought it "played a big role in mediating scientific knowledge. Its publications were a real find. It was difficult to approach Leopoldina members and staff, but their scientists were highly competent and the conferences in Halle were an experience for young scientists." Despite this respect, some Politburo members, including Hager, had attempted to eliminate the academy.[13]

"The Antifascist Wall of Protection"

During the 1950s, then, the Party concentrated its efforts on the political discussions with Mothes, while the MHF performed the actual hands on "care and control" work. A turning point occurred, as it did in all areas of life, after the building of the Wall on August 13, 1961.

In October of that year the Leopoldina had planned to hold its yearly meeting in Schweinfurt, West Germany. The topic of the meeting was energy. The invitations had been sent out in June, and as he had done before, Mothes used the invitation as a forum to advertise the role and function of the Leopoldina in the GDR. He reported that since its reopening in 1952, the academy had experienced a strong "growing together, and its function in a time of division of the fatherland, and of the whole world, is well known."[14]

Two months after the invitations were sent out, the Berlin Wall was built. By the end of August Mothes had sent out another circular canceling the meeting. He reported that he was "forced" to cancel the meeting because it seemed impossible for GDR members to obtain unrestricted permission to travel. He expressed concern that the prohibition of travel would deter scientists from visiting the GDR for lectures and conferences.[15]

The circulars were sent to all members of the academy, including the five socialist members, who immediately reported the cancellation to the prime minister's office. Since this was seen as an act of provocation, Mothes was called to the minister's office to speak to Deputy Minister Alexander Abusch on November 30, 1961. The meeting was obviously important, for lengthy minutes of it exist in Walter Ulbricht's files. The government was disturbed that Mothes had made what it considered a political statement against the regime. Abusch argued that the Leopoldina was on GDR territory, and that the government totally supported it and expected "loyalty" or at least some "understanding" that the Wall and new policies were a reaction to Bonn policies. Abusch noted that no other society or academy had sent out a letter in which it was implied that the German Democratic Republic had done something "unlawful." Abusch defended the building of the Wall, citing "aggressive West German measures against the German Democratic Republic, which were supposed to lead to civil war."[16]

Mothes was upset about the building of the Wall and told the minister that the "limits of humanity have been reached when . . . a mother cannot travel to see her sick child in Frankfurt." The state is there for the people, he told the ministers, and not the "people for the state." In response, Abusch claimed that "we are ready to hear about the mistakes we have made," but continued to blame and attack the West Germans for luring East Germans away.[17]

Abusch held the view that the GDR had now "secured" itself, and that, after the "measures of August 13," as the building of the Wall was called in official government circles, it had become stronger. He said that scientists' travel would be limited for a time, but would then resume. Abusch made it clear to Mothes that he was "not demanding that Mothes become a Marxist and that he could also not demand that he give up his *Weltanschauung*."[18]

In January 1962 Mothes and Abusch met again to discuss future "cooperation" between the Leopoldina and the government. It is not entirely clear who initiated this meeting, although it seems likely that it was Abusch. Two main issues discussed were the sudden "retirement" of the Leopoldina's vice-president, Erwin Reichenbach, and the politics of two-state status in Germany.[19]

While the Leopoldina had been refounded at a time when reunification was still a possibility, in 1962 Abusch did not think reunification was possible. The two German states had developed opposing systems—one was capitalist; the other was trying to "build socialism." The erection of the Wall became a physical demonstration of an existing state and demarcated the East from the West. Because of these new developments, complications had arisen regarding the role and function of the Leopoldina. With the newfound leverage that came with the building of the Wall, the state attempted to assert its ideology more aggressively. Mothes stressed the role that he and the Leopoldina had played in reconstructing postwar science by retaining scientists who might otherwise have fled to the West.[20]

The representatives of the state demanded that the Leopoldina present a plan on how the academy fit into the new political constellations. They even threatened that the Leopoldina would not get very far if it held fast to traditions. They justified their control over certain institutions by emphasizing that they received financial support from the state. Mothes reiterated the ideology of the unity of science in times of division but was vague on how the academy fit into the new system.

There is no doubt that the building of the Wall marked a chronological break in the historical development of the GDR. Revisionists in the future will have a hard time reinterpreting or denying its importance as a turning point. The above discussion shows the newfound confidence of the state in asserting itself. Although only hinted at in this official discussion, the unity of science was also threatened more seriously than before by the new physical barrier and the accompanying restrictions on travel. There were renewed discussions in which scientists voiced concerns about the destruction of the unity of science through the political, and now physical, division of Germany.

Although the SED demanded a new "conception" from the Leopoldina, one never really emerged during the 1960s. More pressure, however, was exerted on the academy to rewrite its statutes. This pressure came from the president of the German Academy of Sciences, H. Klare, as well as from government quarters. A copy of a 1969 draft of the statutes exists in party files.

During the early years of the GDR, the Leopoldina benefited from the SED's policy of promoting and supporting science and technology. Unlike National Socialism (which is often interpreted as hostile to modern science and technology), communism glorified science and technology and saw it as the engine for socialist change. This, together with thoughts of unification in the early 1950s, paved the way for the Leopoldina's reopening in 1952.

Oversight and Control: The Ministry for Universities

While the SED formulated policies toward science and scientists and even had close contact with the leaders of scientific institutions, the Ministry for Universities and Technical Colleges was responsible for daily control. The MHF, as its title indicates, was primarily responsible for overseeing the universities, and during the early GDR years, for transforming existing universities into socialist institutions as well as creating new ones. The placement of the Leopoldina under its jurisdiction was therefore to some extent inappropriate and the result of tradition and personalities. It seems that Gerhard Harig, the historian of science, then at the MHF, played a role in reigniting the Leopoldina's flame, and thereafter it remained under the MHF's supervision. Most of the other academies were placed under the jurisdiction of the minister president.

Another reason for placing the academy under the Ministry for Universities may have been the close ties all the leaders at the Leopoldina had with the University of Halle; they were all professors there. But because the Leopoldina was not directly tied to teaching and maintained no research institutes, it was not important to attempts at central control during the early years.

The relationship between the MHF and the Leopoldina began on the occasion of the latter's reopening on its 300th birthday in 1952. The ministry sent representatives to that meeting and to all its biyearly conferences thereafter. The presence of a ministry representative was not always met with enthusiasm by Leopoldina members, and usually there was no clapping after the representative's opening speech. Once the entire West German delegation left the room after the MHF representative's talk.

During the whole period of the GDR, the ministry wrote reports on the Leopoldina and laid out strategies on how to proceed with it. The MHF tended to become more active in its relations with the Leopoldina immediately before, during, and after yearly conferences. A pattern developed over the decades, with variations as the GDR policies and politics changed or developed.

During the 1950s, the reports concentrated on the history of the institution and the MHF's attempt to have it rewrite its statutes. The MHF pushed the issue throughout the period of the GDR, and the Leopoldina managed to postpone the process so that new statutes were never finalized. The Party and the MHF also attempted to increase the number of GDR and East Bloc scientists, especially "progressive scientists" (party members), in the academy's membership list.[21]

From the 1950s through the 1970s, the MHF tried to settle the issue of the statutes without much success. The only existing internal statutes were written in 1944, and the others dated from decades before that. It is thus no surprise that the statute issue came up at least every two years. By the end of the 1950s, the MHF turned to the Ministry of Justice and a lawyer to settle this issue in their favor.[22] (A common feature of totalitarian states is the attempt to control all institutions, but especially educational and scientific ones, by influencing the content of the bylaws.) It was only in 1967 that the Leopoldina finally created a statute draft to be approved at the yearly meeting in 1971. The Leopoldina presidium members had become expert escape artists, however, and these drafts never solidified into official statutes.

In 1952, shortly after the 300-year celebration, in addition to the resolution of the statutes issue, the MHF wanted to increase the representation of "progressive" scientists (party members) from the GDR and other socialist countries, to change the Leopoldina's tasks and methods for the new period, and to influence the election of the president and vice-president. Schlüter, however, did not acquiesce to the MHF and thus created a precedence for his successor, Mothes.[23]

By 1953 Schlüter was appealing to the notion of an all-German community and the historically conditioned structure of the Leopoldina. He differentiated the Leopoldina from other academies through its lack of institutes and the fact that it was not tied to a specific region; he did not want it to be a provincial learned society in Halle.[24]

In the fall of 1955, after the first postwar annual conference, the MHF complained that it had been difficult to influence the organization and course of the conference for two reasons. First, for many years there had been no personal or organizational relationship between the MHF and the Leopoldina; the only connection was financial. Second, the MHF saw a deeper problem in the attitude of the presidium, with its "conservative tradition" and attempt to remain independent from any state interference in its scientific research. Specifically, the Leopoldina did not want to be suspected by West Germans and foreign members of being closely tied to the GDR government.[25] The MHF thought the "relationship between the state organ and the academy reached a head" at the big yearly meeting in November 1955. At that meeting Mothes "rejected" every suggestion of the MHF. It was only with great difficulty that the "role of our state as sponsor of scientific institutions" could be documented.[26]

The MHF representatives came away from the conference believing that "patient influence and constant contact with the Leopoldina, and especially a better relationship with the leaders, need to be created."[27] There was some disagreement among MHF personnel on how to proceed with the Leopoldina. Franz Dahlem, the head of the ministry in 1955, scolded a department head, Dr. Müller, for the tone of his report to the Central Committee, which he characterized as "I came, I saw, I conquered."[28]

During the 1950s political unity was still a possibility. The Leopoldina's position was that German unity was the chief task of all Germans and that the academy had adhered to an "all-German" tradition since its founding.[29]

During the late 1950s, the MHF made concrete plans for how to bring the Leopoldina into its orbit. Although it had reported to Walter Ulbricht in 1957 that more scientists from the East had joined the academy, it was still determined to "take the Leopoldina out of its isolation" and centrally control it, as it did the other academies. In 1959 most of the members still came from the West. Despite the state's yearly grant of 100,000 marks, the MHF had no influence on the Leopoldina. That year it began to toy with the idea of founding an institute, perhaps in the history of science, to increase its influence. It also wanted to influence the cadre policies and increase the number of socialist members. In 1959 the MHF drew up an "action plan" toward these ends. It wanted to open an office in Halle for organizational work staffed with people from the ministry. As mentioned, the MHF already tried to influence the Leopoldina through control of its conferences; this included translating, answering questions, and writing daily reports.[30] From these activities it would appear that it was not only the MfS that kept life and science under surveillance in the GDR.

The Leopoldina continued its all-German rhetoric, and in the invitation to members for the spring 1959 conference announced that

in a time of imminent estrangement of the separated part of our fatherland, we have attempted to emphasize the unity of our culture in the area of science and to intercede on its behalf and to cultivate . . . the humanitarian character of science at every opportunity.[31]

In his opening address at the meeting, Mothes added a twist to the theme with a reference to Galileo and the continual struggle for the freedom of research: "Galileo is a symbol. Every time has its unnamed Galileo."[32]

During the early 1960s the relationship between the Leopoldina and the MHF did not change very much: the Leopoldina continued to adhere to its credo and the MHF continued its plan of attack as outlined above for the 1950s. After the Wall was built in 1961, state control shifted to the Party and the office of the Minister's Council. Following the government's official line, the MHF sanctioned the existence of two separate states and attempted to force the Leopoldina to conduct its affairs within that political context.

During the late 1960s the GDR planned to prohibit all-German societies, and there was discussion in the Politburo of banning the Leopoldina, although I have found no reference to this in the MHF files. On the contrary, by the early 1970s the emphasis was on keeping the Leopoldina in the GDR.

By the late 1960s and throughout the 1970s, the MHF saw the future of the Leopoldina as sitting squarely in an internationalist ethos. The Leopoldina, like the GDR generally, emphasized its international character. It was during this time that the GDR gained international recognition. This is reflected in the creation of embassies worldwide in 1972. In 1967 the MHF wrote that the further development of the Leopoldina should lie in its character as an international society with its headquarters in the GDR. It is thus no surprise that responsibility for the Leopoldina first shifted in the ministry to the Foreign Department of the MHF and then to the department for international relations.[33]

Plans on how to proceed with the Leopoldina increased during the 1970s, as the MHF came to realize that it had been too much on the defensive during the 1960s: "We were on the defensive, and mostly just reacted only shortly before a certain activity. We did not . . . develop an active policy." Since there had not been an "offensive, goal-directed policy," the MHF designed some "necessary measures" to be taken. Although the MHF continued to pursue its initial goals, there were some variations.[34]

In anticipation of the yearly conference on computer science in October 1971, the MHF in February 1971 had already begun to consider how it could influence the further development of the Leopoldina. It claimed that the Leopoldina had remained "bourgeois" and considered itself to be above the "worldwide class struggle between capitalism and socialism": "It cultivates a value-free science and does not feel tied to state directives in such questions." It "embodies the typical bourgeois-conservative learned society," they complained. The Leopoldina continued to consider itself a "German" scientific society, where "German" meant all German-speaking countries in central Europe (Germany, Austria, and Switzerland).[35]

Although the MHF acknowledged that the Leopoldina consisted of internationally recognized scholars, it was not happy with the abundance of West German members and their politics, past or present. It found that many older West German members were ex-NSDAP members who had filled important positions during the Nazi period at the universities or the Kaiser Wilhelm Society. The MHF accused many of these same West Germans of having influential positions in "monopolistic university life and science in West Germany." Their background led to a "continuity of imperialistic university and science policy." As an example, the MHF used Adolf Butenandt, then president of the Max Planck Society and past vice-president of the Leopoldina. He was "closely tied to the West German state science organization" and was a member of the Leopoldina senate. Although there had been a small increase in the number of scholars from socialist countries, the MHF found that the academy still had not improved the political composition of its membership list—the majority of members were Westerners who held "bourgeois, petty-bourgeois conceptions, which included anticommunist and antisocialist ideas." An official at the MHF found that none of the presidium members was a party "comrade," and only three GDR scientists were party members.[36]

To combat this situation, the MHF proposed creating a working group stacked with "progressive" Leopoldina members and an MHF representative. The group would be charged with continually analyzing the development of the Leopoldina, providing suggestions for "improving" its work, and nominating members from socialist countries.[37]

The Leopoldina was valuable to the GDR, according to the MHF, because it brought together scientists from all over the world in the GDR. It considered the yearly meetings, colloquia, lectures, and publications to be of high caliber. The MHF intended to use the scientific events for the "socialist community" by discussing basic issues of science and society from the standpoint of dialectical materialism, and appointing scholars members of the Leopoldina if they adhered to dialectical and historical materialism. (They also wanted to secure the international character of the Leopoldina by incorporating a clause to this effect into its statutes.) The international modus vivendi would be used to limit the West German imperialist influence, while strengthening the representation of the "progressive" scientists whom they wanted to facilitate the "adapted development" of socialism in the GDR.[38]

In 1972 the department for international relations of the MHF described its "principal line" as follows: to keep the Leopoldina in Halle; to repress the West German influence; to increase its own influence step by step; and to use the Leopoldina's events for the GDR's scientific development.[39]

In order to "politically safeguard" the upcoming conference in October 1972, the international relations department planned to use "every opportunity . . . to oppose every attempt to demonstrate the unity of German science."[40] By 1974 the MHF complained that the Leopoldina was increasing its efforts to undertake all-German activities and to use itself as an "all-German clip." As an example, the MHS referred to a couple of lectures at which 50 percent of the attendees were West Germans.[41]

On the surface the relationship between the Leopoldina and the MHF appeared to change when Heinz Bethge became president in 1975. In a confidential conversation with an official from the MHF, Bethge claimed that he intended to shape the work of the Leopoldina so that it would develop an open and trusting relationship with the MHF and the Academy of Sciences— all in the interests of the GDR. Further, he stated that he "rejected" Mothes's style as too "ornate" and overdone, and did not always agree with his science policy evaluations. (This, of course, can be seen as strategy on Bethge's part.)[42]

By 1979, the MHF had noted some progress at the Leopoldina: a GDR scientist was going to give the opening lecture at the 1980 yearly meeting, and there were six SED members on the academy's membership list. The MHF still thought, however, that the West German presence at the scientific meeting was too strong; of the 1010 members present, 718 came from capitalist countries and 292 from socialist countries. There were 356 West Germans and 143 East Germans.[43]

During the 1980s, relations between the MHF and the Leopoldina seem to have normalized, and most of the correspondence relates to organizational matters, the building of a new lecture hall, and the submission of "work plans" each year. Bethge often initiated meetings with the MHF. There were no "action plans" or offensive measures, talk of politics decreased, and the issue of the statutes apparently did not come up again. There continued to be a great deal of respect for the caliber of scientists who belonged to the academy, as well as for the high level of science they practiced.[44]

In 1986, the year of the Eleventh Party Congress, in which science and technology figured prominently, the MHF formulated some expectations about the future work of the Leopoldina, reflecting the Party's vague emphasis on university-scientific life: securing freedom, intensification of science and production, and higher effectiveness in research and engineering education. Within the framework of the Leopoldina's dialogue between scientists from the East and the West, the MHF pressed the academy to foster the "spirit of humanism" and a clear belief in the responsibility of the scientists for "securing freedom and the advancement of societal progress." An official from the MHF also discouraged the Leopoldina from having contact with those who supported German unity: with its "historically conditioned special anchoring in German-speaking lands," the Leopoldina should stay away from all "bourgeois politicians and ideologues who speak of the 'unity of German science' and of the continuation of a 'unified German cultural nation.'"[45]

"COMET": The Ministry for State Security

In November 1958 an operational file was opened at the Ministry for State Security's Halle district office under the code name "Comet." Unfortunately, no explanation is given for the selection of the code name, but we can surmise that the MfS must have been thinking of a foreign object that follows an eccentric orbit, as comets do. The occasion for opening a file and observing the Leopoldina was the suspicion that a group of scientists—nicknamed the "Halle Center" by Mothes himself—existed there which "constitutes the center of resistance against the measures of the Party and government in its creation of socialist universities." In 1958, the MfS had already "smashed" the illegal group "Spiritus Ring," which consisted of twelve university professors, but now found this other group in the "foreground." It therefore began intelligence work on the character of the organization and its most important members, concentrating on the leaders, ten professors at the University of Halle.[46] In addition to resisting the efforts of the government to create socialist universities, the group, according to the MfS, "openly propagated bourgeois ideology."[47] The Leopoldina was also accused of breaking four GDR laws, including those on treason, propaganda and agitation that threatened the state, state slander, and harmful activity and sabotage.[48]

In the course of ten years, the MfS Halle office created more than 18 volumes of material on the Leopoldina, running some 2,700 pages. The files consist of informant reports, summaries of transcripts of telephone conversations, copies of Kurt Mothes's letters, and, most important, summary reports by the case officer on operational activity. Only copies of the substantial summary reports were given to "MfS friends," as the KGB was usually called. Operational files were usually opened if a person or group was suspected of wrong-doing.

It is not clear how the attention of the MfS was drawn to the Leopoldina, but it appears to have something to do with the members' contacts with the University of Halle. Several of the informants (called *Gesellschaftlicher Mitarbeiter,* or "societal staff members" in the 1950s as opposed to *Inoffizielle Mitarbeiter,* or "unofficial staff members," later) used by the case officer had already written reports on a few Leopoldina members before the file was officially opened in 1958. The first informant report from "Gisela" dates from the 1955 annual meeting. It is difficult to determine the real identity and names of the informants by reading the volumes, partly closed and blackened by a staff member at the archive, but it appears that one of the four regular reporters was a professor and another a secretary. The code names of the initial six informants—all in "key positions"—were Egon, Pohl, Fink, Sternheim, Paul, and Förster. Many of the reports were about faculty meetings at the University of Halle in the medical and mathematical-scientific faculty.[49]

These informants were given the job of spying on members of the Leopoldina and their contacts. They were also told to do research on the cadre policies of the institutes headed by Leopoldina members, and to determine the effect of the academy's policies on the University of Halle and on university development in general in the GDR.[50]

Alfred Trautsch was the operational officer who gave instructions to three main informants—Fink, Sternheim, and Egon—on what to look for soon after the operational file was opened. Common tasks for the informants included character sketches of the well-known members, including their political views, a cadre policy analysis, a political assessment of the Leopoldina and its importance, and its role in and influence on university developments in the GDR. Trautsch was interested in identifying members who were "progressive." At the same time, he also wanted to know if there were individuals who were "reactionary" and if there was a "reactionary center."[51]

In addition to using informants—the chief operational method in espionage work—Trautsch looked up all members of the Leopoldina in the "incrimination card file," which meant the NSDAP archive. This card file was used, especially in the 1950s, to blackmail people with politically sensitive pasts, such as NSDAP memberships.[52] The result of the research by the MfS uncovered at least eighty-eight "Nazis" as Leopoldina members. The report broke down these "Nazis" (one of whom simply belonged to the Teacher's Union) into membership in other National Socialist organizations, which means that many members were in several organizations: officers or professors who were especially active, twenty-five; members of the NSDAP, fifty-eight; Members of SA (Sturm-Abteilung), thirty-three; members of the SS, nine; NS-Dozentenbund etc., sixty-three; higher officers, nineteen; and Free Corps, German Nationalists, thirty. Most of the twelve university professors constituting the Halle Center had some affiliation with a National Socialist organization. The MfS even found a letter Mothes had written to Adolf Hitler expressing his gratitude for being named a professor in Königsberg; Mothes also expressed his admiration for Hitler.[53]

Although not as often as some would suspect, the MfS occasionally planted microphones—using Department O, as it was then called—in order to find out more about the "plans and goals of the leading members" of particular groups. This was not easy to do with the Leopoldina, because the MfS determined that meetings took place both at Mothes's house and at the Leopoldina building.[54]

By April 1959, Trautsch had written a summary report on the Halle Center, and five copies were made for various departments and for the "friends." After introductory material on the reasons for the file, a summary of the history and structure of the Leopoldina, and a list of the goals of the Center, the 100-page report became a lawyers' brief in which quotes from informant files and reports were used to illustrate "forms of enemy activity" and "influence in the GDR." According to Trautsch's interpretation, the goal of the Halle Center was to "undermine" and eventually to "abolish" the "constitutional state and societal order of the German Democratic Republic." Members of the center took advantage of the GDR's intelligentsia policies in order to draw attention to themselves, he wrote.[55]

Trautsch made several preliminary recommendations on how to "change the political course of the Center." He thought the Leopoldina could be made to adapt to societal conditions in the GDR, and stressed that it would be a great benefit and an international coup for the GDR if the Leopoldina could "represent" the GDR's political, economic, and scientific interests. Trautsch noted that the Leopoldina's political influence and contacts were extensive, and might be employed in the gathering of scientific-technical intelligence through the foreign espionage units (such as Department XV, the district foreign espionage unit). Specifically, the Leopoldina had many Western contacts that could be used to obtain scientific-technical information.

One way to influence the political course of the Leopoldina was to install a state-selected "director" with the justification that the institution was state-supported. Like the MHF, Trautsch found that it was necessary to change the statutes so that the Leopoldina would "adapt to societal relations in the GDR." He proposed influencing the Leopoldina through leading functionaries and by the creation of opposition or division within the reactionary center itself, for which task he would use rivals like Hans Stubbe, for instance. The final goal was to "isolate" members of the Center. He concluded that the "complicated process" of transforming the Leopoldina into an "instrument of the workers' and farmers' power" was only going to be successful if the MfS employed all its resources.[56]

A little over a year later, in August 1960, another operational plan to "smash" the negative influence of the "reactionary Halle Center" was proposed by the MfS. The Leopoldina was the only academy to have evaded any attempts to be placed under the supervision of the government. The goal of the operational plan was to create mistrust within the Center using "covert means." The MfS also wanted to "expose," "isolate," and "liquidate" the "active enemy" by fabricating material to prove their enemy activity. In order to prevent "resistance" by Mothes, the MfS planned to bestow state medals on him for his work as a botanist![57]

In order to create mistrust among members of the Halle Center, the MfS planned to have the Halle district SED secretary, Comrade Bruck, hold discussions with members, during which he would drop remarks from internal Leopoldina meetings. Bruck was to give as a source a professor in the medical faculty. This type of method was to continue with some variations. Needless to say, Bruck, with whom the MfS later consulted, was not happy with this measure and asked the MfS to hold back. He thought such measures would only increase the problem of scholars "fleeing from the Republic."[58] The fact that the SED secretary was consulted and listened to by the MfS supports the interpretation of the latter's task as sword and shield of the Party; the Party and the MfS certainly worked together and coordinated their tasks.

Soon after the building of the Wall, operational activity relating to "Comet" increased. The MfS made plans to keep the circle of leaders under "tighter control" and to decrease the influence of the active members. They did this through increased work with informants in key positions. The MfS Halle office consulted again with Bruck and reported on the newest developments in "enemy activity." The recommendations of the MfS were eventually to reach the first secretary of the Central Committee of the SED. The MfS sought Hager's support for "active and offensive" action.[59]

In January 1962 a report was written on the "enemy activity of the Leopoldina Center" for the SED leaders in Halle, Comrades Koenen and Bruck, to be used in consultation with Kurt Hager. Interestingly, two versions of this report exist, one for the SED leadership and one internal version for the "Comet" file. The main difference in the texts is that the sources are left out of the Party's version. For example, if the source of information came from "GM Egon," this was omitted. In addition, the code name and registration number for the "Comet" file is left out.

The report focused on activity after the building of the Wall, and this pointed to the circular Mothes wrote canceling the Schweinfurt meeting. The MfS also found that the Center's attempt to have the laws changed in connection with the "security measures" of August 13, 1961, constituted organized enemy activity. There were also measures against the state's attempts to make the economy "free of disturbances," especially in the area of science and the importing of medicine, and "enemy activity against the socialist world system." Other activities that existed before the building of the Wall, such as working against the intelligentsia policies of the Party, continued. In conclusion, the report noted that "NATO secret service" agencies were interested in Mothes, and that the MfS thought he might be recruited during one of his trips to the West.[60]

As a result of the meeting between Bruck and Hager, further measures were taken to increase operational activity around the Halle Center. Future plans were made for closer cooperation between the MfS and the Party and for increased control of the Leopoldina's activities. The department also planned to increase the activities of the telephone-tapping and mail-opening departments.[61]

MfS operatives planted bugs in the rooms of Leopoldina buildings on several occasions, but had to abort one mission when they heard someone in an adjacent room. In August 1962, they decided to try a new method for installing technical devices and planned a break-in at the Leopoldina head-quarters. The break-in was announced in a letter to the district office of state security in Halle and signed by "N." In this mysterious letter, N announced a planned break-in on the night of August 22 in the rooms of a scientific society on August-Bebel Street. He stated that "for personal reasons I would like to remain anonymous."[62]

During the rest of the 1960s, operational plans continued and reports were written. Since it never did find conclusive proof that the Halle Center had broken any GDR laws, and since many of the professors had retired, the MfS closed the operational file on "Comet" on September 4, 1969, and opened an object file. Unlike an operational file, which is used in an investigation to gather material and information to prove suspected wrong-doing, the object file is used when the object is under surveillance but not for the purpose of legal prosecution.[63]

Comparison with Other All-German Societies

By the 1950s, and especially the 1960s, there were very few institutions on East German soil that could maintain activity in both the East and the West. There were several reasons for this, the first and most obvious of which were the physical barriers—first the sectors and difficulties in passing the border, and then the Wall. Second, in its attempt to transform institutions into social-ist organs by changing the composition of the personnel into committed communists, the SED did not encourage contact between East and West. Unless an institution already had a strong all-German sentiment, the evolu-tion of general policies, which tended to demarcate the East from the West, naturally isolated institutions from the West.

Immediately after the Wall was built there was great concern that the two scientific communities would be cut off from each other; indeed, I have found frequent references to the impending loss of the unity of science with the new physical division. For example, at the Berlin-Buch Institute of the Academy of Sciences, there were many discussions in the days after August 13, 1961, on the unity of German science. Many scientists thought that the unity of German science would be destroyed because conferences held in West Germany, for example, could not be attended by East Germans.[64]

A few other institutions in Germany either had had dual headquarters (like the Leopoldina) or had always emphasized an all-German ideology or propa-gated one during the division of Germany. The Goethe Society in Weimar (with Western headquarters in Frankfurt) was one of these few all-German societies. Although not associated with the natural sciences (except for the fact that Goethe himself had also dabbled in the sciences), the society experienced a pattern of development similar, though not identical, to that of the Leopoldina. Like the Leopoldina, the Goethe Society attempted to preserve the unity of the society during a time of national division. At its general meetings it, too, used all-German rhetoric in an attempt to keep the concept of German unity alive.[65]

Although the Soviet occupation forces allowed the society to reopen in 1946, it did not have its first yearly meeting until 1954. A break in continuity occurred when the president, a West Berliner named Andreas Wachsmuth, died. Wachsmuth, who had become president in 1951, has been praised for keeping a sinking ship afloat and for his aplomb at mediating East-West tensions. He was not a Marxist, and therefore it is not surprising that the SED saw the society as a Trojan Horse that was importing bourgeois ideology into the GDR. Kurt Hager's stance toward the society was similar to his feelings about the Leopoldina, although he favored the Leopoldina because it helped support East German science. In 1957 Hager stated that all-German institutions were "agents of Western ideology."[66]

Although a systematic analysis of the relationship between the SED, MHF, and MfS and the Goethe Society has not been undertaken, it is clear from the existing material that similarities exist between it and the Leopoldina's relationship with those same organs. Like the Leopoldina, the Goethe Society stood under the MHF's jurisdiction. Figures like Alexander Abusch commented on the society and its role in cultural policies. There was an ideological battle between the socialist and the capitalist world on who should inherit the spirit of Goethe. Goethe strangely became a model in the fight for a higher productive force in the economic, moral, and intellectual arena. A similar chronological development took place from the 1950s to the 1970s for the Goethe Society. By the 1970s, the Goethe Society, like the Leopoldina, began to emphasize its international character.[67]

Conclusion

There certainly is truth to the statement that the Leopoldina remained free from communist influence. The story of its relationship with the state organs shows how they attempted, but mostly failed, to exert their control. The Leopoldina survived, in part, because of its tenacious hold on tradition and its ability to play for delays. Since the Party's goal was to promote science and technology, it was in its interest to keep the Leopoldina in the GDR, even if it did not conform to its image of a scientific institution.

Several historical developments also protected the Leopoldina and made it valuable to the regime. The academy was refounded at a time when German unity was still a possibility, and so its identity as a clip or a clamp holding the two German scientific communities together was not at odds with state policy. After the building of the Wall, it became increasingly difficult for the Leopoldina to hold fast to this tradition; thus it was lucky when the state set forth an internationalist policy in the late 1960s and early 1970s. Then the Leopoldina fit right in and could be used as a signboard.

If the Leopoldina was able to resist the measures of the Party and the state, could others have done the same? The GAS (as Peter Nötzoldt has shown in Chapter 6) was changed from a Learned Society to a major research academy along the lines of the academy in the Soviet Union. Many universities were transformed, but there were pockets of resistance, especially of the old intelligentsia during the early years. The state was more intent on smashing resistance at the universities, since, as educational centers, they are tools for implementing a new world view. The Leopoldina was a somewhat eccentric, relatively small academy with no research institutes and no students. It therefore was not of prime interest in the transformation process, although later there was great interest in supporting it because it added international prestige to the GDR.

A similar pattern of transformation occurred during the early years of the Third Reich after the National Socialists seized power. The National Socialists had concentrated their attention on transforming the universities according to their ideology while neglecting the Kaiser Wilhelm Society because of its nature as a research institution.

The Leopoldina also had other elements protecting it: it could have moved to its headquarters in Schweinfurt, West Germany, a move the Party and state did not want. When there were tensions between the Leopoldina and the state, the former threatened to leave. Mothes and other academy leaders were internationally known and highly respected scientists even though they were politically conservative and were part of the older generation. In the end, their all-German rhetoric was a highly political statement even if the goal was to adhere to value-free science.

The role of each angle of the power triangle illuminates both the intentions of the state and the way in which it implemented its goals. The Party was the general policy maker, while the MHF stayed in direct contact with the Leopoldina. Kurt Hager, first a member of the Central Committee and then of the Politburo, was in charge of the Leopoldina and had many "conversations" with Mothes and Bethge. In a time of crisis even the prime minister was called in. It should be noted here that the Party never wanted to, and never tried to, influence the science done by Leopoldina members or censor the topics of the conferences. In the early years the Party was primarily interested in controlling personnel issues and persuading the leadership to come to its side; this was part of the cadre and intelligentsia policy. This interest in controlling and monitoring members continued throughout the GDR period. After the Wall was built, the government pressured the Leopoldina to conform politically to the physical fact of the division of Germany into two social and political camps. In the end the academy fit into the GDR's new internationalist policy.

Of all the state organs, the MHF exerted its control the most. Although during the 1950s and 1960s it characterized itself as too passive and thus made action plans in the 1970s, it still attempted to change the statutes and make the Leopoldina admit more "progressive" scientists. The MHF "retired" Erwin Reichenbach.

Finally, the MfS was the sword and shield of the Party and was responsible for "smashing" enemy activity. The fact that the SED branch office in Halle could veto some of the MfS's operational ideas shows the extent to which the MfS was an arm of the Party and not an independent state within a state during the first two decades of the GDR's existence. In fact, the MfS closely cooperated with the SED during critical periods. The Party even received a copy of an MfS internal file report on the Leopoldina so that plans could be made on how to proceed with state measures against the activity of the Halle Center. Although to some extent an exaggeration, the MfS rightly characterized the group as "reactionary" and against the socialist transformation. They even found "dirt" on the leaders' "Nazi" past, although it was never used against them. In the end, the operational file on the Center was closed in 1969 because most of the leaders under surveillance had retired from the university (most willingly) and the MfS thought this removed them from the picture.

Finally, there were very few all-German institutions in the GDR. The comparison with the Goethe Society shows the extent to which state policy was structurally similar for the existing all-German institutions.

Surely the above narrative telling the story of the power triangle's effort to control, influence, and steer the direction of the Leopoldina has implications for the interpretation of the GDR as a totalitarian state during the 1950s. There is no doubt that the story also indicates that the state attempted to control institutions and life to a high degree as the GDR was attempting to build socialism. Because the Party and state had encountered resistance they needed to keep science under surveillance. Of more interest than labeling the phenomena, however, is understanding how a new political ideology and movement in power attempt to carry through a program. A common feature of most nascent regimes with extreme political ideologies (from the left or right—and often labeled "totalitarian") is precisely this state control, which is used to implement a vision.

III

Disciplines and Professions

8

Frustrated Technocrats: Engineers in the Ulbricht Era

Dolores L. Augustine

In the Ulbricht era, technological progress was proclaimed to be the motor of socialist change.[1] It was to be carried forward by the "new technical intelligentsia," a new class of technical specialists recruited largely from the proletariat and the peasantry and open to women, but lacking professional autonomy. Did engineers in fact enjoy the kind of power and prestige that the central role assigned to them would seem to entail? Or were they the passive recipients of a culture of technology based on a Soviet model? In this chapter I will focus on the history of the engineering profession in the GDR up until 1971. In particular, I will look both at the ways in which the state remolded the profession and sought to use it to pursue its technological and economic goals, and at the professional ethos of engineers.

The engineering profession was radically transformed after 1945, as the Western model of autonomous professions was replaced by the Soviet model of state-run professions.[2] Many professions continued to exist after 1945, and in fact were allowed to continue to be an organizing principle of society, but they were subjected to state control, cut loose from their historical connections with the bourgeoisie, and robbed of their ability to regulate themselves and to organize themselves as interest groups vis-à-vis the state and industry.[3] The work of engineers was also profoundly transformed by state economic policy, notably the institution of the planned economy and the New Economic System, or NES (introduced in 1963), which promised to decentralize economic decision-making, as well as to promote technological progress. This chapter will attempt to reveal the inconsistencies of state policies, as well as the ways in which they worked at cross-purposes or brought about social and economic changes that had unintended consequences.

Though revolutionary changes took place in the engineering profession in the GDR, there were also continuities with the precommunist past. First, an "engineering mentality" persisted that could be termed technocratic in the sense of dedicated to a narrowly technical, non-ideological view of problem-solving. Second, perceptions of the history of the engineering profession in the precommunist era continued to provide important points of reference for the professional identity of engineers. In comparison with some professions, such as medicine and law, engineering experienced only incomplete professionalization up to 1933.[4] According to Konrad Jarausch, however, the Nazi state won the loyalty of many engineers by reversing these trends and providing engineers with professional opportunities and prestige in a way that the Weimar Republic had not. This collective memory of pseudo-professionalization (reinforced, perhaps, by comparisons with the West German engineering profession) framed the relationship between engineers and the East German system in ways that could not be openly expressed.

It should be noted that use of the term "engineer" as the central concept of this chapter is not entirely unproblematic. Theoretically, the engineering profession in the GDR became part of the technical intelligentsia (encompassing all cadres in technical and scientific fields who had advanced formal education), though in practice engineering continued to be regarded by many as a distinct profession, as can be seen in state, party, and industry records, as well as in literature and film, East German sociological studies, and the memories of many former citizens of the GDR.[5] In the 1950 census, engineers were placed together with *Techniker* (technicians) and related professions, whereas in the 1964 and 1971 censuses they formed a separate category. The term "engineer" is also useful in studying continuities with the pre-1945 era.[6] Another terminological difficulty lies in the fact that the term *Ingenieur* long lacked an exact definition in the German language, including as it did all technical occupations above that of skilled worker or foreman. In the GDR, the title of engineer was generally reserved for graduates of the Technische Fachschule, or technical college (whose engineering graduates were called *Ingenieur*), and the Technische Hochschule, or technical university (whose engineering graduates were called *Diplom-Ingenieur*), though in exceptional cases, the title was granted on the basis of professional experience and expertise alone (presumably mainly to engineers of the precommunist era).[7]

The first section of this chapter deals with state educational policy and the recruitment of engineers, looking at how well the state bureaucracy planned recruitment, the impact of the exodus of engineers to the West, and state attempts to change the social and gender mix of engineering. The second section addresses the professional concerns of engineers, particularly pay, perquisites, and power structures in the factory, asking how these aspects of professional life affected engineers' attitudes toward the state, as well as their decisions to stay or to move to the West. The third section examines the relationship of engineers to the Socialist Unity Party. The fourth section looks at the actual work of engineers, particularly in research and development, focusing on a case study of one socialist enterprise in Berlin.

State Policy on Advanced Technical Education and Engineer Recruitment

Up to the building of the Wall, GDR policy makers were faced with the exodus of technical specialists. In 1950, engineers and members of related technical professions constituted 106,777 of the GDR's 7.2 million workers.[8] At the height of the brain drain, in 1958–1959, about 100 engineers a month were fleeing the GDR.[9] The impact of this exodus was far greater than the mere figures indicate. The loss of several technical specialists at once created a crisis situation in some industries and firms. For example, in the third quarter of 1950, 11 engineers working at the Electrical Appliance Works of Treptow (Berlin) quit and took jobs in West Berlin. The firm director fled around the same time.[10] A vital research project at the VEB Farbenwerk Wolfen was disrupted in 1958 when three members of the research team left the GDR.[11] Machine construction was the hardest hit state-run industry, reporting in 1958–1960 losses each month of about 25–30 employees with university-level education.[12] Moreover, GDR policy makers felt that shortages of engineers would put the brakes on technological advances, particularly ambitious programs such as the Chemistry Program. In 1958, authorities expressed the fear that the GDR would be lacking 30,000 engineers in the future.[13]

The state responded with a two-pronged strategy, offering experienced engineers, who were mostly educated before 1945, special inducements to stay in the GDR, while at the same time initiating a boom in advanced technical education that had parallels in other East Bloc countries (see Chapter 5). From 1951 to 1955, enrollments of day students at technical colleges (most of whom were engineering majors) grew 41.4 percent, from 34,737 to 49,132.[14] The number of graduates from technical colleges in the six most important engineering majors doubled between 1957 and 1961, rising from 8,884 to 17,744, while admissions rose 78.3 percent, from 15,648 to 27,898.[15] University enrollments in engineering (day students only) jumped from 14,860 in 1956 to 18,670 in 1961 (a 25.6 percent increase), while overall university enrollments rose only 17.3 percent in that period.[16] By the late 1950s, there were 9 engineering students in the GDR for every 10,000 inhabitants, while there were only 4 engineering students for every 10,000 inhabitants in West Germany.[17] In the era of the NES, university engineering enrollments expanded at a particularly rapid rate, rising 51.8 percent (from 18,670 to 28,344) between 1960 and 1965, and 62.2 percent between 1965 and 1970 (to 45,967). By contrast, between 1965 and 1970, the number of university students in all other majors grew only 16.8 percent.[18] The number of employed persons with engineering degrees rose from 189,604 in 1964 to 281,219 in 1971.[19]

Several factors discouraged the hiring of these engineering graduates, however, leading to a job shortage in engineering after the building of the Berlin Wall.[20] Many engineering positions were still held by employees without degrees, and factory administrators were reluctant to force them out, partly because they believed, as did many in GDR industry, that experience counted for more than academic training.[21] This belief grew out of the state planners' emphasis on production rather than on research and development. In addition, particularly during the NES, factory administrators were fearful of losing their jobs to academically trained experts, and thus were reluctant to hire and promote personnel with engineering degrees.[22] Gradually, these problems were overcome, as the number of engineers in industry rose from 31 for every 1,000 employees in 1961 to 68 per 1,000 in 1970.[23]

In the 1950s and early 1960s, at a time when the GDR was expanding the number of engineers in industry, the state was also trying to change the social mix of the higher professions, including engineering. The GDR provided opportunities in engineering to young people of working-class or peasant origins that had never before existed in German history and still did not exist in West Germany. The state required that 60 percent of all engineering students be of peasant or proletarian origins, and it encouraged young skilled workers to pursue engineering degrees. (Workers under twenty-five were often "delegated" to regular degree programs or to the workers' and peasants' faculties, whereas older workers generally enrolled in part-time, correspondence, or night programs.) Socialist enterprises undermined this policy to a certain extent because they wanted to keep highly skilled workers in production.[24] It was important to overcome this resistance to the state mandate, not only to promote the recruitment of workers to higher professions, but also to deal with an overall decline in applications, due in part to the low birth rate in 1940–1945.[25] A partial solution was sought in the expansion of correspondence and night programs.[26] However, the class-based quota system, with its emphasis on class origins over ability, generated controversy within the Party from the beginning, and it was dropped during the NES.[27]

The state also encouraged women to go into engineering, partly for ideological reasons (socialism stressed the equality of men and women), and partly because of the shortage of engineers. Female students who became pregnant before graduation received special assistance, and female factory workers were provided with special child care and other benefits if they took up university studies.[28] Women were given preferential treatment in the admissions process, but there were no quotas. Women were particularly attracted to technical college programs because they were shorter than university programs. By contrast, women—who carried a heavy burden of household and family duties—seldom took advantage of night and correspondence programs in engineering.[29] The NES increased opportunities for women in engineering as their percentages among those admitted to engineering programs at technical colleges rose from 12 percent in 1961 to 17 percent in 1964, and at universities from 6 percent in 1961 to 8 percent in 1964 and 9 percent in 1966.[30] Whereas in 1950 only 3.3 percent of all engineers and technicians were women, in 1964 women made up 7.5 percent of all engineers, and in 1971, 8.6 percent of all employed persons with engineering degrees.[31] Though some female engineers—particularly those who were young and had children—were relegated to jobs that did not correspond to their training, most were placed in true engineering positions.[32]

Historically, educational booms have often brought about expanded opportunities for the lower tier of society, in some cases for women as well (a good example would be the Federal Republic of Germany in the 1970s). The story is somewhat more complicated in the GDR. Because of difficulties in recruiting enough students of the appropriate class background, the quota system was actually a drag on educational expansion. The biggest rise in enrollments in engineering programs came *after* the quotas were dropped (as part of the NES). Nonetheless, the *absolute* number of students from working-class backgrounds continued to grow, as did both the absolute and the relative numbers of women in engineering programs.

Pay, Perks, and Power

The engineering profession in the GDR was characterized by a profound generational divide between those who entered the profession in the precommunist era and the "new technical intelligentsia." This divide was exacerbated by a policy of giving special contracts *(Einzelverträge)* involving higher pay, auxiliary pensions, and superior housing to older technical experts as an inducement to stay in the GDR, while engineers and *Techniker* on a regular pay scale earned only one-quarter to one-third more than factory foremen (according to a 1958 study).[33] In certain socialist enterprises, such as VEB Bergmann-Borsig, some engineers actually earned less than foremen.[34] Before the NES, engineers could not count on bonuses for special individual achievements, either.[35]

Through these policies, as well as through professional experience and an advantageous position on the job market in the West, older engineers gained a power base that they used to the disadvantage of younger engineers. Many recent engineering graduates were not given positions that befitted their degrees, but were used as substitutes or as errand boys or girls. Their employers often kept them in limbo for years, not letting them know when they could expect to advance to more responsible and interesting work. Moreover, older engineers displayed a lack of confidence in young engineers, denying them professional autonomy. For example, older engineers and even foremen kept a lot of knowledge to themselves, with the result that the young engineers could not work when the former were absent due to illness.[36]

This discrimination against young engineers was disadvantageous for the state in many respects. Discontent over low pay was a common complaint of engineers who fled to the West. Low pay also discouraged skilled workers from pursuing engineering degrees.[37] In addition, by stunting the professional development of young engineers, the GDR wasted human capital. Moreover, it was politically counterproductive to favor Nazi-era engineers (many of whom were of middle-class origins) over the "new technical intelligentsia," which supposedly represented the future of socialism.[38] Neither the westward exodus of engineers nor the NES brought significant change. As late as 1970, the 1952–1953 pay scale for engineers was still in place in some industries, while skilled workers' wages had risen to the point that many of them were earning as much as engineers. This was exacerbated by the fact that engineers were taxed at a higher rate than workers in production.[39] The state appealed to factory administrators to do more to promote the professional development of young engineers, and cooperation between young and old engineers was fostered in work and research units *(Arbeits- und Forschungsgemeinschaften),* but without much apparent success.[40]

Ultimately, demographics probably played a greater role than state policies in the waning of this generational conflict. Census data show how the balance between the generations of engineers changed during the first twenty years of the GDR. In 1950, roughly 30 percent of all engineers and *Techniker* were fifty or older and about 60 percent were forty or older.[41] Presumably, most of these older engineers would have entered their profession in the Nazi period or earlier. By contrast, in 1964, about two-thirds of all engineers were *under* forty, and thus were almost certainly educated after 1945, and a further 13.2 percent were in their forties, the great majority of whom were post-1945 graduates.[42] Just under one-fifth were fifty or older, of which perhaps 60 percent were educated before 1946.[43] Thus by 1964 members of the "old technical intelligentsia" probably did not occupy much more than one-tenth of all engineering positions, though their power probably exceeded their numbers.[44] By the end of the Ulbricht period, the old guard had virtually disappeared from the scene, and the age structure of the engineering profession had again changed, with engineers fifty and over making up only 7.7 percent of all engineers.[45] Thus, while in the 1950s the "old intelligentsia" was able to defend a position based on special contracts, expertise, professional possibilities in the West, and authoritarian structures, the changing of the guard in the 1960s was reinforced by a partial takeover of the engineering profession by young people.

The Relationship between Engineers and the Party

It is remarkable that in the late 1940s and early 1950s, party officials at the highest levels and the party "grass roots" seem to have been working at cross purposes. Many party officials at the factory level treated engineers with special contracts as representatives of the hated bourgeoisie, an attitude that seems to have had an impact on the way some workers saw engineers. In discussions with party envoys at the Transformer Works Karl Liebknecht in 1951, workers complained that while new collective contracts led to lower wages for workers, the intelligentsia was given special contracts and higher salaries and pensions. One worker expressed the suspicion that if it were not for these incentives, all members of the old intelligentsia would flee to the West.[46] At a 1950 district party meeting in Berlin, one delegate was greeted with resounding applause when he criticized the special privileges of the intelligentsia.[47]

Resentment was not, however, confined to the privileged members of the "old intelligentsia." At times, party fanatics treated *all* engineers as "class enemies." A party secretary in a Dresden steel works told party members to gather material on members of the technical intelligentsia that could be used as evidence that the latter were reactionaries; according to one source, "he created in the entire factory an atmosphere of fear and distrust." He was relieved of his duties.[48] Such fanatics were to be found in other factories as well. One party member reported at a district party meeting in 1949, "We had in our plant a comrade [that is, a party member] who thought himself to be the most radical person around . . . His attitude toward the intelligentsia was completely wrongheaded."[49] A trouble maker (probably a worker) in another factory complained about managers who were former members of the Nazi Party. They were reassigned, but the man continued to complain, "Our leadership has become fascist." Sending him to party "schooling" did no good—he could not be silenced.[50]

While such openly aggressive behavior was not condoned, a more subtle sort of distrust was allowed to simmer among the party officials at the factory level, as can be seen in the assertion of a party member at the same 1949 party meeting in Köpenick: "The Socialist Unity Party is a party of the workers and . . . we are in danger of letting this working-class element in the party fall under the table to a certain extent . . . A large part of the intelligentsia and white-collar workers are completely and wholly our comrades. But a large number are just painted red on the outside."[51] Some engineers were intimidated by this atmosphere. At one factory, the technical intelligentsia remained largely passive at a meeting on its behalf. When asked why he had not participated in the discussion, one member of the audience replied, "I have a wife and children!"[52]

Though these attitudes probably just reflect a lag in the reception of changes made by the Party, it is not surprising that they had a negative impact on the relationship between the Party and engineers. Party and union records of this period are filled with exhortations to make more of an effort to win over the technical intelligentsia, but successes were modest.[53] At the socialist electronics firm Elpro, forty out of sixty newly hired young engineers left the *FDJ* (*Freie Deutsche Jugend,* the communist youth organization) in their first two weeks on the job in 1959.[54] At a party conference of the Berlin district in 1956, it was noted that many members of the scientific and technical intelligentsia in industry were "politically unstable" and prone to ideologically "false" points of view.[55] Nonetheless, from 1946 to 1957 just under 18 percent of all engineers and *Techniker* were party members, while 13.8 percent of the workforce were party members in 1952. This reflects the degree to which those engineers interested in professional advancement joined the Party, and is in line with the total figures for white-collar workers (a category that included engineers): about 18 percent were party members in 1950–1952.[56]

Engineers were far more likely to join the communist-affiliated engineering organization *Kammer der Technik,* or Chamber of Technology, whose membership rose from 44,707 in 1956 to 110,733 in 1961 and 192,996 in 1971.[57] Not a true professional organization, it hardly dealt with the professional concerns of engineers, and in fact attracted a membership that included not only the technical intelligentsia, but also *Techniker,* skilled workers, and other professional groups. But neither did the *Kammer der Technik* concern itself particularly with political indoctrination. On the national level, the main focus of the *Kammer der Technik* was technology, and it conducted conferences on new technologies and published the most important technical journal in the GDR. This work was very important to engineers, though some considered Western technical journals and conferences to be superior to those in the East. Unable to address professional concerns and unwilling to promote an ideologized vision of technology, this organization defined its mission in purely technical terms, thus re-creating the engineer as the apolitical caretaker of technology in modern industrial society.

Research and Development

We now turn to the actual work of engineers in research and development (R & D) and the impact of working conditions on the way engineers saw their profession and the system. According to a 1953 report, "Securing conditions that will allow them to devote their full energies to their specialized work without any bothersome obstacles—that is more important to the creative scientist and engineer than the question of their pay."[58] Changes in the scope and nature of their professional activities bred considerable discontent among engineers in the early years of the GDR, leading some to flee to the West. They had to conform to the timetables and goals laid out by state planners; they no longer had influence over hiring and firing practices; and they were forced to work together with workers in research units and production meetings. It was noted in a district party meeting in Köpenick in 1949 that at a recently nationalized firm, some workers "were against planning in general, reasoning that scientists, researchers, and engineers cannot be pressed into a narrow mold."[59] Engineers were frustrated by the enormous amount of paper work that kept them from technical work—particularly reports and red tape connected with ordering parts—as well as by the many meetings held during working hours.[60]

Though only one aspect of engineering, research and development was essential in maintaining the international competitiveness of the GDR economy.[61] In the words of Walter Ulbricht, "The outcome of the peaceful competition between the two world systems will mainly be determined by which societal order best promotes science and technology, and makes best use of the opportunities created."[62] In the GDR, the most important research and development for industry was conducted at five central institutes (run by the National Economic Council until 1965, thereafter by the industrial ministries) and at institutes administered by the socialist combines, or VVBs—industrial branch institutes, central development and design bureaus, central laboratories, and large enterprise R & D units designated as "scientific-technical centers." In addition, there was a proliferation of much smaller factory R & D units in VVBs and the regionally managed sector of the economy.[63]

The development of industrial R & D is best illustrated by the case study of the Berlin socialist enterprise Karl Liebknecht. Part of the AEG Company before the war, Karl Liebknecht was one of the three largest transformer and electrical equipment works in Europe in the mid-1960s (with 4,000 employees in 1966), and the sole GDR producer of high-voltage electrical transformers (up to 400 kv), switches, and other high-voltage equipment.[64] Was the situation at Karl Liebknecht typical of East German industry? A basis of comparison is to be found in reports of central government agencies, particularly a 1962 study that systematically analyzes the reports of 138 major R & D divisions and research institutes and a 1966 study on fulfillment of the state plan for science and technology.[65]

One can distinguish six issues.[66] First, engineers complained that Karl Liebknecht concentrated too much on day-to-day production, particularly special orders—many of which were for other East Bloc countries—to the detriment of the development of new products. The R & D staff was caught up in short-term projects not involving technological advances, and they neglected basic technical research and product development with broad applicability until 1966.[67] This was fairly typical for GDR industry at the time. On the national level, enterprises tended to underestimate the importance of research and development relative to production, according to the 1962 study. Even the VVBs forced the R & D institutes under their control to do administrative work, write reports, compile and publish information, and deal with problems arising in production. The 1966 report mentions that R & D personnel were sometimes even forced to work in production, probably to help meet plan deadlines.[68]

Second, reports on Karl Liebknecht complain of shortages of engineers and other specialists with advanced degrees, also an endemic problem of the GDR.[69] In 1963–1965, 13 percent of R & D specialists in the GDR and in West Germany were university graduates, while 43 percent of the West German R & D specialists but only 26 percent of their GDR counterparts had degrees from technical colleges. In a 1971 survey, R & D workers stated that the principal impediment to effective research and development was the lack of technical personnel.[70]

Third, Karl Liebknecht faced serious structural problems owing to the inadequacy of its laboratories, which resulted from a shortage of space and funding.[71] Lacking the necessary production facilities, the R & D division had to have test models built by technology departments (which were concerned with production technology) or by the regular factory facilities. This situation resulted in considerable delays. With the opening of a GDR institute for the testing of electrical equipment stalled for years, Karl Liebknecht had to use testing facilities in Czechoslovakia. The scarcity of testing facilities had a serious impact on the morale of some engineers. One applied for a job in West Berlin because he was concerned that he would not be able to continue to work in his specialization in the GDR because the lack of testing facilities at Karl Liebknecht made the development of new high-voltage switches very difficult.[72] Similar problems existed on the national level. Of the 138 R & D divisions and institutes profiled in the 1962 study, 25 complained about a lack of space, measuring instruments, or other facilities. Many additional complaints had been registered through other channels.[73]

Fourth, at Karl Liebknecht and in the GDR in general, the quality and assortment of components and other materials were well below international standards, and there were frequent delays in the delivery of materials needed in research and development. In 1962, 35 out of 138 R & D divisions and institutes reported that they did not have the materials and tools needed for research and development. The 1966 report points to one cause of such bottlenecks: firms that needed components did not inform component producers in time, and component producers refused to make contractual agreements with firms using their components. This resulted from a lack of coordination between socialist combines, or even between enterprises in the same combine. In other cases, components (such as semiconductors) were scarce owing to the Western embargo. Other equipment was unavailable because it was exported. For example, Carl-Zeiss-Jena exported about 95 percent of its analytical measuring instruments for physical optics in 1966. The remaining 5 percent of production covered only one-third of the needs of GDR industry. Engineers often wasted a fair amount of time trying to "organize" components, instruments, and the like. However, a real effort was made to improve the quality of R & D materials in the late 1960s and 1970s.[74]

Fifth, a 1966 report at Karl Liebknecht pointed to deficits in the system of information dissemination, charging specifically that GDR industry was not able to keep up technologically because of a lack of state guidance, as well as a lack of information on the latest technology, costs, products offered by competitors, and production methods used by the competition. Cooperation with universities and research institutes was also deemed woefully inadequate. It took a very long time to procure much-needed technical literature, and Western publications were almost unobtainable, owing to the shortage of foreign exchange. Articles on new technological developments (including some translations of American and Soviet articles) were published in an internal newsletter. Up until 1962, however, technical journals themselves were circulated only among division heads.[75]

In the GDR, the lack of a civil society and a market economy impeded the flow of information on technological developments and production methods, as well as on costs, demand, and markets, and the state failed to make up for these deficits of the system. Despite the work of the Chamber of Technology *(Kammer der Technik),* many seem to have found it difficult to keep up with technical advances in East Germany. According to reports on "abandonment of the republic," engineers and chemists were resentful over being denied permission to attend conferences in the West. This issue, which contributed to the decision of some technical specialists to flee westward, was largely "resolved" by the building of the Wall. However, according to the 1962 study on research and development, R & D personnel were still on occasion denied permission to visit trade fairs or attend conferences (generally in the East Bloc), though less often than in the past. The difficulties that the staff of Karl Liebknecht had in getting technical literature seem to have been typical, particularly in the 1950s, though according to the 1962 study on R & D in the GDR, complaints on this subject had greatly decreased. The flow of technical knowledge was also impeded by the rather tenuous links between industry and the universities and academies and between GDR institutions and those in other East Bloc countries.[76]

State planning of research and development—which promised greater rationality in project selection than was the case under capitalism—could have served as a substitute for information. However, topics were largely selected by R & D personnel rather than by central agencies, particularly in the era before the New Economic System. To the extent that research and development was truly centrally planned, bureaucratic involvement in R & D injected into the process a strong element of irrationality, which grew out of the planners' lack of technical expertise and lack of knowledge of conditions in specific industries, as well as the proliferation of red tape. Economic usefulness and costs of projects were hardly taken into account. Market research was neglected and consumer and industrial needs ignored.[77]

Sixth, delays in getting newly developed products into production were a perennial problem of GDR industry. The causes were many: enterprises were not rewarded for getting a new design into production within a short time period; state investment funds were not always sufficient to start up production of all completed projects, particularly before the New Economic System; a lack of production capacity resulted from poor coordination of R & D and production planning, as well as from the poor state of equipment and machinery, scarcity of factory floor space, and lack of personnel; and the number of R & D projects was too great in some enterprises and combines.[78]

With the advent of the New Economic System, studies were conducted at Karl Liebknecht to find ways of speeding up research and development, and bonuses were introduced to reward R & D personnel for completing projects on time or early. According to an internal study, however, the rate of increase of production declined between 1960 and 1968, reportedly owing to weaknesses in research and development. In 1953, plan fulfillment reached 95 percent, but by 1962, it was reported that the enterprise had not been able to fulfill the plan for years. It was found that Karl Liebknecht spent less on R & D than the typical Western corporation.[79] By 1966, all three R & D sections had fallen behind the *Weltniveau* (that is, the international technological standard). At a time when the top Western corporations were working on equipment for 750- and 1000-kv transmission, Karl Liebknecht was still struggling to complete its program for 380-kv transmission, which had been established as the enterprise's top priority under the Central Plan (*Z-Plan*) in the late 1950s. Nonetheless, Karl Liebknecht did achieve some tangible successes, such as its 125-mv power transformer. Also, a new generation of labor- and material-saving high voltage switches were developed and went into production at the end of the 1950s. In 1961–1962, 15–20 percent of production consisted of new or improved products.[80]

How effective was the R & D establishment of the GDR overall? Raymond Bentley contends that though the GDR had one of the top-ten economies and was third in the world in research and development (in terms of man hours), the GDR was technologically not on the forefront, even in its strongest areas, was outperformed by West Germany, and was falling behind in labor productivity. He believes that under NES, the state largely failed in its attempts to push research in areas that were of high priority in terms of technological advance and user needs. The decentralization of project financing and bonuses, as well as the awarding of bonuses on the basis of individual performance, was largely unsuccessful.[81]

Conditions in industrial R & D clearly had a major impact on the way engineers saw their profession and the system under which they worked. At Karl Liebknecht there was widespread discontent among the enterprise's technical intelligentsia owing to what they perceived as the weakness of the enterprise in research and development, resulting in the loss of 90 employees out of 233 who were allowed to pursue part-time university or college degrees between 1952 and 1961, as well as personnel hired after graduation.[82] The reports on the exodus of engineers to the West specifically point to problematic working conditions in R & D as a major motivation.

Despite these conditions, GDR engineers identified strongly with their work. Their professional pride and sense of injustice at conditions that prevented them from doing "their job," which they obviously saw largely in technical terms, are striking. Typical of the less successful sectors of the GDR economy in the 1950s was the attitude of engineers in television and radio R & D: "They take it as a personal insult that their research work is in decline relative to that in the West."[83]

Conclusion

It was shown in this chapter that there were considerable contradictions in GDR engineering policies. Attempts to provide skilled factory workers with the opportunity to pursue engineering degrees met with the resistance of industrial administrators who were concerned with the fulfillment of production goals. The boom in technical education could not take off until the 60 percent quota for students of peasant or proletarian origins was dropped. The policy of offering special privileges to precommunist-era engineers undercut the position of the "new technical intelligentsia," creating large pay differentials, as well as a power base that older engineers could use to the disadvantage of younger engineers. State planning failed to provide the funds and incentives needed in research and development.

The New Economic System ironed out some of the inconsistencies of the system and brought about changes: an acceleration of growth in advanced technical education, a greater emphasis on scholastic ability in the admissions process, an end to class-based quotas in advanced technical education (leading to a significant decline in the percentage of engineering students of proletarian origins), and greater opportunities for women in engineering. But there were also limitations to change: factories continued to resist hiring engineering school graduates; merit-based bonuses were instituted, but at least in some industries, the basic pay scale for engineers hardly changed, and the gap between the earnings of engineers and the earnings of skilled workers narrowed. Research and development was given a more solid financial foundation and was centralized to a certain extent, but problems persisted: the use of R & D staffs in production; the shortage of engineers on R & D staffs; problems in procuring components and other materials; the export of instruments urgently needed in GDR industry; a lack of coordination in R & D between enterprises; and the state's inability to identify high-priority areas of research.

Did the balance between ideology and the demands of technological and economic progress shift to the advantage of the latter in the 1960s? Certainly, there are no reports of engineers being treated as the class enemy by party fanatics in the 1960s, and class-based quotas were dismantled during the NES. But in some ways, the orientation of the 1950s certainly was pragmatic, for example, in the privileges accorded to older engineers. And in other ways, the 1960s do not seem particularly pragmatic. For example, the resistance of enterprise administrators to the hiring or training of engineers persisted, an expression of the sense of rivalry that political appointees felt toward highly educated technical experts, as well as the belief (antibourgeois in nature as well as counterproductive from the point of view of technological progress) that experience was more important than professional credentials. After the building of the Wall, problems in acquiring Western technical literature worsened, as did difficulties in obtaining permission to attend conferences abroad.

Another major finding of this chapter was that an unofficial "culture of technology" existed that deviated from the officially propagated one, though it could not organize itself into a subsystem capable of challenging state authority.[84] There is good evidence that GDR engineers were offended by the lack of competitiveness of some sectors of the GDR economy, as well as by working conditions and job definitions that kept them away from what they saw as the work of the engineer, which they usually defined in technical terms. Closely related was a sense of professional identity that at times rebelled at the reduction of professional autonomy, low status, and low pay of engineers after 1945. These attitudes were "proto-professional" in that they contained an unspoken claim to high professional status, a claim that was denied the opportunity of collective or public expression. Framed by this outlook, experiences of generation conflict and rejection as "class enemy" by some elements in the Socialist Unity Party created a sense of alienation that contributed to the long-term technological, economic, social, and political problems of the GDR.

These conclusions are relevant to the larger conceptual framework of this book. First, the "engineering mentality" in the GDR did not fit in at all with the Soviet model of the "new technical intelligentsia," whose implementation was also impeded by the inadequacies and inconsistencies of the system. Some of the most promising aspects of the Soviet system were not effectively put into practice, such as planning so that the number of graduates matches the number of job openings, or were partially dismantled, as in the case of opening engineering to women and those outside the bourgeoisie. The material presented here has pointed to the limitations of change under the New Economic System. In general, policy shifts (a second major theme of this study) failed to bring about an amelioration of what engineers saw as the basic problems of engineering in the GDR in the Ulbricht era. Third, engineers were nonetheless not less likely to join the Party than were members of other professional groups.

Fourth, it was shown in this chapter that the creation of the planned economy had a major impact on engineering, reducing the professional autonomy and authority of the engineer with regard to the selection of R & D projects, the ordering of materials, influence on hiring and firing, influence on the funding of R & D, and access to information (procurement of technical literature, attendance at conferences, and so on). Engineers in research and development were forced into administrative positions, and sometimes even into production. In addition, in the planned economy (particularly before NES) production was given priority over research and development, leading to resistance to the hiring of engineers and the undermining of their work.

No fullscale comparison with a market-economy system is possible here. However, this chapter provides some points of departure for comparisons between the two Germanys (the fifth theme of this volume). In fleeing west, some engineers drew obvious conclusions from their own tallying of the pros and cons of the two systems, though of course many stayed. The rivalry between East and West Germany also extended to engineering. For example, the GDR, whose research and development capacity was smaller than that of the Federal Republic, expanded advanced technical education in part so as to be able to catch up with its Western rival, but ended up with a larger number of engineers. The two Germanys also had much in common, owing to a common past—for example, the narrowly technical view that engineers had of their profession. One could also point to the two-tiered system of *Fach-*

schulen (*Technische Fachhochschulen* in the West) and universities, though I have found no indication of an East German equivalent to the superior status of the *Diplom-Ingenieur* so typical of the Federal Republic. On the other hand, educational opportunities for the lower strata of society became more similar in East and West Germany by the late 1960s, when the GDR dropped the quota system. Opportunities for women in engineering proved to be a more enduring advantage of the East German system.

This chapter also provided material for a consideration of the continuities between Nazi Germany and the GDR and the similarities of the two systems. Though the professional ethos described here did not originate in the Nazi period, it certainly flourished in those years. A partial continuity in personnel between the pre- and the post-1945 eras was exacerbated by policies that favored the "old intelligentsia." However, the introduction of the planned economy and the "new technical intelligentsia" brought about enormous discontinuities. Engineers found the changes that accompanied socialism to be intrusive and disruptive, probably more so than those brought about by National Socialism. Looking at the GDR as a "modern dictatorship" (analogous to Nazi Germany), one can focus on the tension between ideology and the demands of technological and economic progress. Even in the phases in which technological progress was given high priority, notably the New Economic System, ideology and party rule ultimately triumphed.

It was this subservient role of science and technology that prevented engineers from whole-heartedly supporting the system. Critical observers of East German technology and economic policy, they nonetheless could not develop political resistance or alternate models of technological development. They were prevented from doing so, not only by their lack of professional autonomy and the general deficiencies of civil society in the GDR, but also by a professional ethos that declared them to be apolitical custodians of technology.

9

Chemistry and the Chemical Industry under Socialism

Raymond G. Stokes

Throughout its existence, the German Democratic Republic was in a quandary.[1] Poorly endowed with natural resources and with a small domestic market, the GDR had to trade abroad to survive, but the Cold War and other factors limited foreign trade and foreign-exchange earnings. Unable, therefore, to import sufficient raw materials, the GDR was forced to search for domestic substitutes for items that might be available more cheaply (or of higher quality) abroad.

The chemical industry could offer such substitutes and was therefore critical to the political and economic welfare of the regime. What is more, since the industry is research-intensive and since Germany had a long tradition of excellence in this area, it had the potential to demonstrate the GDR's scientific and technological prowess and thus to symbolize the success of GDR-style socialism.

In this chapter I will focus on the development of GDR chemical science and technology in political and economic context. My main concern is with industrial application of chemical science and technology, rather than with the university training of scientists and engineers or basic research in chemistry. The bulk of the chapter deals with the emergence of the main characteristics of the GDR chemical industry during the period 1945–1965, although I consider briefly the period 1966–1990.

The organization of the chapter is straightforward. I first examine the period 1945–1965, looking both at continuities from previous periods in German history into the GDR period, and at changes through 1965. Along the way, I touch on a number of themes, including German technological tradition; divergence of GDR industrial culture from that in the Federal Republic of Germany; and technology as an expression of cultural values, political realities, and economic constraints. The final sections consider the further development of these themes between 1966 and 1990 and offer conclusions based on this case study.

Chemistry in the GDR, 1945–1965

Continuities

Despite the GDR's initial rejection not only of the Nazi period but also of most of the conventional German past, its chemical industry displayed a pronounced affinity for long-held traditions within the German organic chemical industry.[2] These traditions for the most part antedated the Nazi period, although some were intensified during it; the traditions thus ran very deep. What is more, the GDR chemical industry shared many of these traditions with its counterpart in the FRG.

There were numerous continuities in traditions from before 1945 into the GDR period, but I focus on three major aspects: machines and techniques; traditions of excellence in German chemistry; and basic assumptions about coal as the most important raw material in chemical production.

MACHINES AND TECHNIQUES   There were, of course, major discontinuities in this area from before 1945 to the postwar period. I mention them here only briefly before turning to the more essential continuities. Much of the machinery and equipment of the chemical industry in the area was destroyed during the war. I. G. Farben's giant Leuna Works, for instance, by far the largest chemical plant in postwar East Germany, suffered a destruction rate of about 50 percent.[3]

Moreover, beginning in the summer of 1945, the Soviet Occupation Authorities pursued punitive policies with significant effects. Dismantling assumed dimensions unknown in the West. For the Leuna Works alone, approximately 120,000 tons of machines and equipment were slated for dismantling, most of which was carted away by the end of 1946. At the height of activity, beginning in March 1946, Leuna deployed more than 38,000 employees to assist the Red Army in removing machines and structural materials from the works. Included were eight functioning compressors for synthetic gas, large-scale installations for methanol synthesis, and equipment for synthetic gasoline production.[4]

Losses from bombing and dismantling at the Leuna Works were substantial, but they do not tell the whole story. Soviet policy changed significantly in mid-1946. Instead of removing machines and factories from its zone of occupation to the Soviet Union, the Soviets seized the actual property of key firms, transforming them on site into Soviet Corporations (sowjetische Aktiengesellschaften, or SAGs) for more effective exploitation for the Soviet economy. The new policy changed the conditions under which German firms operated. The Soviets, for example, returned some artifacts seized previously "so that the production targets that the Soviet military administration had set in the meantime could be fulfilled."[5] Under Soviet control, East German chemical plants reattained most of their previous production lines and began reconstructing damaged facilities, something that continued when the SAGs were turned over to the East German government in 1954.

In other words, despite bombing damage and dismantling, the GDR chemical industry retained much of its capacity from the prewar period. And although the damage and dismantling represented important aspects of the legacy of the National Socialist (NS) period for the GDR, it was not what had been destroyed or carted away, but what *remained* that represented the most concrete and lasting legacy of Nazism. Construction and design of East German chemical facilities had taken place primarily during the NS period; much of this capacity was either intact or fairly easy to repair. In other words, the layout of the plants and their hardware represented a physical embodiment of NS-era assumptions, objectives, and policies.[6]

What is more, most chemical engineers and industrialists who operated the plants, especially during the 1950s, came of age professionally during the NS period and shared many of the assumptions that contributed to the plants' design. An important segment of this group left the GDR for West Germany during the 1950s, and this generation was later replaced by another, trained under the GDR regime. Still, East German planners, chemical engineers, and industrialists continued to negotiate through 1990, at least in part, with the physical embodiment of the ideas and assumptions of those who had designed the facilities, men who had in many cases been active during the National Socialist period.[7] One constellation of ideas centered on the promotion of excellence in chemical production. Another assumption was that coal was the preferred raw material for producing feedstocks, or starting materials, for chemical production.

TRADITIONS OF EXCELLENCE By virtue of inheriting a significant portion of the physical and human legacy of pre-1945 large-scale German chemistry, the GDR also inherited a set of ideas that lay at the heart of the industry's technological and commercial success: ideas centering on the promotion and pursuit of excellence. Along with their counterparts in the Federal Republic, GDR chemical engineers and managers retained two key components of this tradition of excellence. First of all, well into the 1960s at least, they maintained a steady commitment to research and development, with R & D efforts centered mostly on industrial laboratories.[8] Despite limited means at their disposal for R & D as a result of the plight of the GDR economy in the 1950s, emigration to the West, and the pressures of plan fulfillment, researchers in the GDR chemical industry were often able to develop world-class technologies in plastics, synthetic fibers, and synthetic rubber.

Second, GDR chemical engineers and managers judged their success and gained inspiration for future directions of chemical technology primarily by looking at international best practice, something that today would be termed "benchmarking."[9] This is not to say that the GDR chemical industry successfully maintained international standards across the board and throughout its history—in fact this was certainly not the case, as the development of petrochemical technology in the GDR indicates. Still, GDR industrialist personnel, much like their pre-1945 German counterparts, maintained as their ultimate goal the attainment of internationally recognized technological best practice.

COAL-BASED CHEMISTRY Politically and economically, chemical industrialists were some of the most liberal businessmen in pre-1945 Germany.[10] Heading an industry that had always been heavily dependent on export markets for its commercial viability, German chemical industrialists championed free trade more than those in other industries. At the same time, however, they thought in terms of economic self-sufficiency in raw-materials procurement and in manufacturing. In both these areas, the German organic chemical industry from its beginnings emphasized the importance of securing a domestic supply of raw materials (most important, coal) and tended to favor maximizing manufacturing at home. The industry's emphasis on autarky thus predated the Nazi Period, but it also constituted an area of common ground between I. G. Farben and the Nazi party in the period 1933–1945.

After World War II and well into the 1950s—and in both the GDR *and* the FRG—the argument that domestic production of coal represented the most reliable, secure, and self-evident source of chemical raw materials continued to hold sway.[11] The perseverance of this argument affected technological decision-making during the 1950s to a large degree and manifested itself, not only in the continued emphasis on coal-based production technologies even as international best practice began to favor petroleum-based chemistry, but also in the tendency to choose technologies that permitted maximum *flexibility* in feedstock input.

What accounted for this continuity in thinking in terms of coal-based chemistry, especially when U.S. chemical and petroleum firms had already demonstrated that petroleum-based chemistry could be both cheaper and more productive in certain areas? In the GDR, of course, one reason was the shortage of foreign exchange mentioned earlier, and petroleum, too, was in short supply. But chemists and chemical industrialists in both the GDR and the FRG made a number of assumptions about how to design chemical plants, and they changed only slowly. For instance, the optimal scale of petrochemical plants was much larger than that of coal-based ones, entailing differences in design. The massive scale of petrochemical facilities required new methods of handling large volumes of materials smoothly and without interruption. Moreover, designers in the chemical industry—who were, in Germany, primarily trained as chemists—had to abandon the traditional value assigned to "elegance" in coal based chemistry, which favored indirect processes with high yields that could be described well theoretically. To produce petrochemicals, they had instead to concentrate on the most direct path to the final product regardless of how messy, theoretically obscure, or low in yield, relying on recycling part of the reactant to increase yields.[12] Moving away from such long-standing traditions was a slow—and often painful—process in both German successor states.

In sum, then, the GDR chemical industry featured major continuities in machines, equipment, and fundamental design and production assumptions from pre-1945 into the postwar period. In this, it was not particularly different from its counterpart in the FRG. In both the GDR and the FRG, the assumptions (more so than the machines and equipment) predated the National Socialist period, although the industry's autarkic stance with regard to raw materials was intensified under National Socialism. The major impact of the Nazi period on the chemical industry in both the GDR and the FRG, however, was that the personnel who ran the plants in both cases had come of age professionally during the period and thus were influenced (in terms of selection of raw materials, production processes, and products) by the heightened attention to autarky.

Still, by the late 1950s, West German chemical industrialists began to discard many of these ideas, especially in petroleum-based chemistry; East Germans, on the other hand, intensified their commitment to autarky, although the bases for and meaning of the concept changed considerably in the postwar period.

Changes

Two major sets of changes took place during the 1950s and early 1960s within the GDR chemical industry. First, the industry fell behind its counterpart in West Germany technologically and commercially; second, it committed itself more intensively to coal-based chemistry at precisely the same time that firms in the West moved decisively away from it.

FALLING BEHIND THE WEST   East German chemists, sharing the tradition of emulating international best practice, recognized the importance of petroleum-based chemistry at the same time as their counterparts in the West.

Researchers at the Leuna Works had begun in 1951 to develop processes for producing polyethylene—a key plastic of the postwar period and one that is generally produced from petroleum-based feedstocks—in other words, at about the same time as West German firms. From the start, however, the East German program was plagued by problems, and at the beginning they were especially severe. As an economic planning group put it in 1957, the "forward movement [of development work on polyethylene at Leuna] was catastrophically hampered in the initial years by completely inadequate opportunities for doing engineering-technical work."[13] Construction of a small-scale experimental apparatus began in 1953, and it started operation in 1954, but the initial reactor could not maintain a sufficient temperature. At the same time, seals on the high-pressure apparatus were "inadequate." Researchers at Leuna thus could not sustain a continuous reaction even for a couple of hours. At the end of 1954, after completely reconstructing the apparatus, researchers produced white polyethylene for the first time. Unfortunately, the product came out "not as a viscous mass [which was polyethylene's normal form], but rather in the form of white, asbestos-like flakes." What is more, reaction conditions allowed only 4 percent of the ethylene feedstocks to be polymerized into polyethylene.[14]

Leuna researchers had made enough improvements by 1955 to build a pilot plant, but it produced only contaminated polyethylene. Information exchanges with Soviet researchers in Leningrad made the major problem clear: Soviet chemical engineers had discovered that the minimum throughput of ethylene gas at the pilot plant had to be higher than in the original design if uncontaminated polyethylene was to be produced.[15]

The path through which this particular difficulty in East German polyethylene development was solved highlighted a key characteristic of GDR research culture. Although the country's scientists and engineers were aware of and influenced by developments in the West, their most regular and influential professional contacts were with their counterparts in the Eastern Bloc. In fact, by the second half of the 1950s, there was regular exchange of know-how and information and technical cooperation in organic chemicals technology between representatives of socialist countries, and especially in the early 1960s, East German planners anticipated relying heavily on the Soviet Union for technical assistance in developing a domestic petrochemical industry.[16] Technological change became a vehicle for increased technological cooperation between and integration of the countries of the Soviet Bloc. Closer integration into the Soviet bloc also had the effect of reorienting research and development emphases within East Germany.[17]

Information exchanges were clearly useful, but they also mirrored the limits and weaknesses of the individual countries in the Soviet sphere. When Soviet and East German researchers met to discuss cooperation in polyethylene research, development, and production in mid-December 1955, for example, the Soviets reported that, although they were already producing high-pressure polyethylene at the rate of 1,000 tons per year, they desperately needed additional high-pressure compressors to increase production. Their own machine-building industry was not capable of supplying them, and they therefore asked Leuna for help. The Soviet visitors had the impression that Leuna might provide them with the necessary technology. A Dr. Geiseler, who headed Leuna's polyethylene program, admitted that the know-how for manufacturing the necessary compressors existed in his plant, but stressed that he could not translate this potential into practice: at that moment, Leuna possessed neither the materials, machine tools, nor the engineering personnel necessary for the job. He suggested—and the other East German representatives agreed—that the Soviets look to West German suppliers instead.[18] This anecdote indicates widespread problems in East German industry, including insufficient cooperation between firms, especially in different industries; poor flow of information, materials, and equipment in the Soviet Bloc; and continued technological dependence on the West.[19] Such dependence was especially striking in the chemical industry, where, despite some success in efforts to refocus technical cooperation eastward, there were also negotiations with the French and, as will become clear shortly, much more important, with the British on transfer of petrochemical technology.[20]

In spite of attempts to search out and deploy foreign expertise, the East German polyethylene development program proceeded at a snail's pace. During 1958, for instance, the polyethylene production facility was "still very much subject to disturbances." After six years of intensive research activity, Leuna engineers could produce only a single ton of high-pressure polyethylene in the entire year of 1958.[21] In contrast, already in 1956 a single plant in West Germany, the Rheinische Olefinwerke (jointly owned by BASF and Shell), featured polyethylene production capacity of 35,000 tons per year.[22]

Research and production efforts continued, with only sporadic success, into the 1960s. By April 1961, the country's leadership finally agreed to remedy this situation by purchasing know-how and key materials for building a high-pressure polyethylene production plant from two British firms, ICI and the engineering company Simon-Carves. The decision is especially significant since it cost the GDR government 781,250 pounds sterling in precious hard-currency reserves during the severe political and economic crisis that preceded the construction of the Berlin Wall in August 1961. By November 30, 1965, with outside help, three high-pressure polyethylene production lines were in place, but technical problems remained. Only on December 30, 1965, was ethylene of sufficient purity produced for the first time. Production finally commenced in January 1966.[23]

By that time, however, East Germany's development of high-pressure polyethylene production had taken fourteen years. The country's technology was at least a decade behind that of the West. In a rapidly changing industry, such delays could be devastating; indeed, although the East German chemical industry continued to grow and perform well into the 1980s—especially in relation to its counterparts in the Soviet bloc—it never fully recovered from falling behind so significantly during the 1950s and early 1960s.[24]

THE GROWING COMMITMENT TO COAL-BASED CHEMISTRY
In 1957 and 1958, the GDR abandoned its Second Five-Year Plan (which was supposed to have covered the years 1956–1960) and replaced it with the First Seven-Year Plan to cover the years 1959 to 1965. The new plan jettisoned Stalinist promotion of heavy industry to the exclusion of all else. Instead, East German planners stressed lighter industry, consumer-goods production, and precision manufacturing. They also planned to refocus production patterns in the GDR so as to complement the production palette of other states in the Soviet-dominated Council for Mutual Economic Assistance.[25]

The chemical industry would, of course, play a vital role in realizing these plans. Accordingly, people in the highest positions of authority in the GDR participated actively in making them: the Fifth Party Congress of the ruling Socialist Unity Party, held from July 10 to July 16, 1958, decided to establish a petrochemical industry in the GDR.[26] Walter Ulbricht himself sketched out the details of the new plans for the chemical industry in a lengthy address in Leuna on November 3 and 4, 1958.

Essentially, Ulbricht made three key points. First, highest priority in the new program would be given "to rapid development of the production of modern plastic materials and synthetic fibers and, in order to do this, to tapping into the raw material that has of late become available to us—namely, petroleum."[27] Second, modernizing the chemical industry would require considerable research and development expenditures.[28] Finally, departing to some extent from the conventional Western conceptions of technological progress in the chemical industry of the 1950s that were evident in the rest of his talk, Ulbricht announced a sort of "two-track decision" with regard to feedstocks and associated technologies. Petroleum-based chemistry would indeed start in earnest in the GDR as a result of "the generous help of the Soviets, which they will demonstrate with the delivery of 4.8 million tons of petroleum in 1965." Still, Ulbricht announced with some fanfare that "the establishment of modern, high-molecular chemistry will continue to take place on the basis of acetylene made from carbide [in other words, on the basis of older, coal-based technology!]. *Thus the VEB Chemische Werke Buna will become the largest producer of carbide in the world.*"[29]

The ambiguity in Ulbricht's long speech was unmistakable. The East German chemical industry was capable of participating in the most modern developments in organic chemistry. By contrast, the plan called for the GDR to rely to an extent unknown elsewhere by the late 1950s on acetylene made from carbide—produced in turn from domestic coal—as a key feedstock for producing modern plastics and synthetic fibers. The tension between modern and traditional technologies was reflected in more detailed planning for the chemical industry, and indeed emerged as a salient characteristic of GDR industry as a whole: as a history of GDR technology puts it, "Old next to the new, that was a general characteristic of the state of technology. Scientific and technical achievements of the highest level . . . stood as island solutions right next to backwardness, gaps, and deficits."[30] Rather than commit wholeheartedly to petrochemicals, GDR planners continued to speak about "adaptation" of technology to maximize use of domestic resources.

At this juncture, it will be useful to highlight both a fundamental similarity and a crucial difference between developments in the chemical industry of the GDR and that of the FRG in the late 1950s and early 1960s. First, the similarity: the GDR was not alone during the 1950s in featuring an industry in which old and new technologies stood side by side. In fact, this is a characteristic of all industries in all societies. It was certainly true of the FRG during this period, and old technologies served as the basis for the "economic miracle."[31] Even in the advanced sectors of West German chemical technology in the late 1950s and early 1960s, there was considerable talk about "adaptation" of old technology.[32]

The key difference between the two industries had two dimensions. One was the acceleration of coal-chemical technological development in the GDR at precisely the time this was in decline in the FRG and elsewhere and despite the promise of Soviet oil supplies for the nascent GDR petrochemical industry. The other was the GDR's unwillingness and/or inability to depart from tried and true technological traditions when the time had come to do so, in other words, its inflexibility.

These differences grew larger, not smaller, over time. Although official groundbreaking at the first major petrochemical production facility in the GDR, Leuna II, took place on October 8 1959, less than a year after Ulbricht's announcement of the chemical industry's new course in November 1958, the GDR continued to rely heavily on coal-based feedstocks, in part because oil imports from the Soviet Union did not reach the level promised in the late 1950s. A 1965 projection of the "demand of the chemical industry for the most important organic basic chemicals" for the years 1966 through 1970 indicated that the biggest single source of feedstocks would continue to be "acetylene for chemical uses." In addition, "the proportion of organic basic chemicals made from petroleum in the GDR, reckoned on the basis of carbon content, will reach 35 percent of total consumption of the chemical industry in 1970."[33] In contrast, the West German organic chemicals industry had reached this level already by 1958–1959; by 1970, the proportion there was around 90 percent.

In the late 1950s and early 1960s, chemical industrialists in the FRG were still generally hedging their bets by emphasizing flexible technologies that would allow chemical production on the basis of petroleum, coal, or natural gas, but they were also—as is clear from the above figures—by this time in the process of discarding acetylene chemistry and moving inexorably into petrochemical production.[34] Despite lingering similarities between the GDR and the FRG, as a result of much slower technological change in the GDR and its growing commitment to production on the basis of coal, there were by the mid-1960s two very different technological cultures of chemical production in the German area.

Chemicals in the GDR, 1966–1990

As a result of a variety of factors, the GDR chemical industry performed well into the 1970s, especially in the context of the Eastern Bloc. Building the Wall in August 1961 stabilized the GDR regime and prevented the outflow of scientists and engineers to the West. Petrochemical production in the GDR may have been delayed seriously, but it was nevertheless functioning and expanding by the late 1960s. The New Economic System, introduced by Ulbricht in 1963, promised economic reforms that would help rationalize production, stimulate innovation, and introduce a new, dynamic price system. Finally, worldwide demand for chemical products remained strong. Taken together, all of this meant that the GDR chemical industry was competitive on international markets. As Raymond Bentley has shown in an analysis of exports of the FRG and the GDR to Yugoslavia, in 1968 the GDR was able to obtain on average 94 percent of the purchase price per kilogram of chemical products compared with that obtained by the FRG. For fully one-fourth of all product groups considered in Bentley's analysis, the GDR could demand the same or a greater price compared with the FRG in 1968.[35]

The situation did not last. East German success in the 1950s was primarily a function of further development of prewar technologies (and often usage of prewar machines), a practice that could only lead eventually to diminishing returns. The limits of this almost exclusive reliance on prewar technologies became evident by the late 1950s: even as the older technologies reached their peak performance, the GDR chemical industry performed poorly in new technologies.[36]

In addition, the political-economic framework within which the GDR developed its chemical technology did not prove conducive to successful and timely innovation. The NES did not live up to its promises, and the fall of Ulbricht in 1971 marked its complete abandonment. Worldwide demand for all sorts of products slowed considerably in the midst of recession even as capacity (especially in plastics) continued to increase. The oil crisis of 1973–1974, followed by that in 1979–1980, slowed further the GDR's tentative moves toward petrochemical production: the Soviets, who had previously supplied virtually all of the GDR's petroleum, could now sell their product for hard currency in the West. According to Raymond Bentley, "In the 1980s, as a result of higher world prices for mineral oil and limitations of supply from the Soviet Union, the GDR felt compelled to intensify its use of brown coal not only as an energy source but also as a raw material for the chemical industry." The back-pedaling on feedstocks was accompanied by declining numbers of personnel devoted to R & D, slower increases in R & D expenditures, and low rates of investment. The result was decreased competitiveness compared with the FRG. By 1987, the GDR could obtain on average only 9 percent of the kilogram-price charged for equivalent chemicals by FRG firms (this compared with 94 percent in 1968). The GDR could command a price equal to or greater than that charged by the FRG in only 5 percent of all product areas.[37]

Conclusion

In 1945, there was a single German organic chemical industry. By the mid-1960s, there were two very different ones. The history of the GDR chemical industry is suggestive because it provides the basis for an explanation of *how* GDR society grew apart from that of the FRG and of *why* the GDR's chemical industry lagged behind that of the West.

The process of engaging in a major technological transition—in this case, the transition from coal-based to petroleum-based chemistry—provided a "window of opportunity" and a locus within which fundamental redirection of technological culture occurred, which in turn had far-reaching economic and social implications. Preparing for and beginning a technological transition of these proportions required active and intensive engagement in technology transfer, which helped integrate GDR industry more fully into the Soviet Bloc and distinguish it more clearly from industry in the West. GDR industry looked to the West for inspiration for developing new technologies. But in the realm of everyday, hands-on development of technology, GDR industry turned to the Soviets for know-how, for instance, with regard to ethylene production; the Soviets, by contrast, looked to the East Germans for chemical process machinery. The East Germans depended on the Soviets for supplies of petroleum and also for aid in designing and building a refinery dedicated to production of petrochemical feedstocks. A parallel process took place in the West in dealing with the same technological transition, and the result was increased integration of FRG industry into the US-dominated world economy and more pronounced differences between the FRG and the GDR. Changes in chemical technology represented only one of many areas of intensive technological change in the 1950s and 1960s, and therefore only one of many paths through which the two German successor states grew apart technologically.[38]

Technological changes in the 1950s and beyond helped create differences between the two German successor states. What is more, in this case, the difference, although it had many dimensions, was especially striking in that one state plainly performed *better* than the other: GDR industry lagged far behind its FRG counterpart. What explains the lag? All other things being equal, one might expect that the GDR would not have fallen so far behind the FRG, at least in chemical technology. Immediately after 1945, the basis for successful technological change existed in both German states, in terms of equipment, personnel, and reputation. Potential GDR demand for the products of the chemical industry—many of which could substitute for materials in short supply—was, if anything, greater than that of the FRG. GDR chemists, like their western counterparts, were well aware of the latest technological developments around the world and oriented their research and development programs around that awareness.

There are three major reasons for the GDR's relative inferiority. First, the GDR's main partner in its technological transition to petrochemicals, the USSR, was not especially capable. This was in marked contrast to the FRG's major partners, corporations in the United States and Great Britain. Second, owing to unreliable sources of petrochemical starting materials, GDR industrialists were forced to delay the transition to petrochemical technologies longer than their counterparts in the FRG. Third, GDR chemical firms proved to be poor innovators, and even worse at applying innovations effectively: they did not cooperate well with one another or with factories in other related industries; they frequently had difficulty obtaining materials and equipment vital to research and development; and, even when research was effective (which it often was), they could not translate it into a tangible production process.

These three reasons for East Germany's relative inferiority in turn require some elaboration. After all, the fact that the GDR's major partner was wanting and that the GDR did not innovate effectively seem to point to a more basic problem: there was perhaps something fundamentally flawed—from beginning to end—with the system of "real existing socialism."[39] This may be true, and seems intuitively obvious in the wake of the collapse of the East Bloc. But I would caution against accepting this thesis without further examination. Regardless of its many shortcomings, after all, the GDR delivered substantial technological sophistication, a relatively affluent standard of living, and an economy that many envied. This performance may have been in spite of—rather than because of—state planning and other aspects of the "socialist system." The point is that we just do not yet know the primary causes of the GDR's relatively poor performance, systemic or otherwise. Furthermore, even if the "system" was ultimately to blame for the GDR's performance, it is not clear that it could not have been changed, especially in light of the failure of the New Economic System in the 1960s.[40]

Perhaps the most important point here is that the question of *why* the chemical industry of the GDR performed poorly compared with that of the FRG is extremely difficult to answer because the list of plausible explanatory factors is long and defies causal judgment. Some of the possible factors were systemic, including centralized state planning, ideological rigidity, and consistently poor research and development performance. At the same time, though, many factors were contingent and largely beyond the control of GDR authorities, including deficient oil reserves; lack of foreign exchange in a world economy increasingly dominated by Western countries with hard-currency reserves; and loss of key personnel to the West. Establishing more certainly the relationship between the systemic and the contingent factors that contributed to the relatively poor performance in innovation in the GDR will be possible only after many more case studies are completed. For the chemical industry, it appears that over time these factors interacted with and amplified one another, hindering East Germany's development of a petrochemical industry, and ensuring that GDR chemicals could not compete on world markets.

10

Nuclear Research and Technology in Comparative Perspective

Burghard Weiss

The subject of nuclear research and technology is a significant but complex topic in contemporary history: its scientific-technological, economic, ecological, military, and political importance, and the implicit ambivalence surrounding its use, have made it the object of persistent social disputes and debates. The aim of this chapter is to provide an overview of the history of nuclear research and technology in East Germany by defining the various phases in its development.

The approach is comparative with an eye on developments in the West (particularly West Germany) and in the East, for example, the Soviet Union, as well. While the development of nuclear research and technology in those countries has been examined, no studies for East Germany are based on extensive archival work.[1] Joachim Kahlert's 1988 study discusses the use of nuclear energy against the background of questions concerning its social use, and it is based on published sources only.[2] More recent publications have been chiefly interested in the effects of German unification on East German nuclear research—a process that has led to large-scale layoffs, downsizing, and shutdowns of institutions.[3] A recent historical research project under the direction of Reinhard Koch is focusing on safeguarding the historical sources; publications are being prepared.

Prohibition

In 1945 law number 25 of the Allied Control Council for Germany prohibited research and technology in applied fields such as aeronautics and nuclear research. These restrictions were not abolished until 1955 with the ratification of the "Paris treaties," which gave the Federal Republic sovereign status as a state. An analogous development could be observed in East Germany. The postwar period, that is, the decade between 1945 and 1955, could thus be described as a phase of prohibition.

Prohibitions stimulate attempts to circumvent them, and this was the case in the field of nuclear research. The application of nuclear energy to military as well as to civilian uses was a very dynamic process, and scientists were motivated to get involved. But involvement was not possible as long as the allied restrictions stayed in force. The result was a growing impatience among scientists, which eventually expressed itself in attempts to revive at least basic nuclear research. Since the establishment of new institutions devoted specifically to nuclear research was prohibited by law, such actions could be pursued only under the "camouflage" of existing institutions such as universities or academies.

Institutions that had already been involved with nuclear research before 1945 were predestined to be revived. Since many research institutes had been transferred westward during the final years of the war, however, the territory of the Soviet Occupation Zone, the future GDR, had few institutional remnants that could be used as "nuclei" for these efforts. One such site was the former research institute of the German Reich Post Office in Miersdorf near Zeuthen (on the outskirts of Berlin). Here, financed by the National Socialist Reich Postal Minister Ohnesorge, scientists had been able to pursue nuclear research under relatively favorable conditions. Although the large apparatuses dating from that period (a cascade generator built by C. H. F. Müller in Hamburg and a cyclotron developed by the institute) had been dismantled by the Soviets, the institute itself had survived. Shortly after the establishment of the GDR, the plenum of the Academy of Sciences resolved to turn Miersdorf into the academy's "Institute for Nuclear and Atomic Physics." For "obvious reasons," it was to operate "for the time being" under the innocuous name "Institute Miersdorf." At the end of 1951, a committee was appointed to plan the development of the institute: it consisted of Friedrich Möglich, Georg Otterbein, Robert Rompe, and Rudolf Seeliger—the elite of East German physics.[4]

The remnants of the Kaiser Wilhelm Institute for Brain Research in Berlin-Buch were a second, though less important, "nucleus" for work in the field of nuclear research. The Russian-born scientist Nicolai Timoféeff-Ressovsky had founded radiation genetics prior to 1945 at this institute's department of genetics. An accelerator (cascade) had been available for radiation genetics and isotope production, but it was dismantled after the war. Like all the former Kaiser Wilhelm institutes remaining on East German territory, this institute was absorbed into the Academy of Sciences. It was turned into the academy's Institutes for Biology and Medicine (under the direction of W. Friedrich) and Solid State Physics (under the direction of F. Möglich).

The work that was carried on at the universities at this time was negligible and limited to low-cost "string physics." An internal SED review of the state of nuclear physics research in the GDR mentioned Professors Paul Kunze

(Rostock), Wilhelm Messerschmidt (Halle), and Alfred Eckardt (Jena), who were involved with cosmic radiation research. The SED considered all three to be politically unreliable—"not suited for sensitive work." This political criterion was to gain increasing importance during the 1950s in the wake of the SED's policy on university-level education, and it further narrowed down what was already a thin layer of personnel.[5]

Basic research was thus revived at the academy prior to 1955 on a very modest basis. Efforts were limited to restoring what had once existed: for example, a replacement for the neutron generator dismantled by the Soviets was procured from the "Transformer and X-Ray Union Dresden" (formerly Koch and Sterzel). This apparatus was unreliable owing to the low quality of the materials. Although nuclear reactors were forbidden, planning was even carried on in this field. In a confidential discussion in July 1954, experts from the SED explored the possibility of expanding nuclear physics research in the GDR. At the center of the discussion stood the building of a nuclear reactor.[6]

Establishing Nuclear Research and Technology

The prohibition phase in both German states ended in 1955. Restrictions were lifted in the wake of the signing of the Paris treaties, and applied nuclear research and development was permitted. The Soviet Union adjusted its policy accordingly: it allowed "specialists" to return to East Germany, and this gave a considerable boost to the country's pool of intellectual talent.[7]

Although nuclear research was limited exclusively to civilian applications, the geopolitical situation had played a role in the lifting of the restrictions. The controversy over the legitimacy of nuclear armament had become increasingly heated from the early 1950s on. In the face of worldwide criticism of the nuclear arms race, the superpowers believed that the proliferation of civilian technology was an instrument for diverting attention away from the dominance of the military uses of nuclear technology. In December 1953, President Eisenhower had launched the program "Atoms for Peace." Aimed at freeing up nuclear research for civilian uses, it was intended as a propaganda antidote to criticism of the nuclear arms race. The climax of this campaign was the First Conference on the Peaceful Uses of Atomic Energy. Held in Geneva in August 1955 under the auspices of the United Nations, it introduced the tremendous possibilities of nuclear technology to a broader public. Here the superpowers turned the proliferation of nuclear technology into a question of prestige and an instrument for exerting concrete political influence. The Soviet Union tried to counteract the American initiative with even more impressive apparatuses and gadgets.[8]

As part of the "Atoms for Peace" program, the United States had proposed that the supervision of fissile material be transferred to an international "atomic bank." In 1955, however, this suggestion was already dead, for the Soviet Union had made such a step contingent on the outlawing of all nuclear weapons of mass destruction. This ran counter to the defensive doctrine of the United States, which was based on increasing nuclearization. The idea of an atomic bank was replaced by bilateral agreements on the delivery of research reactors and nuclear fuel, which both superpowers signed with their satellites (GDR/USSR in 1955, FRG/USA in 1956).

The simultaneous introduction of nuclear technology in the GDR and the FRG also produced innovations in political institutions, with similar structures emerging in both German states. The new technology—because of its presumed importance to the national economy—called for a specialized department in the executive branch: West Germany created the Federal Ministry for Atomic Affairs, East Germany the Office for Nuclear Research and Technology. These organizations, set up at the level of departmental ministries, had to confine themselves to administrative tasks. Professional expertise was located outside of the executive in the field of research. The West German government set up an advisory committee, the German Atomic Commission, to make systematic use of the knowledge of experts. Its structural counterpart in the East was the Scientific Council for the Peaceful Use of Atomic Energy associated with the Council of Ministers.[9]

Given the newness and complexity of nuclear technology, research and development could no longer be carried out within the framework of existing institutions such as universities and academies. Consequently, "research centers" or "central institutes" were set up. In the West they constituted the first, nuclear generation of what were called, after 1970, large-scale research establishments (Großforschungseinrichtungen). Given the federal structure of the FRG, decisions where to site these installations had to take into consideration the ambitions of various Länder (states) wishing to participate.[10] The GDR, geographically smaller and economically weaker than West Germany, and organized on strict centralist lines, concentrated on setting up a Central Institute for Nuclear Physics, the later Central Institute for Nuclear Research (Zentralinstitut für Kernforschung, or ZfK).[11]

The Central Institute, one of whose tasks would be to run the large apparatuses supplied by the Soviet Union, would be set up not within the framework of the Academy of Sciences, but as a separate entity. It was given this special status not only because of the importance accorded to nuclear research and technology at this time, but also out of consideration for the financial burden it was expected to impose. Within the Academy of Sciences, a structural change in favor of the natural and technical sciences had begun in the mid-1950s and had progressed at a rapid pace: by 1957, fully 66 percent of the 1,044 scientists in the academy were working in scientific-technical institutes and using up 68 percent of the budget. Large-scale nuclear research and the financial burden it involved would have strained the precarious balance of disciplines in the academy.[12] In addition, the founding of a special institute would secure the political influence of the SED vis-à-vis an academy that was still promoting "bourgeois science."

A forested plot of land in Rossendorf near Dresden was chosen as the location for the Central Institute for Nuclear Research. This decision was influenced by the scientific-technical infrastructure of Dresden (home to the Dresden Technical University) and the wishes of specialists who had returned from the Soviet Union. The large apparatuses—the research reactor and the cyclotron—were supplied by the Soviet Union and assembled with the help of Soviet experts. The institute's building for reactor technology and nuclear physics also followed Soviet models down to the details. By 1958, the institute already comprised five sections, namely, reactor technology and neutron physics, physics of atomic nuclei, radiochemistry, materials and solid bodies, and technology.[13]

The available equipment was very good, if we base our assessment on the specifics of the large apparatuses. The research reactor was a swimming pool reactor with two megawatt thermal power, a so-called Material Test Reactor (MTR). MTRs were characterized by a high neutron flux and simple construction: the light water in the pool served simultaneously as moderator, shield, and coolant. The research program, using the high neutron fluxes of the reactor, was aimed at applications in neutron physics, a field of basic research that was promising but still in its infancy at that time. The reactor and the program were very similar to those of Heinz Maier-Leibnitz at the Munich Technical University in Garching, West Germany.[14]

On the occasion of the start-up of the Rossendorf research reactor in December 1957, Heinz Barwich, the director of the Central Institute, made the following assessment: "Our German Democratic Republic now has the very real possibility of winning, within the foreseeable future, a respected place in the ranks of the states that are currently leaders in nuclear research and technology."[15] This was certainly a realistic assessment in 1957. That the Central Institute was set up so quickly and furnished with up-to-date equipment shows that the Soviet Union at the time still had an interest in helping for its German comrades to enter the nuclear age. In the wake of the Geneva conference and in view of the "struggle of the systems," the propaganda success this entailed was probably more than welcome. During congressional deliberations in the spring of 1957 in preparation for amending U.S. atomic energy law in favor of West Berlin, Admiral Lewis L. Strauss, the head of the

US Atomic Energy Commission, cited the extent and speed of the work in Rossendorf as evidence of the Soviet Union's political resolve.[16]

In the area of applied nuclear research and development, scientists at the Central Institute concerned themselves with developing reactor technology. As early as July 1956, East Germany had concluded a top secret agreement with the Soviet Union concerning the "granting of technical assistance . . . in building a nuclear power station."[17] The nuclear power station Rheinsberg was to become a pressurized-water reactor (PWR) of Soviet construction. It wasn't operational until 1966 owing to delays in delivery, the need for technical backfitting, and economic circumstances that will be discussed below. The decision to use a Soviet reactor determined the work of the Central Institute in the area of nuclear technology for years to come. For the period 1958 to 1965, the mission of the Central Institute included, in addition to basic research and the production and application of radioactive isotopes, "involvement in the development, design, and construction of power reactors," that is to say, the optimization of Soviet reactor technology.[18]

Crisis

The early stages of nuclear research in both German states developed along similar lines. This changed at the beginning of the 1960s. The initial euphoria triggered by the Geneva conference in August 1955 had by the end of the 1950s given way to a more realistic assessment. The reason was the realization that the boom would be less dramatic and not as rapid as initially expected: atomic energy was still much too expensive. While the power companies, given the range of applications of relatively cheap and abundant coal, held back with options for nuclear power stations, companies in the nuclear industry were forced to realize that the costs of developing these stations were far greater than originally projected. Calls for state subsidies grew louder and louder. In 1963, West German industry withdrew its funds from the nuclear research facility Karlsruhe, a move that "sent a signal" to the state.[19]

Calls for state subsidies created considerable ideological problems for the West German government, which followed a neoliberalist economic doctrine that largely rejected subsidies. However, concern for the competitiveness of export-oriented German industry and its share of the future market for nuclear energy made assistance programs seem justified. The development of nuclear energy thus became part of the West German government's policy of providing the basic necessities of life. East Germany, by contrast, offered a diametrically opposed picture.

In the late 1950s, the economic situation in the GDR, difficult from the beginning, deteriorated to the point of crisis. The economy was suffering from withdrawal symptoms. Despite additional Soviet deliveries, urgently requested again and again, a dramatic shortage of supplies and raw materials developed, and production chains were broken. The supply of necessities to the population was also getting worse. Above all, there was a shortage of high-quality industrial products and food. The Party was confronted with signs of growing discontent among the population; some cities saw open unrest. At the same time, the stream of escapees from the GDR swelled dramatically: in 1959, about 145,000 East German citizens left the country; in 1960 the figure rose to just under 200,000; and by the middle of August 1961, when the Wall was built, an additional 160,000 escaped. It was not only the political framework in East Germany that motivated these citizens to turn their backs on the GDR, but also the palpable effects of economic blunders by the SED. For instance, the "socialist spring," during which the compulsory collectivization of agriculture was carried out in 1960, had led to a clear decline of living standards.[20]

Added to this was the inefficiency of the planning system, which became increasingly apparent. That inefficiency can be read from the ratio of investments to domestic national income: while investments in the GDR between 1950 and 1963 rose by 380 percent, national income during the same period grew, in terms of percentage, by less than half. The industrial growth rate fell from 12 percent to 6 percent between 1949 and 1961.[21] The SED, too, had located the causes of the wretched situation in the economic sphere.[22] The result was a reform of the state planning sector, leading to the "New Economic System for Planning and Managing the People's Economy." The reform was preceded by an economizing along the lines of cost-benefit analyses.

This "systemic crisis of the GDR," as M. Lemke called it, was bound to have repercussions in the field of nuclear research and technology, since this field had shown itself to be particularly capital-intensive. In the period from 1955 to 1962, a total of 470 million marks had been invested in nuclear technology in East Germany, 395 million of which was under the jurisdiction of the Office for Nuclear Research and Technology. The lion's share of the latter sum, 280 million marks, had gone to the nuclear power plant in Rheinsberg. Doubts soon arose as to the economic usefulness of that installation. As late as 1960, the State Planning Commission (SPK) had asked the Scientific Council to examine whether it would be possible to establish 600 megawatts of nuclear capacity in East Germany by 1972. The study revealed that while this goal was attainable with a considerable technological risk, it was economically not justified in view of the burden it would place on East Germany's national economy. Owing to the costs already incurred, the study cast doubt on the Rheinsberg nuclear power plant as a supplier of energy.

However, the government was understandably reluctant to draw the immediate consequence of halting ongoing construction and leaving the plant standing as a shell. As Max Steenbeck wrote, "The only sensible way of operating the installation is as a large-scale experimental and testing site," that is, as a contribution to the further development of the pressurized-water reactor in East Germany.[23]

The SED's Central Committee no longer shared the view of Steenbeck, Robert Rompe, and other scientists that the nuclear power plant had to be seen as a costly but indispensable contribution of the GDR to reactor development. The Central Committee had come to believe that the energy needs of the GDR could be met in the next ten to fifteen years much more cheaply and quickly by the production of conventional energy from fossil materials such as coal and oil and "its sensible distribution in the socialist camp." Preliminary work for the expansion of the nuclear power plant and preparations for the "600-megawatt program," including research and development, should therefore cease "immediately and in their entirety."[24]

The renunciation of reactor development by the GDR became the official line of the SED-state apparatus. This meant that economizing, radical cost-benefit thinking took hold in the sphere of research funding. So-called proportionalization became the maxim, that is, the process of measuring expenditures for a field of research against its presumed utility for the national economy. The basis for this was laid by the resolutions of the Fourteenth Plenum of the Central Committee of the SED in November 1961. These resolutions called for a stronger concentration of resources, in other words, massive savings as well as a strengthening of central planning.

Nuclear research now lost the special status it had held up to that time; the result was a far-reaching restructuring of the way research was organized. The first victim was the faculty for nuclear technology at the Dresden Technical University, which was dissolved in 1962 by the State Secretariat for Universities and Technical Colleges. Simultaneously, the Scientific Council's Commission on Nuclear Energy—at the urging of the Central Committee of the SED—recommended that the research institutes and enterprises be removed from the jurisdiction and supervision of the Office of Nuclear Research and Technology, and that the latter's responsibilities be transferred to a secretariat of the Scientific Council created for that very purpose.[25] Largely gutted by these steps, the office was eventually abolished altogether. Its original director, Karl Rambusch, had already been suspended long before that. The Scientific Council itself was attached to the Research Council of the GDR as the "Section for Nuclear Research and Technology."[26] Low-energy physics came to a halt entirely at the Academy of Sciences' Institute for Nuclear Physics in Miersdorf; its director, Gustav Richter, had to leave the institute. The Central

Institute for Nuclear Research in Rossendorf was hit especially hard: theoretical work for the fast breeder reactor had to be terminated, investment funds were cut, and consumption and personnel funds were frozen beyond 1970. Pits that had already been excavated for expansion construction were filled in again. The Central Institute was to focus on basic and "exploratory" research and, in the applied field, on supplementary and complementary research. Beyond that, there were discussions about requiring the institute to pursue work that was directly "production-effective." In 1963, finally, the Central Institute was integrated into the Research Association of the Academy of Sciences, over the emphatic protest of Klaus Fuchs.[27] The end of its special role could not have been made more obvious.

Stagnation

The Third Atomic Conference in Geneva in September 1964 marked a low point for nuclear research . The GDR was represented by an unofficial delegation of ten members, only nine of whom returned to East Germany. The defection of Rossendorf's former director, Heinz Barwich, was indicative of the severity of the crisis. The summary written by the East German delegation after the conference was clear and to the point:

> It must be said that, in the scientific-technological field, West Germany
> has caught up with the leading countries in the area of nuclear energy . . .
> The GDR has continued to lose ground in the development of nuclear
> technology during the past years. In particular, a very rapid development
> can be observed in West Germany, while de facto stagnation has existed
> in the GDR for a few years.[28]

The state-decreed moratorium on nuclear research was not accepted by scientists without dissent. Klaus Fuchs, who always gave priority to party discipline, accepted the need for savings but maintained that the decision about where and how these savings were made should be left to scientists. There were many complaints that decisions were being made without prior consultation of those affected.[29] Bertram Winde of the SPK justified the cuts to Helmut Faulstich, the new director of the Central Institute, by saying that nuclear energy would not be needed in the GDR until ten to twenty years later than projected, that East Germany could not afford nuclear physics, and that it would be better to send specialists to Novosibirsk to let them do nuclear physics there. These statements drew a sharp protest from the party leadership of the Central Institute, which complained to the Central Committee about Winde's "false and subjective views."[30]

Thus the moratorium that the SED decreed in the field of nuclear research had varied consequences: it not only caused the GDR to continue to lose ground in nuclear research and technology, but also led to an estrangement or distrust between science and the state apparatus. What was left behind were "frustrated technocrats."[31]

The reform and reorganization of the economy, the New Economic System, gave special importance to "prospect-planning" (planning for the future), and this invariably affected the area of research and development.[32] The champions of nuclear research raised their voices once again. Given the extreme scarcity of resources, however, there was no longer a consensus on which path to pursue.

The struggle over the drafting of a directive for prospect-planning lasted nearly two years and ended largely in paralysis. There are two main reasons for the duration and intensity of the conflict: first, the resources that were available to the GDR for its own research activities were scarce and compelled a kind of prioritizing that would have been unthinkable in West Germany; second, there was a continuing lack of clarity about the line of the Soviet Union, which, protestations to the contrary notwithstanding, lagged behind the West in technological development. In the West the pressurized-water reactor was developed to the point where it was ready for serial production and could be purchased "off the rack" by the middle of the 1960s. COMECON, by contrast, had nothing like this available at the time.[33]

The Soviet Union had entered the stage of commercial use of nuclear energy at about the same time as the United States, and in both countries light water reactors were at the forefront of development from the outset. However, unlike the West, where the pressurized-water reactor quickly developed into a production type, the Soviet Union pursued a variant that was never realized in the West, the pressure pipe reactor. In this design, the fuel cells, composed of thin rods of slightly enriched uranium-dioxide, were contained inside large pipes in which the coolant, light water, was circulated under pressure. The moderator was graphite, which, arranged in blocks and furnished with vertical ducts, housed the pressure pipes. The decisive factor in the choice of this design was that one could do without the large-volume pool of the pressurized-water reactor. The production of such containers had posed problems for heavy industry in the Soviet Union. Like the first nuclear power plant in the Soviet Union, Obninsk (1954), numerous other nuclear power plants (Belojarsk I and II, Leningrad, Chernobyl) were of the pressure pipe type, which had been developed to a capacity of 1000-megawatt blocks (RBMK-1000).[34]

The development of the pressurized-water reactor was delayed considerably. The PWR was used for the first time in 1964 in Novovoronesh as an experimental power station. It was developed further by the beginning of the

1970s. During the 1970s, reactors of this type (WWER-440, WWER-1000) were exported to the countries of COMECON, among them the GDR. Its development had not been without technical problems. A serious accident had occurred in 1969 in Novovoronesh, when the reactor of block 1 (which had been put into operation in 1964) dropped to the floor of the pressure vessel after a rupture of the thermal shield. The cause was material fatigue.[35]

Another factor that slowed development was the skepticism of leading Soviet energy scientists, who cast doubt on the economic efficiency of nuclear energy, in particular its ability to compete with Soviet pit coal. That attitude is perfectly understandable, especially against the backdrop of the energy-policy situation in which the Soviet Union found itself: according to a 1962 estimate, its territory contained one-third of the world's energy reserves. These riches grew substantially year by year in absolute numbers as considerably improved methods of prospecting led to the discovery of new deposits: reserves rose far more quickly than output. In 1967, the potential pit coal reserves of the Soviet Union were estimated at more than 4 trillion tons. The known reserves of oil and gas at the beginning of the 1970s were four times what they had been in 1950. Given this situation, it would take a special kind of dialectic to justify the rapid development of nuclear energy.[36] For the Soviet Union, more than abundantly endowed with natural primary energy sources, developing an industrially manufactured nuclear power plant was not an urgent economic necessity.[37]

The fact that Soviet development of a pressurized-water reactor was lagging behind put in doubt the SED's party line of procuring complete nuclear power plants from the Soviet Union. Engineers understood this much more quickly than did the party bureaucracy: because of its special construction, the pressure pipe reactor would have made it easier to divert fissile materials like plutonium. This feature prevented its export to and installation in the GDR. Added to this was the experience of erecting the nuclear power station Rheinsberg, where technical problems with the pressurized-water reactor components supplied by the Soviets had delayed its completion by several years.[38] In view of all this, it seemed necessary for the GDR to cooperate in drawing up and implementing plans for nuclear power station projects, to make an independent contribution to reactor development. But in which direction could this be pursued? While engineers of the state-owned nuclear power plant, such as Rambusch, favored the development of a pressurized-water reactor and its use in the long run, Fuchs advocated the view that the Soviet Union would take the pressurized-water reactor merely as an intermediate technological stage, a transitional phenomenon, in order to catch up with and surpass the West with a breeder that would be production-ready immediately. For that reason it would make sense to concentrate resources in this promising field, the breeder reactor.[39] It was a fight with the gloves off: while Rambusch accused Fuchs of pursuing his "private philosophy" with the breeder, Fuchs countered with the charge that Rambusch was listening to the "gospel from Karlsruhe."[40] The Nuclear Research Center Karlsruhe, the headquarters of breeder-reactor development in West Germany, had backed off from overly optimistic prognoses about this type of reactor. Rambusch was very well informed about relevant developments in the Federal Republic, thanks not least to the work of the agents of the Ministry for State Security.[41] The economic situation in East Germany hardly permitted an independent contribution to reactor development. The consequence was a growing dependency on Soviet technology.

Dependency

Complaints about the Soviet Union's reluctance to cooperate were voiced from the very beginning. In fact, the substance of the complaints changed very little. As a follow-up to the Second Atomic Conference in Geneva in 1958, a discussion was held in the office of Bruno Leuschner, the chairman of the state planning commission.[42] A review of what was learned in Geneva concluded that East Germany "has not sufficiently kept up with other countries, in particular the Federal Republic." In addition to the emigration of specialists to the Federal Republic, there was a lack of cooperation from the Soviet Union. Four points were outlined in the report in concrete terms:

1. The supply of specialized materials either took an unacceptably long time or appeared entirely impossible. For example, no uranium metal or zircon could be obtained from the Soviet Union. In 1958 the GDR had not a single gram of uranium metal of its own, not an ounce of heavy water, nor any reactor graphite or other specialized materials. The production of fuel cells—which was already being done in Karlsruhe (West Germany), for example—was impossible in the GDR.
2. The authors of the report next complained about obstacles to the sharing of experience among socialist countries: "We can get access more easily—and more quickly—to institutes or companies in capitalist countries than in socialist countries."
3. Training facilities for reactor technicians were inadequate compared with those of Western countries. The critics therefore called for more intensive travel and exchange of experiences to raise the level of cooperation between the USSR and the GDR.
4. Lastly, there were psychological barriers impeding progress. Particularly dangerous was the "rose-colored reporting" (Schönfärberei) that was

taking hold in the East German press. It posed the risk of "attributing to the Soviet Union successes that do not actually exist." The account of Rossendorf in the East German press was "a typical example of damaging rose-colored reporting, which cannot be helpful to progress in our field." There was a tendency to hush up gaps that were not embarrassing at all, for example, the use of research reactors for study purposes, "and this frequently results in promises that for purely material reasons cannot be kept." According to the critics, a radical course correction was soon needed.

The protocol went on to say that cooperation required that everything be fixed contractually. The Soviet Union was not very flexible when unforeseen problems arose, especially when it came to the kind of complaints that occurred with large projects:

> Requested specialists usually arrive contrary to plans, that is, either nobody shows up at the specified time or a specialist with the wrong qualifications, or someone who was not requested, appears out of the blue. In short, there was a complete lack of "operationalism," which drove up expenses and cost time, something that would surely have led to serious business complications with capitalist partners. We also feel that the letters exchanged about this incident were not very objective. This created ill feeling on our side, which was not in keeping with what has so far been friendly cooperation.[43]

The problems that appeared during the construction of the Rossendorf research reactor were repeated at the nuclear power station Rheinsberg.

The technical and bureaucratic difficulties were not infrequently joined by a lack of openness, especially with regard to the latest production technologies. For instance, the electronics division of the SPK complained in the summer of 1959 that it had proved impossible to visit Soviet institutes and factories in the field of semiconductor technology. Such visits were urgently required, however, to gain experience for drawing up plans for the East German semiconductor plant in Frankfurt-Oder.[44] Apparently complaints of this nature could be resolved only at the highest level, that is, between general secretaries.

It was entirely in line with a restrictive Soviet information policy that East Germany did not receive any information about the Soviet reactor program, which the Russians revealed only as part of the Third Geneva Conference in 1964. The situation was no better within the framework of COMECON, leading first to the paralysis and then the failure of talks about atomic cooperation within the organization. No trace here of collaborating as partners;

rather, one could make out tendencies toward control and hegemony on the part of the Soviet Union.

This emerges clearly in the example of Czechoslovakia. Czechoslovakia was the only country within COMECON that tried to develop an independent line of reactors, namely, the heavy water-moderated natural uranium reactor. This design had the advantage of not requiring uranium-enrichment technology, and this made it particularly attractive for countries that, like Czechoslovakia, had extensive reserves of natural uranium but lacked the technology to enrich it. (Incidentally, in the West, Canada initially pursued the same path.) However, the Soviet Union showed no inclination of supporting this development, which would promote Czechoslovakia's economic self-sufficiency.

In 1963, the Soviet Union demanded that all member states of COMECON submit their ideas for the development of nuclear energy to the USSR. However, the member states either did not respond at all or provided only limited information. Czechoslovakia sent its plans for a reactor to the secretariat of COMECON, but with the condition that they not be passed on to the member states (including the Soviet Union). This was a sign of the climate of suspicion that existed within COMECON at that time. Czechoslovakia tried to avoid going it alone by winning over other COMECON member states to its conception of a nuclear reactor. Those states, the GDR included, responded passively, for the Soviet Union had made it clear via unofficial channels that no support could be expected from its side for "this special version of a reactor."[45] Czechoslovakia therefore had no alternative but to switch over to COMECON's reactor type, the pressurized-water reactor.[46]

The project for a joint 800-megawatt–capacity nuclear power plant among COMECON states, discussed within the framework of the "Permanent Commission of COMECON for the Peaceful Use of Nuclear Energy," never progressed beyond the stage of ideas; the GDR had made it clear from the beginning that it would not and could not share the investment costs. East Germany, which early on had criticized insufficient cooperation within the framework of the Permanent Commission, chose the path of direct cooperation with the Soviet Union.[47]

The desire for control and hegemony dominated even when no economic issues were at stake. It showed itself also in the field of basic research, concretely in the Unified Institute for Nuclear Research, which had been set up in Dubna in 1956 in response to the founding of the European Organization for Nuclear Research (CERN). In the mid-1960s, negotiations were held about reorganizing the institute in Dubna. Their goal was to strengthen the influence member states had on the work being done in the laboratories. The spokesman for the critics was the People's Republic of China, whose representatives used the talks to settle scores with the Soviet style of leadership. It was not enough that all important positions at the institute were held by Russians. The Soviets were monopolizing the choice of research topics, the use of funds, and scientific successes. In short, so went the criticism of the Chinese, the Soviet Union was using money and collaborators from member states for selfish national purposes.[48] Even taking into account that the criticism from the Chinese came at a time when the ideological conflict between the communist superpowers was approaching its climax, there was some truth to what they said: Dubna was a Soviet institute shaped in accordance with the Soviet style of research. While the Chinese terminated their membership as of July 1, 1965, the GDR, of course, remained a member. It paid approximately 5 to 7 percent of the total budget, and in return was allowed to send roughly twenty collaborators to Dubna.

East Germany's share of the research capacities in Dubna became a growing problem as domestic research installations were cut back and the country became increasingly dependent on installations in the Soviet Union. In the wake of this "externalization" of research, East Germany became more vocal in asking for greater influence at the institute in Dubna, a request to which the Soviet Union acceded only up to a point. For instance, the Soviet Union was successful in a number of highly innovative fields of research: high-energy physics (physics of elementary particles), plasma physics (nuclear fusion), and the synthesis of superheavy elements. East German participation in these fields was minor or nonexistent.

Another conflict concerned the supply of nuclear fuels. The Soviet Union's desire for total control clashed with East Germany's desire for self-sufficiency in raw materials. The Central Committee chose the option (which proved illusory in the end) of bringing nuclear fuel under national control as soon as possible. As early as 1956, Ulbricht asked Khrushchev what kind of help the GDR could expect from the Soviet Union in setting up production installations with a capacity of twenty-five tons of pure metal uranium and twenty kilograms of plutonium a year—a question that strikes us as naive today. In his answer, Khrushchev dismissed his East German comrades with the smug comment that East Germany now had the specialists who used to work in the Soviet Union, and they were "capable of accomplishing this task successfully." In addition, Soviet organizations were willing "to give the necessary aid by granting consultations."[49]

In plain language, this meant that delivery of Soviet installations for reprocessing nuclear fuels was out of the question. Subsequently, the GDR had to obtain prefabricated nuclear fuel cells from the Soviet Union, at a price that was three times what it would be on the world market. Spent fuel cells had to be returned. This painful state of dependence was a strong incentive to setting up a production facility for fuel cells in conjunction with the nuclear power plant Rheinsberg in East Germany. Estimates on how long it would take for the investment to pay for itself ranged from one to three years. At the instigation of Fuchs, however, this plan was not carried out. The option he pursued, that of bringing the plutonium cycle of the fast breeder under national control, suffered the same fate: the hopes placed in the fast breeder were not fulfilled.

Application

If there was a country in the industrial world in which the use of nuclear energy made economic sense, it was East Germany. After the war, there was essentially only one primary energy source available on the territory of the Soviet Occupation Zone, later the GDR: brown coal. Since the energy policy of the SED was built on "national" self-sufficiency—that is, it largely did without the import of alternative energy sources such as pit coal, oil, natural gas, and so on—the country became almost completely dependent on brown coal: in 1950, brown coal accounted for 92.8 percent of primary energy production in East Germany; in 1958 the figure was 95.8 percent.[50]

Another factor specific to East Germany aggravated the energy situation: the inefficient use and waste of energy. Consumption of primary energy per capita in East Germany was 20 percent higher than in West Germany. This was caused, on the one hand, by a low, state-subsidized consumer price, which gave no incentive for energy conservation, and, on the other, by poor heat insulation and the use of inefficient and increasingly obsolete machinery. Added to this were the high losses that result when brown coal is converted into electricity. Repeated energy shortfalls had to be made up, like it or not, by importing oil and natural gas from the Soviet Union. Judging simply by demand, a large-scale expansion of nuclear energy in East Germany thus would have been justified. However, this expansion occurred only slowly because it was done exclusively through the import of pressurized-water reactors from the Soviet Union (see Table 10.1). The share of electricity from nuclear energy in the GDR never exceeded 12 percent compared with roughly 34 percent in West Germany.

Thus at no time was the use of atomic energy a realistic way for East Germany to break free from its fatal dependence on brown coal. This became especially apparent in the face of the supply crisis at the beginning of the 1970s. When the Soviet Union sharply raised the purchase price for oil and natural gas within COMECON, the only thing the GDR could do was turn away from oil and toward indigenous brown coal. The maximum yearly output of about 260 million tons of coal had been reached in 1970–1971,

Table 10.1 Capacity of GDR nuclear power plants in 1988

Power plant	Type	Capacity	Operational since
KKW Rheinsberg	WWER	70 MWe	1966
KKW Lubmin b. Greifswald			
Block I	WWER	440 MWe	1973
Block II	WWER	440 MWe	1974
Block III	WWER	440 MWe	1977
Block IV	WWER	440 MWe	1979
Blocks V-VIII	WWER	4x440 MWe	under construction
KKW Stendal	WWER	2x1000 MWe	planned

Source: Joachim Kahlert, Die Kernenergiepolitik in der DDR. Zur Geschichte uneingelöster Fortschrittshoffnungen (Cologne: Verlag Wissenschaft und Politik, 1988), p. 131.

Note: East German state nuclear power plants have been shut down since 1990.

however, and it was not surpassed until 1980. To be sure, considerable efforts were made to maintain this output or even surpass it. However, the increasing exhaustion of the deposits drove up extraction costs, with the result that the efficiency of energy production steadily declined. In 1988, Kahlert, without knowing it, reached a conclusion that now seems the final historical verdict on the GDR:

> The GDR is caught in a dilemma in terms of energy policy: on the one hand, it has to fall back on its brown coal reserves for reasons of resource economy and foreign policy; on the other hand, the high degree of self-sufficiency in energy policy comes at a high price. Investments in the energy and fuel economy account for about 25 percent of total investments every year, and thus tie up a considerable portion of the funds that would be urgently needed for innovations in other industrial sectors.[51]

The situation in the energy sector and the historical dependence on Soviet nuclear technology played a significant role in the economic downfall of the GDR.

Conclusion

The way nuclear technology was used and how it began were similar in both German states; at the start, then, the two Germanys were equal. The race for nuclear science and technology started even before the Allied research restrictions had been lifted. The real take off, however, came after the August 1955 Geneva conference. In the East, efforts to reestablish nuclear science were initially hampered by a lack of manpower, know-how, and instrumentation.

Although the German atomic specialists who had worked in the Soviet Union during the occupation and early postwar period returned in 1955, there continued to be a lack of equipment, and this need had to be filled by imports. Complaints about insufficient quantity and the poor quality of imported technology could be heard in the East and in the West.

East German state policy centered on education, but basic research was soon undertaken as well, and before long it reached international standards. The industrial development of nuclear power technology, however, met with criticism focusing on the relatively high expenses needed to produce electrical energy by nuclear reactors in comparison with traditional coal-burning power plants. In dealing with these objections, marked differences emerged concerning the state of "national" economies and the kinds of decision-making processes that developed. While in the West, the so-called economic miracle made West Germany competitive internationally, the East sunk into a deep economic depression aggravated by severe losses of personnel through emigration to the West ("Republikflucht"). While in the West the federal government decided to increase its budget for, and engagement in, nuclear research and development, in the East, just the opposite took place: internal discussions of the Socialist Union Party in 1962–1963 led the GDR government to freeze expenditures on nuclear research and technology. The financial cut back in the East was followed by the decision to suspend nuclear reactor development entirely. As a result, nuclear power plants had to be imported from the USSR.

Debates centered on whether future East German nuclear research should focus on the backfitting of Soviet light-water technology or on bypassing this technological step by "jumping" to fast-breeder technology. This discussion split the GDR nuclear science community. What made the breeder so attractive to GDR scientists like Fuchs was its image of being the most modern and advanced technology, as well as the Promethean idea that stood behind it: the breeder whose production rate of nuclear fuel was supposed to exceed its nuclear fuel consumption seemed to guarantee national energy autarky, a policy that had motivated all GDR industrial policy from the beginning. East Germany's desire for self-sufficiency, however, clashed with the Soviet Union's policy of control, making the idea of national control over the nuclear fuel cycle pure moonshine. What was left for the GDR was a rather modest use of Soviet light-water technology.

—Translated by Thomas Dunlap

11

Politics and Computers in the Honecker Era

Gary L. Geipel

The history of the German Democratic Republic's final years includes the dramatic story of a powerful, centralized leadership that invested great resources and rhetoric in the development of advanced technology only to be undone, at least in part, by its failure to keep pace with competing systems. The very intensity of the GDR's efforts to overcome technological backwardness and the very fact of its demise make East Germany an excellent case study with which to test broader understandings of the relationship between politics and the development of technology.

Chief among the advanced technologies pursued in East Germany was information technology (IT), including computers, microelectronic components (microchips), software, and data-communications equipment. In the past two decades, the IT field has produced data-transmission networks, plant automation systems, educational and entertainment devices, and many other products now synonymous with the so-called information revolution in the noncommunist, industrialized world. That revolution fueled unprecedented growth throughout much of the global economy, and set in motion transformations of industry in the most advanced countries, away from heavy manufacturing and toward the service sector.

Beginning at a special Central Committee session of the East German Communist Party in 1977 and continuing through the end of the Erich Honecker regime in 1989, East Germany invested billions of marks in, and allocated a disproportionate share of its research and development resources to, the promotion of independent development and production across almost the entire spectrum of IT. Microchips, mainframe computers and later personal computers, printers and other "peripheral" devices, software and a host of more mundane computer components all were to be manufactured in the GDR itself. For a country of seventeen million people, the attempt to achieve technological autarky across such a wide spectrum of components and products was indefensible from an economic standpoint. With few links to a global division of labor, East Germany managed to achieve only small production runs of IT components and finished systems, which invariably were several years behind the international state-of-the-art. The one-sided allocation of resources to IT and a small set of other high-technology industries exacerbated investment shortfalls in other, more basic sectors. In the aftermath of German unification, a number of scholars and journalists inside and outside the GDR speculated that the pursuit of autarky in IT was evidence for the incompetence of the country's leadership and a primary cause of East Germany's economic collapse. Since economic analysis would have dictated another course, the observer is drawn to political explanations in attempting to understand East Germany's autarkic strategies for the development of high technology.

The Context for East German Technology Policy

"Every country's images of its own capabilities and its surrounding environment influence the choices it perceives and makes," Bruce Parrott argued in his study of Soviet technology policy.[1] An assessment of capabilities surely must enter into the rational formulation of policy. The case of the GDR, however, seems to demonstrate that decision-makers' images of their "surrounding environment" are the dominant factors influencing the choice of technology policies. Five aspects of the international environment perceived by East Germany's leaders are particularly important: the imperative of technology, the fear of dependence, isolation in the socialist community, trade with the Soviet Union, and the quest for legitimacy.

The Imperative of Technology

In the words of an American scholar, "technology is the new 'holy Grail' of international economic competition . . . It is seen as the essential life blood of the modern state and world economic prowess."[2] Few top GDR leaders would have disagreed. "The key to the significant improvement of economic effectiveness in the 1980s lies in a considerably higher effectiveness of science and technology," General Secretary Erich Honecker told the tenth Party Congress of the SED in 1981.[3] Kurt Hager, the chief ideologue of the SED, cited nothing less than a West German business journal in claiming that "without chips from Japan, the Americans cannot produce computers, the Germans cannot produce cars, the Siemens company cannot construct a telephone network, and Boeing cannot build airplanes."[4]

The widespread application of IT appeared to East German leaders to be the ideal response to their diagnoses of the GDR's economic problems in the 1970s and 1980s. Faced with rising energy prices, perceived absolute limits in securing raw materials, and a declining supply of labor and thus the end of "extensive" growth strategies, they pursued a policy of *Intensivierung* (intensification), in which greater productivity was to result from constant or declining inputs of labor and materials.[5] Automation on the basis of IT was to promote "a new level of *Intensivierung*," according to East Germany's minister for electronics in 1976 and many others thereafter.[6] "The quality of final products and the competitiveness of exporters depend increasingly on microelectronics. The opportunities here are worth gold for our country," Honecker argued in 1977.[7]

The Fear of Dependence

The GDR's attempt to avoid any dependence on the capitalist world economy was a frequent theme in the statements of the country's leaders, and was an attitude of fundamental importance in the development of East German technology policies. A speech by Honecker in 1986 was typical, in which he linked the 1986–1990 Five-Year Plan to the need for "expanding the GDR's maneuvering room in trade policy and ensuring that the country cannot be threatened economically."[8] "Avoiding critical dependencies" was described by the GDR economist Hans-Jürgen Wunderlich and many others as one of the key factors determining the GDR's relationship with the capitalist countries.[9] East German rhetoric attributed the risk of dependence on the West to a U.S.–NATO desire for military superiority. Often, the capitalist world was described as engaging in a deliberate "war of technology" against socialism. "Ultimately, the object is to attain military superiority over socialism, and from that position to apply strong-arm politics to roll back socialism," according to one tract that reflected a prevailing line.[10]

The paranoia of East German leaders had some grounding in reality. The GDR clearly did not enjoy legal access to all types of Western technology or full membership in the Western scientific establishment. Its exaggerated fear of dependence, however, probably was based as much on the shortcomings of the East German economy and perceived social instability as on the actual policies of Western countries. Parrott suggested that regimes are "wary of widespread technological contacts when leaders doubt their society's internal cohesion and ability to withstand the disintegrating influences of foreign cultures."[11] GDR leaders almost certainly realized that it is difficult to transfer foreign technology without transferring the foreign ideas that gave rise to that technology.[12] In reference to the developing countries but with no less relevance to the GDR, Christopher Freeman wrote that "heavy reliance on imported technology is an inescapable necessity for most countries in the world. The economic consequences of this situation are perhaps not too serious, but the political and cultural consequences are very great."[13]

In addition to the potential for political and cultural contamination, it is likely that one of the primary reasons for the East German leadership's fear of dependence on the West was that it lacked the ability to sustain large-scale hard-currency expenditures for Western technology or the willingness to take on the additional debt that a Western shopping spree would have required. The East German economics minister admitted as much in a book after the country's collapse, in which he wrote that "unfavorable economic conditions hindered the import [of high technology]."[14]

One of the great ironies of East Germany's technology policies is that while the country produced the vast majority of its microchips and computers at home, the GDR regime's dependence on the capitalist world's basic research, prototypes, and specialized equipment in the IT field was not overcome but actually increased over time. As a result, East Germany's espionage apparatus devoted ever-growing attention to illegal and costly covert acquisitions of IT for purposes of reverse engineering and equipping the GDR's R & D labs.[15]

Isolation in the Socialist Community

Whether the GDR's fears of dependence on the West were driven by the potential for outside political manipulation, by East Germany's limited ability to afford hard currency purchases, or by the regime's concern over cultural contamination, the logical alternative would seem to have been cooperation and trade with other socialist countries in the Council for Mutual Economic Assistance (CMEA). It is important to understand the reasons East Germany instead felt increasingly isolated among its ostensible partners.

The success of CMEA in promoting joint R & D and trade in the computer field was mixed at the outset and declined steadily. In the 1970s, when computers had longer lives, CMEA's ES (unified systems) and SM (small systems) programs divided tasks among manufacturers in most of the Eastern European countries and the Soviet Union, and were somewhat successful in producing computers that functionally duplicated Western machines.[16] In the 1980s, however, as economic crises worsened in Eastern Europe and the pace of Western IT development increased, CMEA became increasingly ineffective in fostering mutually beneficial trade relationships involving IT or in promoting joint computer development. The CMEA countries pursued independent strategies, or descended into technological chaos; incompatible capabilities and goals were the rule.[17]

In this environment, the GDR emerged with the most advanced IT industry in Eastern Europe and usually ran high-technology trade surpluses with most of its allies.[18] The GDR could count on little in return. East German ministry officials and enterprise managers complained frequently of long delivery times, failures to meet promised levels of quality or quantity, the unavailability of spare parts, and poor customer service in their dealings with other IT producers in CMEA.[19]

Where joint R & D was concerned, divisions of labor were usually based not on competence but rather on forced marriages arranged by central authorities for political reasons or on the whims of individual enterprises. The frequent result, according to an East German observer, was "partial incompetence on one side and often well-grounded skepticism on the other [GDR] side . . . This working style pre-programmed collapse."[20] Another factor limiting the effectiveness of CMEA cooperation was the intense secrecy of Soviet enterprises and research organizations regarding their most advanced technology. East German officials and industrial managers voiced universal bitterness regarding the perceived one-sided nature of the GDR's scientific-technical contacts with the Soviet Union. Much of this was blamed on the priority of military R & D in the Soviet Union, to which even scientists from other Warsaw Pact countries were said to have had almost no access.[21] By the end of the 1980s, some of the GDR's frustration even spilled over into its press. Referring to East Germany's expensive development of a 1-megabit Dynamic Random Access Memory (DRAM) prototype, the SED ideologue Otto Reinhold said that the GDR was forced to proceed independently since "cooperation with other socialist countries in this field is not yet at the necessary level."[22]

Resource Pressures and Trade with the Soviet Union

As already indicated, the GDR's high-technology sectors were among the strongest in CMEA. Partly as a result, the GDR's trade with the Soviet Union was complementary rather than substitutional, with the GDR providing technology and finished products in exchange for raw materials and lower-level goods.[23] The GDR's leaders felt they had little choice but to make a virtue out of necessity. High technology became a political tool with which East German leaders believed they could manage their country's energy dependence on the Soviet Union.

The dependence of the GDR on the Soviet Union for its energy supplies, and particularly for oil, was large. Through the mid-1970s, CMEA's complicated moving-average formula for the pricing of Soviet oil worked in East Germany's favor, shielding it from the rise in world-market prices. The GDR could pay for its oil and raw materials with manageable exports of finished goods to the Soviet Union. By the late 1970s, however, the price of Soviet oil began to climb to world levels, and the GDR faced a painful squeeze. In order to balance its accounts with the Soviet Union, it had to increase the value of its exports and/or reduce the amount of oil it imported by streamlining industry. The application of advanced technology was vital to both options.[24] Soviet oil prices declined again somewhat in the mid- and late-1980s, but by then the Soviet Union—under its new leader, Mikhail Gorbachev—had begun to pressure its Eastern European allies into providing more and better products in exchange for energy supplies.[25] Evidence will be presented below to suggest that for the East Germans, IT was a vital bargaining chip in dealing with Soviet pressure for finished goods.

The Quest for Legitimacy and Status

The speed with which the regime of Honecker and the rule of the SED ultimately fell indicates that the leaders of the GDR never succeeded in demonstrating to their own people the naturalness of, or the need for, a separate "workers' and peasants' state on German soil." However, that failure was not for want of an effort by East Germany's rulers to develop legitimacy. The GDR's attempt to increase its legitimacy as a separate German state remained at the heart of its international relationships and therefore, in my view, at the heart of its policies to promote the development of advanced technology.

Thomas Baylis provides a useful distinction between two forms of legitimacy: legitimacy that is "wertrational," what Max Weber called a "rational" or "legal" form of rule based on shared values; and legitimacy that is "zweckrational," based on performance in providing the goods, services, and other satisfactions demanded by a population.[26] As a young state that was believed by most of its own population to have been invented by an outside power, the GDR initially could not inspire a value-based legitimacy. Perhaps accepting this, the East German leadership seemed to focus increasingly on securing legitimacy based on performance.

In pursuit of a performance-based legitimacy designed to raise living standards, the East German regime's record in obtaining and applying new technologies was of obvious importance. Fred Klinger described the effective management of scientific-technical resources as the decisive factor in securing "prosperity, the loyalty of the population, the necessary resources for the execution of domestic and foreign policies, and the rank and influence of a society in the global division of labor."[27] This is true not only because the development and application of technology are keys to economic growth, but also because the very fact of certain technical achievements can lend prestige to the country in which they are realized.[28] Klinger is a West German social scientist, but his observations would not have been disputed by Erich Honecker, who had the following to say about national scientific-technical performance: "How a country behaves in this competition has decisive consequences for its rank as a developed, industrial state—with all the related impact on labor productivity, national income, and the life of individuals."[29]

In 1983 the powerful GDR economics minister, Günter Mittag, used stark ideological terms to cast the economic challenge as "that much more important, since our country is on the dividing line between the Warsaw Pact and imperialist NATO, and must confront the acute international class conflict every day."[30] By early 1989, the leadership's rhetoric took on an almost desperate character, but was still not sacrificed. In an article that appears prophetic in retrospect, the ideologue Otto Reinhold commented on the role of the Federal Republic of Germany in East Germany's quest for international status: "We cannot choose our vis-à-vis [the country to which the GDR relates itself]. And that will remain so. But from this result [come] a number of important conclusions . . . The first is that the SED must carry out an economic policy predicated on high dynamism. Anything else would be a policy of suicide . . . a waiver of dynamic economic development would lead to a dangerous situation."[31]

Like Reinhold, other East German leaders viewed the stakes in the quest for technological achievement and performance-based legitimacy as enormous: the very survival of the country. Even when this was not explicitly stated, it was implied, using the common formulation, "My workplace is my battle station for peace." Peace was understood to be synonymous with the stability and survival of the GDR.[32] The Honecker regime almost constantly invited judgment of its scientific-technical performance on the question, "Who will master the Computer Age—capitalism or socialism?"[33] Going even further, the SED often seemed to cast the question of its legitimacy in terms of mastering IT alone: "Microelectronics, as an important material basis for economic *Intensivierung*, is tied as closely as possible to the purpose of socialism and the strength of its appeal."[34]

Impact of the International Environment on East German Technology Policy

In my view, the East German regime's perceptions of the international environment determined the country's IT strategy. The notion that leaders' perceptions of their country's international environment should drive technology policies is less self-evident than it might seem. It is clear that a responsible national leadership is wise to consider the global strategic and economic challenges that face its industries, in arriving at sound policies to promote the advance of technology. It is far less clear, however, that perceptions of the international environment should be the primary consideration influencing the choice of technology policies. Domestic scientific capabilities, raw materials supplies, the amount of available investments, and the nature of anticipated markets—among other factors—all would play a very significant role in the development of a rational strategy to promote technological advance. The case of East German IT policy suggests that such important factors can be forced by international political considerations into a subordinate if not nonexistent role.

The Pursuit of Autarky

The belief of East German leaders that information technologies were essential to economic health existed side-by-side with their sense of political isolation and fear of dependence on other countries. The result, in my view, was the pursuit of autarky in the development and production of IT. "We proceed under the assumption that significant increase in our own production of microelectronics must be a priority," said Otfried Steger, the minister of electrotechnology and electronics, at a crucial SED Central Committee conference in June 1977 that served as the starting gun for East Germany's subsequent IT strategy.[35] Statements by top East German leaders setting the policy of autarky in IT increased in clarity throughout the 1980s. The drive was focused in particular on the manufacture of microelectronics, the circuits and memory devices so crucial to the modern industry. In 1981 at the SED's tenth Party Congress, Erich Honecker stated that "by 1985 we want to be in a position to meet the greater share of the economic demand for microprocessors through our own production."[36] At a Central Committee meeting in 1984, Honecker demanded increases "in the breadth and in the technological level" of microelectronics production.[37] Calls for increased "breadth" in the mastery of IT, or Bedarfsdeckung (meeting needs), implied movement toward autarky and were repeated throughout the 1980s.[38]

It is important to make a distinction between the pursuit of autarky and the achievement of that goal. The GDR did not achieve all the goals it set for itself in IT production, nor did it seek to produce every type of component or finished system available on the world market. Moreover, as Macrakis demonstrates in Chapter 4, the GDR became increasingly dependent on Western prototypes and crucial IT devices for its own R & D effort. Autarky for the GDR meant autarky in the mass production of IT components and systems that its leaders felt were essential to the East German economy, or that they felt would enhance the country's prestige and the legitimacy of its economic system.

Those qualifiers should not diminish the scale of the East German effort, however. For a country of its size, the GDR remains unique in the annals of IT development for the range of materials, components, and finished systems it sought to manufacture. The components produced from the most basic logic circuits to high-capacity DRAMs using Very-Large-Scale Integration (VLSI) technology. In the course of developing this broad range of circuits, the Carl Zeiss Jena (CZJ) enterprise also became the first socialist firm to manufacture the full spectrum of microchip-production technology.[39] By the mid-1980s a division of labor had developed in which CZJ performed most of the R & D for, and engineered prototypes of, the most advanced micro-chips, while the Microelectronic combine in Erfurt (KME) took charge of the mass production of microelectronic components. Several other com-bines, in particular Ceramic Works Hermsdorf, provided raw materials to the microelectronics enterprises, or undertook production of specific circuits. The GDR's microelectronics effort culminated in the late 1980s in the very expensive and later very controversial effort to develop a 1-megabit DRAM.

In the 1980s, the Robotron enterprise was the CMEA's largest producer of computers. It sought to construct an ever-expanding product line, ranging from 8-bit toy machines to 32-bit mainframe computers by the end of the decade, as well as several models of printers, plotters, and other peripheral devices. Beginning in the mid-1980s, Robotron was heavily involved in sup-plying the personal computers (PCs) and other devices required for Com-puter-Aided Design/Computer-Aided Manufacturing (CAD/CAM) solu-tions, which were promoted by the central economic authorities in East Germany to the point of obsessiveness.[40]

Trade and International Cooperation in IT

The autarky of the East German IT sector is evident not only in the country's declared goals and in its own specific accomplishments; indeed, trade data also show that the GDR depended on foreign suppliers for a declining share of its information-technology needs. This is consistent with the East German leadership's perception of the international situation, in which technology purchases from the nonsocialist countries were seen as limited by embargoes, difficult to afford, and conducive to dependence and political blackmail— while technology suppliers in the CMEA countries were considered inferior, unreliable, and secretive. At the same time, there is evidence that the GDR sought to expand its exports of IT to other CMEA states, in an effort to increase its economic clout.

East Germany's espionage apparatus, aided by a network of shadowy trad-ing firms, managed to obtained embargoed Western information technology in small quantities, and the GDR occasionally placed small, legal orders for less-advanced IT to fill urgent gaps in its own production. There were, however, no instances of large-scale or sustained East German purchases of information technology from the nonsocialist countries. Statistical indicators bear out the GDR's consistently low level of reliance on Western computer equipment and components. East German statistics were not precise. However, even in the broad category of "electro-technical and electronic products"—which probably included very little IT—East German purchases from West Germany in 1988 were valued at about 81 million valuta marks, less than 1 percent of total trade turnover between the two countries. A comprehensive study of East German computer stocks undertaken by a U.S. research firm just prior to German unification found that "imports [of computers] from the West were subject to approval from the authorities and were allowed only in exceptional circumstances."[41] Another outside observer described the GDR's trade with the West in advanced technology as "more the repair shop of plan failures than a genuine technology-transfer channel and a dynamic source of growth."[42]

Two observations are particularly important, and set the GDR apart from almost every other CMEA country. First, there is no single IT product or component category in which the GDR depended on imports from the West for its own automation needs. While it may have addressed short-term production gaps with Western imports, it never sacrificed domestic production capabilities in any area of IT in favor of a Western supplier. Second, East German willingness to import IT products from the capitalist countries did not increase dramatically or consistently during periods of reduced East-West tension. In the consumer sector, where duties and currency-exchange rules would have been relatively easy to calibrate in response to improved East-West relations, East German authorities undertook no changes to ease costs for private citizens eager to obtain personal computers or other IT from the West. In the industrial and scientific sectors, it also remained extraordinarily difficult for all but the highest priority firms and research institutes in the GDR to obtain the required permissions and hard-currency balances needed for the purchase of Western computer equipment, components, or software.[43] This was in contrast to policies in Hungary and Poland, which in the late 1980s permitted significant numbers of Western PCs to be imported. By 1987, for example, almost 70 percent of new computers in Hungary originated outside of CMEA.[44]

Several observers argued that the existence of intra-German trade was largely responsible for East Germany's comparative economic success in CMEA.[45] The clear economic benefits of obtaining goods from the FRG make the GDR's reluctance to do so in the case of IT all the more striking, and lend credence to the notion that East German technology policy was motivated more by international political perceptions than by economic rationality. As Parrott wrote, "The more hostility a regime perceives from other states, the more reluctant it will be to give them diplomatic leverage by seeking and depending on their technology."[46]

Even with CoCom (the Coordinating Committee for Multilateral Export Controls), the GDR could have filled its needs for small-scale educational computers and computer printers in West Germany or secured reliable supplies of simple microchips.[47] Alternatively or additionally, it could have obtained licenses for the manufacture of certain IT products, as did almost all the East European countries at various times.[48] Instead, cooperation with Western firms in the electronics sector was late in coming and restricted to a few final-assembly contracts or small-scale component deliveries.[49] Despite the signing of an accord on scientific-technical cooperation between East and West Germany in 1987,[50] and some neutral-to-positive evaluations of joint ventures in East German scholarly literature,[51] the GDR under Honecker did not permit any IT joint ventures with Western firms, even for the development of new products.[52] This was in marked contrast to the situation in Hungary and Poland—and even Bulgaria and Romania to lesser extents—which benefited in the 1980s from joint ventures with Western firms in the computing field.[53]

In my view, the avoidance of such relationships in the GDR was due to perceptions of the strategic importance of IT, combined with a sense of the GDR's political isolation and susceptibility to blackmail. Autarky was preferred to almost any expansion of IT dependence on the West, and especially on West Germany. Differences between East German IT trade policies and the attitudes of other East European countries were probably due in part to the existential risks perceived by the GDR's leaders.[54] Hungary and Poland, even in a worst-case scenario, could not be blackmailed or cajoled out of existence by a dependence on Western high technology. The specter of unification, however, left East German leaders with the unending fear that too much economic and technological integration with the West would beg the question of the GDR's separate identity.

The GDR's IT policy with regard to the nonsocialist countries can be described as defensive: an attempt to prevent the expansion of dangerous dependencies and counter the effect of an embargo. Where IT trade with the Soviet Union and other CMEA countries was concerned, the GDR appears to have followed a mixed offensive and defensive policy. In the 1980s, the GDR continued to pay lip service to the importance of joint-development schemes with the Soviet Union and other CMEA countries,[55] and could point to a list of programs and protocols it had signed.[56] Those agreements were long on lofty goals but short on specific commitments. In practice, the GDR frequently managed to avoid what it saw as disadvantageous entanglements. Despite what must have been considerable pressure from Moscow, for example, Robotron avoided any formal joint ventures with Soviet enterprises until 1989, when a joint software firm was announced.[57] East Germany's policy of autarky was to some extent a defense against the failures of CMEA and the perceived one-sidedness of R & D cooperation with the Soviet Union.

In its IT trade, however, the GDR attempted to take the offensive in obtaining favorable access to raw materials and finished goods from its partners. Robotron's sales to the Soviet Union averaged more than 3 billion marks per year in the second half of the 1980s.[58] The sales of a single East German IT firm, therefore, represented fully 5 percent of the overall volume of trade between the two countries.[59] Former East German officials maintain that the sale of IT in particular gave the GDR an extraordinary "return on investment" with regard to the other CMEA countries. By virtue of their relative quality and reliability, East German computers at the peak of their appeal afforded the GDR five marks of buying power in CMEA trade for every one mark invested by Robotron in producing them, according to a former top ministry official.[60] It is not surprising, then, that in some years more than half of all PCs produced in the GDR were exported, for the most part to the Soviet Union. The share of exports to the USSR in East German mainframe production was even more dramatic, approaching 90 percent in the case of some models. As general rules, Robotron exported about two-thirds of its production of computers and peripherals, and two-thirds of those exports went to the Soviet Union.[61] Since East Germany's own demand for computers was not even close to being met, the degree to which decision-makers were willing to permit the export of PCs indicates that they played a very significant role in responding to Soviet pressures for high-quality goods and in managing East Germany's raw materials dependence on Moscow. Taken together, the GDR's IT production profile and its willingness to allocate a large share of that production to CMEA trade gave it considerable clout in its relations with Eastern Europe and the Soviet Union.

Findings here support the conclusion that the GDR's disdain for Soviet *perestroika* and for Moscow's opening to the West was due in considerable measure to the fear that East Germany's role as the leading foreign supplier of IT to the Soviet market would be eroded in favor of imports from the West. As Alfred Schüller expressed it in 1988, "[The GDR] would have to worry about its superior economic and technological status in the CMEA, if the USSR were successful in moving its modernization programs forward by passing over the GDR in favor of technologically more advanced partners in the West."[62] Instead, in Steven Popper's view, East Germany wished to continue playing "the part of Japan in a well-defined CMEA division of labor."[63] Given the GDR's level of effort and investment in IT, the worst-case scenario would have been one in which it remained dependent on the Soviet Union for key components but in which Moscow was no longer dependent on the GDR.[64]

The "Propaganda Trick"

The perceived futility or danger of obtaining needed IT from other countries played an important role in determining the breadth of East German production. That is not the only explanation, however. By pursuing a considerable variety of high-profile items rather than focusing on the expansion of a particular niche, the GDR's leaders also appeared to be motivated by the desire to prove that their industry possessed the same prowess as other industrialized economies. It is important to examine how the GDR's top leaders sought to use the breadth of their IT accomplishments to remedy both their international and their domestic legitimacy crises. In addition to their uses in the East German economy, the microchip and the PC became propaganda tools of the highest order.

In the last years of the GDR's existence in particular, the unveiling of new microchips or other IT products invariably coincided with a date of considerable symbolic meaning. The partial realization of a PC production goal in 1986, for example, was described on national television in a tearful speech by a twenty-three-year-old computer assembler at the SED's eleventh Party Congress.[65] The prototype 1-megabit DRAM chip was to be handed over to Erich Honecker just prior to the GDR's thirty-ninth anniversary in October 1988. Development was accelerated, however, so as to permit the prototype to be displayed at a major exhibit of East German technology in Moscow during the summer of 1988.[66] And the GDR's first 32-bit microprocessor was rushed out in time for the country's fortieth (and last) anniversary celebration in 1989.[67] East German sources indicated that crash efforts to prepare prototype devices for unveiling—especially before a Party Congress or other major event—often meant tying up production lines, despite considerable cost in lost output of other items.[68]

Some East German scientists and ministry officials still claim that the propaganda uses of IT breakthroughs were byproducts rather than motivating factors of technology-development efforts. For example, officials of the electronics ministry and the Ministry for Science and Technology argued that command of the 1.5 micrometer VLSI technology inherent in the 1-megabit DRAM was an absolute requirement for East German progress in electronic control systems, and that "the propaganda trick" with the chip was a Politburo game not supported by serious scientists.[69] Since top Politburo officials such as Honecker and Mittag had to approve the megabit project and similar efforts, however, it is quite possible that "the propaganda trick" was a determining factor in their thinking. The stakes involved in the GDR's quest for legitimacy were very high, and it does not seem unreasonable to conclude that the breadth of East German IT efforts had as much to do with their perceived legitimacy benefits as with real or perceived economic necessity.

For the domestic audience, IT breakthroughs were described as affording the GDR membership in a select club of countries. According to Honecker in 1988, the GDR was "among the few developed industrial nations in the world that have conquered the entire complex of development, production, and application of microelectronics."[70] In his speech at the ceremony for the 1-megabit chip, Honecker invoked Marx and Engels and declared the chip proof that "when one considers all aspects of social life, socialism is far superior to capitalism."[71] At one of his last public appearances in 1989, Honecker used another chip as evidence for a maxim he resurrected from the German workers' movement: "Den Sozialismus in seinem Lauf hält weder Ochs noch Esel auf" (Neither ox nor donkey can slow the progress of socialism). The phrase (which rhymes in German) was meant to reassure his people that the "triumphal cries of the Western media" about the death of socialism were wrong.[72]

Few East Germans appear to have been reassured, however. There is widespread anecdotal evidence to suggest that by the late 1980s, most East Germans reacted to the regime's promotion of IT with cynicism, not with respect or confidence. A former official described what he thought to be a common popular reaction to announcements of new microelectronics breakthroughs in the media: "We have trouble filling our shopping baskets. Pretty soon they'll just give us all a microchip instead of something to eat."[73] One observer quoted an East German factory worker as saying that "they [party members] talked their heads off about megabit chips and Computer-Integrated Manufacturing, while we lacked spare parts for the machine tools falling apart around us."[74] Some of the GDR media coverage of microelectronics breakthroughs was almost painfully silly, such as when a scientist at Carl Zeiss Jena wrote that at difficult points in the R & D process, he would "think of what a beautiful country Honecker and the other leaders created out of the ruins of 1945," thus making his task easier.[75] The numerous private citizens I encountered in visits to the GDR since 1985 regarded the regime's claims of IT developments as lies or as meaningless diversions from real problems. It appears that the GDR leadership failed utterly to use its technology policies to gain legitimacy among the population. The depth of this failure was portrayed by an East German commentator in early 1990:

Reports of success in science and key technologies—under the flagship, microelectronics, and its masthead, the 1-megabit chip—could do nothing in the last several years to enhance the already tarnished appearance of East German science at home and abroad. The transformation of what amounted to billion-mark reverse-engineering projects into "super weapons" was regarded with shock rather than as an answer to economic, ecological, and democratic challenges.[76]

The Honecker regime also tried to impress the outside world with its IT accomplishments. Targeting a capitalist audience, Honecker and numerous other top officials took advantage of product unveilings to declare Western efforts such as CoCom a failure. Their rhetoric invariably described the "embargo" as having been "broken through."[77] In the late 1980s, the propaganda surrounding IT achievements also was targeted at an audience in the Soviet Union, implying that the production of various microchips refuted the need for reform. As late as August 1989, at the ceremony surrounding the 32-bit microprocessor, Honecker argued that "all in all the development of microelectronics in the GDR underlines the correctness of the path we have chosen."[78] One year earlier, Honecker had presented a prototype 1-megabit chip to Mikhail Gorbachev on a visit to Moscow, along with best wishes for the success of the "historical process" of change under way in the Soviet Union.[79] The message of the encounter was clear: the Soviet Union may need change, but the GDR's technological achievements demonstrate that it already follows the best course.

Autarkic production profiles, trade policies designed to ward off political risks, and the use of technological breakthroughs for propaganda purposes were the most visible and most fundamental manifestations of East Germany's IT strategy. Those fundamental aspects of IT strategy, strongly influenced by perceptions of the GDR's international environment, determined a range of other policies related to technology development in East Germany. Those policies included the allocation of investments and scientific-technical labor, the structure of industrial enterprises, the establishment of linkages between industry and academia, the role of private firms and individuals in the manufacture and use of IT, and the activities of the GDR's Ministry for State Security that related to the development and use of technology and the control of information flows.[80]

The Perceived Absence of Choice

East Germany did have other options. Like other small industrial countries, it might have focused its resources on developing a small set of advanced technologies, or on the refinement of less sophisticated products in which it enjoyed comparative advantage. By developing industries in which it had a reasonable chance to compete on the world market, the GDR might well have earned the hard currency with which it could have obtained IT from the West. While such Western technology, owing to export controls, would not have represented global state-of-the-art IT, it almost certainly would have allowed the GDR to automate its factories and workplaces at a faster rate than was possible using domestic IT production almost exclusively. With regard to microelectronics, the GDR might have emulated the Scandinavian countries in the West (and to some extent Hungary and Czechoslovakia in CMEA), which imported standard memory chips while concentrating their own capacities on the production of application-specific circuits required by their own industry.[81] In the manner of Hungary, the GDR could have done far more to encourage private-sector development, use, and sale of computers and software and to encourage the decentralization of the IT industry in general.[82] In the manner of Hungary, Poland, and even Romania in the 1970s, the GDR could have encouraged cooperative ventures and licensing in the IT field with Western firms to encourage the infusion of more advanced technology and capital.

That such alternatives were rejected in favor of the risky and economically suspect course of autarky is testimony to the degree of causal influence that perceptions of the international environment had on the thinking of East German leaders. Looking outward, they saw no room for alternatives. Even after the Honecker regime collapsed, for example, the once-powerful head of Carl Zeiss Jena defended the country's massive microelectronics drive as "not a matter of wanting or being able, but a matter of necessity."[83] Mittag also argued after unification that no real alternatives to independence existed, as a result of "the dominating influence of the Cold War."[84] Even in documents that were not intended for publication and in private conversations after the dissolution of the GDR, autarky was described as the only viable option—short of abandoning the advanced electronics industry entirely.[85] "Complicated international conditions," an official of the East German microelectronics firm KME wrote, "force us to develop microelectronic circuits at the world level ourselves or to secure supplies of similar products on the world market."[86] Since the latter course was deemed too risky, only the path of independence appeared to remain.

The Failure to Learn

The GDR leadership's failure to perceive alternatives to its strategy of technological autarky represented a failure to learn—either from experience, from the few subordinates who offered a different perspective, or from the rest of the world. An elite that viewed its task as one of perfecting an already well-functioning economic order was not in a position to recognize the underlying challenge posed by advanced technology. Ultimately, then, a soft form of what is called "technological determinism" operated in the GDR, forcing change in a regime that would not adjust on its own.

In the relationship of politics and technology in the GDR, one paradox stands out. While information technologies brought about stunning change in other societies—accelerating the rise of service industries, permitting the decentralization of management and production, and changing the nature of work for millions of people—IT development in East Germany was seen as a replacement for economic reform and political change. It is no coincidence that almost any time Erich Honecker was asked about the need for reform in the GDR, he found a way to invoke the country's achievements in microelectronics. He told Belgian journalists in 1987, for example, that microelectronics "put us in the position of continuing our tried and trusted course of the unity of economic and social policy, in spite of complicated international conditions."[87] And during his last meeting with Gorbachev, only days before Honecker's ouster, the East German leader listened to a long statement by Gorbachev that culminated in the veiled warning that "he who does not move ahead decisively will be overtaken by history." In his reply, the only substantive defense Honecker offered was to describe his microelectronics program, "which some did not think could be achieved."[88]

Seeking maximum application of IT while continuing "tried and trusted" policies of industrial-era central planning will strike as fanciful those familiar with the epochal transformations brought about by IT. Honecker, Mittag, and the East German Politburo saw no contradiction, however. The proliferation of new technologies in global industry may not have determined the precise nature of political change in the GDR. What can be said with certainty, however, is that in its failure to reap the benefits of technological change, the East German leadership further damaged its legitimacy at home and abroad and increased the likelihood of its fall from power. Far from replacing reform, the GDR's technology-development schemes helped to demonstrate that in the absence of fundamental change in the organization of its political and economic life, the GDR had no future as a separate state.

12

Between Autonomy and State Control: Genetic and Biomedical Research

Rainer Hohlfeld

There are clear indications that the picture of scientific-technological progress that we encounter throughout the history of the German Democratic Republic had two very different faces. On the one hand there was a constant effort to link research to dialectical materialism and social practice. On the other there was always some support for basic research which, by producing innovations, had become a "productive force" in society and thus was considered deserving of creative freedom.

Central to all concepts of modernization is the notion that scientific-technological progress forms the basis for innovation in the economic development of industrial society. This notion was also an integral component of the model of the "developed socialist society" and thus part of the self-image of the East German party and government apparatus.[1] In fact, the SED's conception of research and science policy, and the guiding ideas of "science and the productive force" and "scientific-technological revolution," can be described only within the context of the rhetoric of modernization, regardless of the degree to which it was actually implemented.

It is not my intent, in this chapter, to make a preliminary decision about the viability of any particular theoretical framework. My approach is both eclectic and pragmatic. I will juxtapose the political claim to power, its images of science and its political strategies, with historical reality as I can reconstruct it. The analysis is based on empirical data that were collected in newly accessible archives, in scientific, literary, and journalistic publications from East Germany, and in interviews conducted between 1992 and 1995.[2]

My approach is composed of three strands that will provide a historical and chronological account, as well as a systematic sociological description. One strand tries to reconstruct the scientific development in genetics and biomedicine from within the disciplines themselves. Another strand focuses on changes in the relationship between politics and science. The third strand seeks to capture the beginnings of a discourse between scientists and representatives of society who were neither scientists nor members of the political establishment.

The Development of Genetic and Biomedical Research

My historical and sociological analysis of the development of the life sciences in the GDR is based on two fields that have undergone a stormy development during the last few decades and were particularly relevant to the political system in East Germany: genetics and biomedicine. Genetics played an important role, first, in the ideological conflicts with bourgeois science in the early years of the GDR (Lysenkoism), and second, following its rehabilitation and a theory shift to molecular genetics, as a theoretical and methodological foundation for the new "productive force of biotechnology."

Biomedicine includes those fields of medicine that use scientific approaches, modern technologies, and insights from medically oriented basic and clinical research in an effort to find scientifically grounded solutions for preventing, diagnosing, and treating diseases. Biological theories are applied to medical questions and medical problems are conceptualized using the theories and methodologies of the life sciences. In this way the biological sciences and medicine are joined together in "biomedicine."

I will trace the development by looking at two prominent scientific research institutes at the East German Academy of Sciences: the Central Institute for Genetics and Research on Cultivated Plants, in Gatersleben, and the Central Institute for Molecular Biology, in Berlin-Buch.[3] These two institutes are illustrative of the fluctuation between political attempts to control them and the preservation of scientific autonomy. And since they were privileged institutions, they are ideally suited for a comparison with their international counterparts.

Rebuilding the Science System

The first phase of the rebuilding of the scientific system (1945–1949) focused on efforts by the Soviet Military Administration to reorganize and denazify the universities, colleges, and scientific institutes. Bourgeois and "apolitical" scientists had to be recruited and granted broad autonomy in research and teaching. At the same time, however, a number of émigré scientists and committed communists returned to the GDR, eager to participate in building an antifascist Germany. They set the tone in some of the leading scientific institutes until the revivement of the postwar generation of scientists in the mid-1960s. And because they were authorities respected even by the SED leadership, they were able to pursue research in an international context and on a par with international standards.

The Institute for Research on Cultivated Plants

The Institute for Research on Cultivated Plants opened in Gatersleben/Saxony-Anhalt in 1945. It saw itself, according to the description of its old director, H. Stubbe, as a continuation of the Kaiser Wilhelm Institute that was set up in 1943. During Stubbe's tenure (1945–1969), experimental plant genetics and plant breeding in Gatersleben focused on producing a large assortment of mutant model organisms (snapdragons and *arabidopsis*) and cultivated plants. This mutation research was driven by the notion that it should be possible to simulate the natural diversity of cultivated plant tribes by compiling an assortment of mutations. The results of these efforts were a rich collection of mutations, which to this day have provided a valuable foundation for work at the institute. Research at the institute during the "Stubbe era" was characterized both by its scientific relevance (evolution of cultivated plants, theoretical questions of plant and bacterial genetics) and its practical and social relevance (resources and methods for breeding new strains of cultivated plants). The institute was scientifically at its height and was competitive internationally: Mendel-Morgan genetics, developed at the institute, became the dominant explanatory ideal, and bacterial genetics using *E. coli* as the model organism began to take off in the early 1950s.

Another pillar of the institute's work was an inventory of the genetic diversity of cultivated plants and their primitive and wild strains. The reservoir of genetic material, valuable for breeding purposes, was collected on trips into the world's "diversity centers."[4] The collection of cultivated plants laid the foundation for the institute's reputation and attractiveness to researchers.

The Beginnings of Biomedicine in Berlin-Buch

The beginnings of biomedical research in the Soviet Occupation Zone (later the GDR) go back to the year 1947. The Soviet Military Administration in Germany, by order number 161 on June 27, 1947, transferred the "biomedical institute," which had emerged from the former Kaiser Wilhelm Institute for Brain Research in Berlin-Buch, to the German Academy of Sciences in Berlin, which had been set up in 1946. We read in the order:

> For the purpose of democratizing German medical science and removing the remnants of a false racial doctrine, and in consideration of the request

by the German Academy of Sciences and the German Administration of Public Health, we decree:

1. The medical-biological institute (Berlin-Buch), which the Soviet Military Administration in Germany equipped and restored in the years 1945–1946, will be handed over to the German Academy of Sciences.

2. The president of the German Academy of Sciences shall

a) preserve the nature and direction of the work at the medical-biological institute in Berlin-Buch as a scientific research institute, which will concern itself exclusively with problems of theoretical and clinical medicine;

b) maintain the operating laboratory for biochemistry and biophysics, and set up as soon as possible, but no later than December 1947, experimental laboratories to study cancer, in cooperation with the Clinic for the Study of the Diagnosis and Treatment of Cancer Patients.[5]

The multidisciplinary approach to medical problems and their scientific basis found expression in the establishment of various departments (for example, departments for biological cancer research, chemical cancer research, pharmacology and experimental therapy, biochemistry, genetics, cell physiology, and biophysics) and a cancer clinic. Research focused particularly on chemical carcinogenesis, chemotherapy (especially in conjunction with studies on the metabolism of tumors and nucleic acid antimetabolites), and the viral etiology of tumors. Important work was done in this last area, especially in the 1950s and 1960s, leading to the discovery and characterization of new oncological viruses.

The East German Cancer Registry, set up by the Robert Rössle Clinic in 1953, was also important. It contained essential information on the epidemiology and genetic components of carcinogenesis and on the characterization and prognostication of tumor development.

The Ideologization of Biology

The opportunities for self-directed research changed when the period of rebuilding the most important institutes and facilities gave way to a phase of ideologization (after 1948–1949). Although the authorities continued to court bourgeois scientists to keep them in the Soviet Occupation Zone, they also launched the "assault on the fortress science" to begin building a Marxist-Leninist science. Intervention by the SED was most far-reaching in biology during the Lysenko era, in the early 1950s. Following the Marxist-Leninist conception of science, these efforts amounted to the creation of a science that took political sides, operated in a dialectical-materialist fashion (as opposed to the mechanistic-materialist method of Western science) on the principle that theory and practice were one, and was oriented toward solving social and economic problems ("practicalism").[6] Marxist-Leninist philosophy and ideology asserted the right to provide the *a priori* categories and methodologies for scientific paradigms in specific disciplines.

The main criticism that party philosophers, social theorists, and Lysenko leveled against Mendel-Morgan genetics was that genetics was an antimaterialistic and bourgeois science because it operated with idealistic, immutable concepts such as that of the gene. Such concepts stood in stark contradiction to the principles of dialectical materialism. Added to this was the accusation—philosophically not very sophisticated—that genetics was disconnected from praxis. Beyond that, Lysenko advocated a consistent Lamarckian position, arguing that the phenotypic characteristics that manifest themselves in individual development can be passed on to subsequent generations. Lysenkoism established itself at schools and universities as the Marxist-Leninist ideology of biology. Mendel-Morgan genetics, as an edifice of theory and methodology, was discriminated against or prohibited. Its practitioners were subjected to political pressure and persecution—first in the Soviet Union, then in Czechoslovakia and East Germany. Decisive for this political intervention was the fact that the state used its instruments of power to suppress competition between theories in favor of one doctrine and to prohibit diverging theories from being expressed or taught.

During this dogmatic period in the history of biology, the Gatersleben institute played a crucial role in resisting Lysenko's doctrine. With help from the Soviet Military Administration in Halle—some of the Soviet officers knew Stubbe from his visits to the Vavilov Institute in Leningrad—the Institute for Research on Cultivated Plants was able to secure scientific autonomy, refute Lysenko's claims experimentally, and develop into the East-European "Mecca" for genetics of the Mendel-Morgan school.[7] This circumstance, along with Stubbe's chair for genetics in Halle and research in the genetics department in Berlin-Buch, must surely be seen as an important reason Lysenkoism was not able to gain a foothold in biological research in East Germany.

The Research Landscape after the Academy Reform

The early 1960s saw a short-lived reform policy that was supportive of science.[8] By granting partial autonomy to the economy and science, the reform could have challenged the SED's claim to leadership in the long run. This phase was followed by an intraparty counter-reform by the Stalinist wing of the SED. The tide of counter-reform included the attempt to tie research more closely to industry and the economy, and a more vigorous implementation of the cadre policy through organizational changes. Exemplary efforts in that direction were the academy reform of 1968, the last step in the transformation of the German Academy of Science in Berlin into a socialist East German Academy of Sciences strictly separated from the science system in the West, and the Third University Reform. Common to both reforms was a structural change from faculties and disciplines to new organizational entities such as sections and central institutes. By reducing the number of decision-makers, these new entities were to create clear centralized relationships of subordination. This organizational change was accompanied by a shake-up of personnel: the first generation of qualified and renowned scientific leaders was replaced with a new generation of party-loyal cadres.[9] With these "reforms" the SED was seeking to eliminate the last remnants of academic self-government and autonomy.

After the failed attempt to tie biology to ideological and political premises, the academy reform was the second large-scale political attempt to intervene in the self-regulating mechanism of scientific research. It found expression in the formula "contract research and project-related funding" *(auftragsgebundene Forschung und aufgabenbezogene Finanzierung)* within the framework of the academy reform between 1968 and 1970. The then president of the academy tried to enlist the support of members for the reform and the new guiding idea:

> Fields of research, especially in the natural sciences and in technology and medicine, also have the task of ensuring the implementation of contract research and project-related funding. I would like to emphasize that it would be utterly wrong to read into the system of contract research a subordination of our institutes to the demands of the institutions initiating the projects . . . It would also be completely wrong if the institutes were to adopt a wait-and-see or even a defensive position. Rather, they are called upon—indeed, obligated—to exert active influence on the setting of scientific tasks, and to submit to their contractual partners offers in line with what they recognize as the highest standard of science.[10]

Responding to these demands, the "Socialist Large-Scale Research Project MOGEVUS" (the German acronym stands for Molecular Foundations of Developmental, Hereditary, and Control Processes) was set up in 1970–1971 for research in the biological sciences. The introduction declared: "The purpose of biological research on developmental, hereditary, and control processes is to expand basic knowledge, with the goal of creating the scientific groundwork for broad areas of application in heavy industry, agriculture, and the food industry."[11] It listed some of the key theoretical questions in molecular biology at the time, such as the molecular mechanism of nuclear and cell division in microorganisms, the regulation of the differentiation and maturing processes in plants and animals as the basis for improving yields, and the molecular mechanisms of membrane transport. Although this was the era of the academy reform with its goal of linking research to economic projects, this list of key questions reveals that the formulations of the reform program did not go beyond a rhetoric of practical application. They emphasized basic research, and their legitimation of research projects was one that Western scientists, too, would have readily embraced: the creation of theoretical foundations as the basis for further applications.

We can see all this very clearly by looking at the institute in Gatersleben, renamed the Central Institute for Genetics and Cultivated Plant Research after the academy reform. As part of the academy reform, work at the institute was redirected to focus on somatic cell genetics of plant tissue cultures. The change was made for methodological reasons and followed the lead of bacterial genetics, which was using *E. coli* as the model organism. "Starting in the 1950s, if we wanted to apply the principles of molecular genetics that had developed from work with bacteria to higher organisms, here at this institute especially to plants, we had to try and find the advantages that microorganisms offer for work with plants. These advantages are chiefly their single-cell nature and rapid reproduction."[12]

The larger development that legitimated the Central Institute's research policy was the intensification of East German plant production. That effort involved a definition of the goals of breeding and design: "Plant traits that previously could not be crossed can be now transferred, which means that one can introduce into cultivated plants characteristics of economic interest that previously could not be interbred. Moreover, specific characteristics can be deliberately altered. However, genetic work produces a genetically manipulated plant, for example, a strain of barley, wheat, or horse bean. Breeders must transform it into a strain that can be used economically by applying the classic methods of hybridization and selection."[13]

The foundations for projects in genetic engineering involving animal production were also created in Gatersleben. The introduction of new genes was supposed to create animals with entirely new characteristics, so-called transgenic domestic animals, for example, cows with higher milk production. The whole process was worked out using the mouse as the experimental animal. The biochemical substratum of the gene (the DNA) for the growth hormone was injected into fertilized eggs with the help of a superfine capillary.[14]

Continuity, more than anything else, characterized research work on plant resources in the two departments of "Gene bank" and "Taxonomy and evolution" (previously "Diversity" and "Systematics"). It formed, and still does, the pillar of the institute's international reputation and attraction, and it modernized itself in tandem with research in experimental genetics and molecular biology. "Taxonomy that is on the cutting edge has to operate as much with the diagnoses of Linnaeus . . . and other dead archival material as it does with the determination of karyological and biochemical characteristics . . . If we now include molecular data in ongoing in-house research projects on taxonomy and evolution, these analyses by no means render all other studies on the level of the organism or the organ superfluous, let alone useless. A meaningful application of molecular-genetic data is conceivable only by looking at it in conjunction with these studies."[15]

The Institutes in Berlin-Buch after the Academy Reform

At the Central Institute for Molecular Biology (set up in the wake of the academy reform), the Central Institute for Cancer Research, and the Central Institute for Cardiac and Circulatory Research, the change to molecular biology occurred early on, at least conceptually: concepts for gene therapy were discussed as early as 1972 as an alternative to prenatal diagnostics and were part of biological prognosis.[16] The Institute's Department for Somatic Cell Genetics pursued a project entitled "Creating a system for specific gene transfer," but it was halted after several years because it could not hope to compete with American groups.[17] Still, the project laid the groundwork for inserting genes into mammalian cells; as a result it was possible, after 1980, to set up and develop the various techniques of gene transfer at the Central Institute.

These developments and skills were supplemented by research in human molecular genetics, which focused on the molecular genetic diagnosis of two forms of muscular dystrophy (the Becker and Duchenne varieties). In addition, work in this section laid the groundwork for a molecular genetic diagnosis of cystic fibrosis, a genetic disorder of the mucous glands.[18]

As for cardiac and metabolic research in Buch, it is notable that it was not limited to therapeutic aspects but was also shaped by epidemiological and sociomedical and preventive questions. This research was greatly facilitated by data that were collected as the basis for a national hypertension register; the data covered wide geographical areas and their collection was not much hampered by regulations protecting the privacy of personal data. Finally, there was work that focused chiefly on the holistic aspect of how the central nervous system regulated cardiac and metabolic processes; this research was able to work with 'classic' methods (as opposed to molecular genetic methods).

The picture we had of biomedical and genetic research in East Germany in 1989 does not diverge much from the pattern of research in other Western industrialized nations. East German research programs and goals were—at least on paper—oriented toward international developments in science and the paradigm shift to molecular biology. However, they were greatly handicapped by inadequate access to instruments and materials; the academy institutes were privileged when it came to procuring equipment. East German specialties included the hypertension and cancer registries and the epidemiological and sociomedical research that was based on their data. In a few cases the lack of a research infrastructure was turned into a virtue and "old" paradigms (Mendel-Morgan genetics, morphology, physiology, and so on) were retained longer than in the West. This situation favored theoretical and methodological pluralism at the expense of being widely excluded from international competition: in cardiac-metabolic research, for instance, scientists studied holistic concepts in regulation research. In some instances this led to the creation of scientific niches (for example, the plant collections in Gatersleben), possibly at the price of forging top positions at the cutting edge of international research.

Science and Science Policy Discussions across Disciplines

Scientific Societies in the GDR

After the establishment of the two German states in 1949, all-German scientific societies continued to exist as unifying bonds uniting the two German communities. The division of Germany was largely ignored in these professional associations, and from the very beginning this was a thorn in the SED's side. With the building of the Wall in 1961, however, scientists were forced to break off their contacts with Western colleagues and establish separate East German professional associations. Opposition to this separation was at times considerable, but in the long run it was successful only in the case of the Leopoldina. Willing or not, most scientists went along with the new structures for professional reasons. In some cases they developed new professional associations on their own and brought them into the unification process.

Officially the purpose of these associations was to promote the exchange of professional experience at home, to advance, establish, and evaluate scientific results ("scientific debate"), and to provide professional training. The official description of East German public health policy put it this way: "The most urgent task of the societies is to promote scientific progress, implement new scientific insights, provide for the exchange of scientific experience in a national and international context, promote professional qualifications, and further cooperative socialist work and friendly contacts."[19]

In addition to providing scientific communication within the professions, these societies were also given explicit political functions: for example, organizing medical-scientific functions and meetings, advising the minister of public health, and drafting suggestions for the composition of delegations to international meetings and for East Germany's participation in international scientific organizations. The activities of the societies were handled by the "Coordinating Council of the Medical-Scientific Societies of the GDR." Plans called for an exchange of information and cooperation among the Council for Medical Science, the Academy for Continuing Medical Education, and biological and medical research associations and institutes.[20]

If that was the intended scenario on paper, what became political reality? According to statements by participating scientists, joint meetings of different professional societies promoted interdisciplinary cooperation. In retrospect, East German scientists and physicians see these meetings as positive and regard them as having been their "scientific home," a place where social and professional contacts were promoted over and above the scientific interests shared by the participants. Nearly all participants have emphasized—in conversations or published retrospectives—that the professional associations played a significant role in professional and nonprofessional communication and in initiating cooperation below the "official" level. Moreover, they maintain that the critiques and comments that came from outside particular fields, made possible by the mixed composition of committees, promoted scientific communication and were helpful to their own projects.

According to these statements by participants, which point unanimously in the same direction, scientific societies in the GDR were the equivalent of what in the West is regarded as the national component of science—a component, however, that is increasingly declining in importance owing to the internationalization of science. Professional societies in East Germany, precisely because of the country's isolation from international scientific life, tried to make up for the lack of exchange of scientific ideas, at least in part.

The Discourse about Progress in Biomedicine and Genetic Technology

In a political system with a centralist claim to power, it may seem paradoxical at first to speak of a social discourse separate from what we expect to be a controlled public arena. However, the scientific and technological progress of molecular genetics and biomedicine created potential applications that Marxist-Leninist social sciences and philosophy could not grasp *a priori* either categorically or methodologically. The prescriptive relationship between Marxist-Leninist philosophy and biology in the 1950s and its failure gave rise in East Germany to a probing effort to create ways of discussing social and epistemological problems of modern biology. That effort involved the natural sciences, philosophy, and the social sciences. It gave equality to all participants and involved real debate. This included early discussion—as part of the "Kühlungsborn Colloquia"—about the possible misuse of research in molecular biology for work on biological weapons. Beginning in 1970, these colloquia brought together a group of scientists, philosophers, social scientists, and artists—hand-picked, however—in the North Sea resort of Kühlungsborn for debates *(Meinungsstreit)* about philosophical and ethical problems of the natural sciences.[21] Scientists from "nonsocialist economic areas" *(Nicht-sozialistisches Wirtschaftsgebiet)* were also invited to these meetings. The colloquia thus went beyond a discussion purely within science.

The meetings in Kühlungsborn discussed the issue of the recombination of genetic material almost concurrently with the debate in the West that began in 1975, after the famous conference in Asilomar, California, on the potential risks of recombining genetic material in the test tube. An informal group came together at the colloquia to discuss the need for safety rules for genetic research in East Germany. After a lengthy delay, the suggestions by this group eventually led to the "Guidelines for in-vitro recombination of genetic material," issued by the East German Ministry of Public Health on November 26, 1985.[22] The guidelines were overseen by a commission and met international standards.

This setting was also fertile ground for far-reaching speculations about the possibilities of the new genetics—and for vigorous dissent. After the Seventh Colloquium on "Genetic Engineering and Man" in 1979,[23] two philosophers and the initiator of the colloquia, E. Geißler, summed up the discussions: "Experiments with the goal of intervening in man's genetic inheritance are now possible methodologically . . . It is thus clear that one can envisage experiments to test whether it might not be possible one day to improve man's characteristic individuality."[24] These kinds of fantasies about eugenics and the promises of progress—the one just quoted was neither the first nor the only one—became the topic of a long-running controversy in the cultural journal *Sinn und Form*.[25] The controversy was set off by an interview with Juri Brezan, the author of *Krabat oder die Verwandlung der Welt:* "Here science could prepare a different, nonbloody end to humanity: the end of man, of how we see man. I, at least, am afraid of the biologists, and I fear we must all be."[26]

In addition to Geißler and Brezan as the main protagonists, other writers, biologists, philosophers, and geneticists with many different views on progress in genetic engineering participated in the debate in *Sinn und Form*. Nor did politics hand down an authoritative decision. Instead, it tended to assume a mediating position in search of orientation: "Of course one must not overlook the fact that certain results of modern science and technology also trigger discussions in our republic about the social consequences and utility of some technologies. I remind the reader of the discussion that was carried on in *Sinn und Form* between writers and scientists about the benefits or harm of gene technology, or of the concerns of many citizens about environmental damage from air and water pollution."[27]

The debate in *Sinn und Form* was picked up in the "Gatersleben Conversations." Organized at the initiative of a few scientists, these discussions were held regularly at the Central Institute for Genetics and Research on Cultivated Plants and were sustained by a small circle of people interested in cultural issues.[28] To the initiators of these meetings, the artists who participated were important because they constantly prompted scientists "to think about the goals, methods, and especially the potential consequences of our experiments."[29] Here, as in the previous debate in *Sinn und Form,* they functioned as the *vox populi* in a kind of spokesman role. Topics of discussion were the implications of research in molecular biology and genetics, potential transgressions of ethical boundaries by human genetics and reproductive medicine, and discrepancies between scientific and social ideas of progress and modernity.

Topics in molecular biology and biomedicine were also discussed in the East Germany media (print, television, and radio) and in public lectures (especially those of the Urania, a cultural center) in an open and informed manner. There was special interest in topics dealing with new developments and findings, for instance, the principles and uses of procedures in genetic engineering. One example was a Urania television program entitled "How Does Gene Technology Influence our Lives?" that reported accurately about ongoing projects in the GDR.[30] At the end of the program viewers could phone in questions that were discussed and answered by the program's science reporters.

The historical course of the debate I have just described, the range of participants, the changing forums, the diffusion of the arguments in the different media, the diversity of opinions, the controversies that were openly aired, and the absence of political directives and authoritative decisions all suggest that a public discussion—albeit a limited one—about the problems of advances in genetic engineering and biomedicine was able to develop.[31] Such a discussion does not fit the picture of a strictly managed and controlled public in East Germany. As was characteristic of East Germany, writers and artists assumed in this discussion the role that journalists, critical scientists, parties, or extra-parliamentary initiatives played in the West. There is thus good reason to regard this debate, within its clearly delineated context, as a precursor to and functional equivalent of a democratic and public process of debate and discussion.

The Interaction of Science and Politics and the Politicization of Research

A basic dilemma characterized East Germany's research and science policy vis-à-vis biology and medicine. On the one hand there was the SED's claim to pervade and control all sectors of society in terms of substance, personnel, institutional set-up, and organizational structure. On the other hand, science had a privileged role within socialism's conception of itself as a "scientific worldview" and a motor to "unleash society's productive forces." However, scientific research could live up to these expectations only in a climate that allowed for creativity and autonomy, which were indispensable prerequisites for theoretical and methodological innovations. But such a climate was in direct opposition to efforts to implement the political claim to power, and was highly incompatible with the need of the SED leadership for security and control. The SED's research policy thus charted a perilous course. It manifested itself in the granting of a certain degree of latitude in the cognitive aspect of scientific research, while the Party and the state controlled all other aspects of research with an iron grip during the Honecker era.

The instruments the authorities used to exert direct influence on research programs and on the theories, projects, and goals of science were agencies of research planning and evaluation, provisions in the State Plan for Science and Technology, and goals set by the SED Party Congresses. These agencies and procedures on an intermediary political level controlled interactions between science and politics. We can interpret these interactions in part as interference in the scientific search for knowledge, in part as a negotiating process, and in part as the self-regulation of science. To implement the claim to the centralized exercise of power in the sphere of personnel and organization, the authorities used primarily instruments of communication policy (contacts with foreign scientists) and personnel policy (the system of cadre nomenclature and party leadership).

Guiding the Cognitive Structure of Research

Beginning in about 1970, research projects in biomedicine in East Germany were grouped together into "research associations" *(Forschungsverbände)*. "Main research areas" *(Hauptforschungsrichtungen)* were established on the basis of a decree by the Politburo of the Central Committee of the SED on January 16 and the Council of Ministers on January 24, 1980: "Analysis of Medical Research and Its Development to the Year 1990." These areas were assigned to the Ministry of Public Health and to the Program of Life Sciences of the Academy of Sciences, which was for this purpose subordinated to the

Ministry for Science and Technology. The activities in the main research areas in both institutions were coordinated by program councils *(Programmräte)* that included representatives from various main research areas, the ministries, the universities, and the academy. Neither the Council for the Program in Life Sciences, nor the Council for Planning and Coordination in the Medical Sciences in the Ministry of Public Health, nor the administration of the various main research areas had the authority to issue directives; instead, their function was advisory and evaluative. Authority to issue directives resided with the head of the state institutions, that is, with the president of the academy or the rectors of the universities. Above them in the chain of authority was the Health Policy Department of the Central Committee of the SED.

In spite of the many committees that were involved in coordination and were authorized to issue directives, the tasks of basic research that were grouped together in the main research areas were essentially shaped by the projects pursued at the institutes under the authority of their directors. These projects were financed with funds from the budget of the respective institutions (for example the institutes of the academy).[32] According to one source, "Research projects were suggested by the institutes in so-called plan proposals [*Planangebot*] and discussed and coordinated in the Research Association for Tumor Diseases at so-called presentation defenses *(Eröffnungsverteidigungen)*. Occasionally they were rejected and others proposed in their stead, and after that they were generally approved by the Ministry of Public Health. The strongest regulatory element was the Research Association, whose scientific advisory council was made up of representatives from East German institutions and groups that were focused on oncology."[33] E. Geißler has said the following about political directives: "Decisions about research topics were made by the higher leadership and in the committees composed of experts, specifically the leadership of the main research areas and the Central Working Groups of the Research Council, though as a rule they were not handed down as directives 'from above.'"[34]

Efforts to use political means to direct research in biomedicine were analogous to Western research planning, at least in the fields under discussion: in most cases scientists participated in defining state directives in planning commissions, advisory committees, and research councils. It is thus not correct to say that research was consistently guided by political directives. Instead, science interacted with politics while maintaining elements of self-guidance and autonomy. As a result of this situation, there was no possibility of scientists' using other research institutions or funds from third parties to continue certain projects they deemed promising and scientifically relevant. The loss of "freedom of research" took the form of a loss of options and a compulsion to go along with the new priorities. Once a certain research outline had been fixed in the state plan, it also meant that a decision had been made against all possible alternatives—in other words, pluralism was no longer possible.

Direct Political Intervention in Research

Work that the Health Department of the Central Committee of the SED considered economically urgent was also placed under "party control." For example, the political directives of the department head (October 27, 1976) regarding the development of "Onkotest," a tumor test-kit, read as follows:

> This is a medical-scientific accomplishment that is currently setting international standards. It could become extremely important in providing health care and health protection to our citizens, and it is suitable for further strengthening the GDR's international reputation. A research group from the Central Institute for Cancer Research . . . is currently testing this procedure in clinical trials. I would like to ask you to please support this strictly confidential project in every way possible . . . Moreover, this could give rise to important opportunities to export the instruments or grant licenses. That is why I am coming directly to you. With socialist greetings.[35]

The following comments came from the scientists' side: "The East German government or SED leadership passed on to the leadership of the academy of Sciences the directive that the Academy should not only cost money but earn money. After that the institutes were told that a certain part of the working groups would have to pursue applied—and thus money-earning—research. The institute leadership then 'chose' the relevant research groups." Another scientist gave concrete examples for this shift of research activities in favor of money-earning research and development: "Basic research in oncology was cut back around 1980 in areas whose potential and methods could be used for practical, economical goals: for example, tumor immunology to develop the 'Onkotest.'"[36]

It has been estimated that about 10 percent of the scientific capacity of the Central Institute for Cancer Research was used for research and development that would eliminate the need for imports and earn hard currency. Judged by international standards, the centrally directed demand for practice-oriented research amounted in most cases to nothing more than "reinventing the wheel." That was not all. Given the lack of foreign currency, research tools for preparatory and analytic work, along with the basic materials (chemicals, biomass, small instruments, and so on), had to be created at the institutes under what were for the most part insufficient conditions, and were even supposed to be sold abroad to earn hard currency. This meant that scientists and qualified technical workers were busy with routine and production tasks below the level of their professional training and competence. This observation should not, however, lead to the conclusion that this form of "practicalism" spelled the end of basic, undirected research.

The Control of International Communication

The participation of East German biomedical and biological scientists in research in East and West and in professional scientific associations for the purpose of comparing levels of scientific accomplishment was politically motivated. This was also true of scientific exchange with the goal of keeping scientists informed of international developments and acquiring an international reputation for the GDR. One political purpose was the fulfillment of governmental agreements and contracts within the framework of the Council for Mutual Economic Assistance (COMECON), another was the recruitment and scrutiny of travel cadres by ministries, the Party, and the Ministry for State Security in cases of contact with the "non-socialist economic sphere." When it came to international contacts, biomedical research and its institutes were subject to the usual restrictions, that is, they could be regulated only by higher governmental authorities (the general secretary of the academy, university administrations, or the Ministry for Health), and on these levels the decisions were also chiefly political. These state authorities paid hardly any attention to suggestions made by the executive committees of professional associations. Beginning at about the end of the 1970s, travel to the West for professional purposes was possible only for travel cadres.

COMECON Policy

Cooperation with institutes in socialist countries occurred primarily within the framework of bilateral or multilateral academy and university agreements as well as within the framework of COMECON agreements, and it had significant political-propaganda functions. Many scientists took advantage of these opportunities, though they were often limited to joint conferences or brief laboratory visits and did not lead to intensive work on joint projects. The primary reason for this was the common lack of research funds and materials. However, these contacts did serve the purpose of exchanging information and cultivating human relations. The GDR's preferred partners were the Soviet Union, Czechoslovakia, and Hungary: "Contacts with research in the East Bloc were intense, state-directed, and promoted via bilateral and multilateral treaties on the level of academies and universities. Travel activities and 'scientific tourism' were extensive and out of all proportion to the result (measured by the number of joint publications)." We are told that brief visits were preferred to longer stays, since materials in the laboratories in those countries were in even shorter supply than they were in the GDR. And although the scientific successes were slight, these treaties served to establish personal contacts.[37]

International cooperation with the nonsocialist economic sphere was possible for travel cadres within the framework of state-regulated agreements (for instance, between universities and academies); individual arrangements were for all intents and purposes no longer possible after about 1980. It was easier to participate in events in Western European countries, even in the U.S. or Japan, than in West Germany, and it was easier to get to West Germany—to Munich, for example—than to West Berlin. In the last years before the end of the regime, the East German government also made it increasingly difficult for West German scientists to visit East German institutes. Exceptions were made for GDR institutes with bilateral agreements with West German institutes (for example, as part of the 1987 Treaty on Scientific-Technological Cooperation between the FRG and the GDR).

A former East German scientist described international cooperation in this way:

> It was difficult for an East German scientist to become part of the international scientific community. Contacts with Western scientists were limited to a small circle of select East German scientists, and this circle grew continuously smaller. Some of these scientists could undertake long visits to Western laboratories, which allowed them to pursue modern research and achieve results of international significance. These visits contributed nothing to the development of science in East Germany, however, for they did not improve the working conditions for the great majority of scientists.[38]

Scientists complained that the lack of exchange of information with international professional circles, and the failure to go head-to-head with the competition—which would have made it possible to assess the quality of their research—were highly detrimental to scientific work. The lack of communication and exchange of information led to certain deficiencies in training and skills, which became visible in the battle for funds and positions after reunification: "People's self-confidence, especially in younger scientists, was much less developed than in their colleagues in Western European countries or even the United States, who have been trained very differently . . . This placed them always on the losing track, since they were not able to present themselves properly. Then there were the language barriers . . . The third factor was that they were not part of the insider family . . . It was very difficult to get in."[39] Added to all this, we are told, were obstacles to corresponding with foreign colleagues and restrictions on publishing.

In sum, interference by the Party and the state apparatus in international scientific communication through a policy of selecting travel cadres based on political criteria had two consequences: it split scientists into two classes and prevented the GDR from participating in the international scientific life of biomedicine. The fact that study trips to nonsocialist countries were possible only rarely must be seen—along with the lack of hard currency—as one of the reasons East German science lagged behind that of the West. East German scientists were unable to integrate themselves into international scientific networks or measure their work directly against international competition. As a result, most were, for all intents and purposes, excluded from the competition for publications and status. This situation became particularly apparent after the collapse of the system in 1989, when socialist job guarantees disappeared overnight and the conditions of a meritocratic and competitive society began to take hold.

Political Intervention in Matters of Personnel

Direct political intervention into the institutional, organizational, and personnel infrastructure of research—with indirect repercussions for the quality of research—took the form of appointments to top-level positions and the selection of travel cadres. Membership in the *nomenklatura* was determined both by scientific qualifications and by "socialist" leadership skills. *Nomenklatura* positions were filled after the relevant persons in institutes and research installations made suggestions to the head of the Work and Cadre Section of the Research Association. Those admitted to the cadre reserves had an excellent record of scientific accomplishments, possessed "socialist leadership" and pedagogical qualities, and were supposed to be ideologically reliable.

An application for cadre membership was usually made by the director of an institute (professional institutions had no influence on this). The director submitted a proposal at the "Directors' Consultations" *(Direktorendienstberatungen)*, at which the head of the cadre, the party secretary, and the "Commissioner for Security and Order" (that is, a representative of the Ministry for State Security) of the respective institute or establishment were always present. Following a preselection, a background check on all candidates was set in motion. If the findings were positive, the suggestion was "activated." If the director of the institute simultaneously held the post of party secretary, his influence on this process of scrutiny could be considerable.

Membership in the SED was an important, though not a necessary, criterion for induction into the Academy of Sciences, promotion to the rank of department or section head, and especially appointment as a travel cadre. In the same way, professional qualifications were a necessary though not sufficient condition for attaining cadre status. All else being equal, SED members were given preference. A lack of membership in the SED could be partially compensated for by other "social" activities, for example, membership in a block party or some position in the unions. We are told that in several instances, applications for travel cadre status for nonparty members with outstanding professional qualifications were ignored by the cadre division of the Central Institute for Molecular Biology or failed security clearance by the Ministry for State Security: "In this way the great majority of young scientists were denied the experience of working in leading foreign laboratories. The cadre development plan of the institute included a list for reserve cadres and young cadres who were slated to move into leading positions. This *nomenklatura* status was attainable only with the consent of the Party (and state security)."[40]

A certain basic policy structure thus emerges from interviews and archival documents. Alongside the actual professional committees, which coordinated and approved scientific projects but had no authority to make decisions, real power lay with the local heads of the respective institutions. At their sides were the party leaders and the cadre leaders, who always had to adhere strictly to the directives from higher organs. The activities of these organs often—but not always—ran counter to institute leaders' plans and recommendations for research policy. The supervisory organs of the Ministry for State Security were also actively involved. What emerges is thus a double and triple structure of control over people and communications. This structure was primarily concerned not with the cognitive content of research, but with cadre policy and the control of individuals. It superimposed itself on and counteracted the actual cognitive, social, and institutional infrastructure of scientific work. It was through this centralized-authoritarian and antimodern policy of control and order—and not through political interference in the cognitive structure of science—that the SED and the state apparatus violated the functional imperatives of the world of science.

—Translated by Thomas Dunlap

IV

Biographies and Careers

13

Robert Havemann: Antifascist, Communist, Dissident

Dieter Hoffmann

> R. Havemann became very active at Humboldt University and elsewhere beginning about 1962. At these institutions he sought to revise Marxist-Leninist policy and the philosophy of our party. His attacks were directed against all basic tasks set by the Sixth Party Congress. Under the banner of fighting dogmatism, which he believes we have not even begun to defeat in our republic, he is attacking the economic, cultural, and science policies of our party and their theoretical and scientific foundation.[1]

This quotation comes from the Central Committee's 1964 internal evaluation of the physical chemist Robert Havemann. At that time Havemann not only was at the center of criticism from the Party, but he had also already become a "nonperson" for official East German society: in the spring of 1964, the Central Committee expelled him from the Party and dismissed him from his professorship at Humboldt University. This was the first step in a process that eventually led to the almost complete exclusion of Havemann from public life in the GDR, making him arguably the most important and prominent East German dissident.

East German society was not the first to make an outcast of Havemann. He had already been persecuted under the Nazis. In fact, Hitler's henchmen had sentenced him to death. Havemann's life was shaped by the turning-points and discontinuities of modern German history, and reflects in almost exemplary fashion what could happen to a politically engaged scientist in Germany from the 1930s through the 1960s.

Havemann was born on March 11, 1910, in Munich. His mother was an artist and his father a teacher and later a newspaper editor, and so he grew up in the sheltered but largely apolitical atmosphere of the German educated bourgeoisie. After graduating from high school in Bielefeld in 1929, he went on to study chemistry in Munich and Berlin. The apolitical atmosphere at home did not prevent him from becoming politically active at an early age. Starting in 1931 he worked for the Comintern in Germany, and in 1932, according to his own statements, he joined the KPD (German Communist Party).

When the National Socialists seized power in 1933, Havemann lost his trainee position at the famous Fritz Haber Kaiser Wilhelm Institute and joined the organized antifascist resistance. He was involved with the resistance group "New Beginning" until the mid-1930s. As his resistance activities were not discovered, he was able to complete his training and obtain his Ph.D. at the University of Berlin in the spring of 1935. He continued his research in chemistry at various hospitals in Berlin. In 1937 he was an assistant at the University of Berlin's Institute of Pharmacology. During that time he made a name for himself in the field of blood biochemistry.

In the summer of 1943, Havemann obtained his *Habilitation* at the University of Berlin in the field of blood chemistry. Only a few weeks later he was arrested by the Gestapo and put on trial before Freisler's People's Court for high treason. After the outbreak of the war, Havemann and the physician Georg Groscurth, along with other friends, had founded the resistance group "European Union." It sought out contact with foreign workers and Jews in hiding and furnished them with forged documents—from passports to food ration cards. The group saw itself not merely as a humanitarian but primarily as a political organization, as is clear from leaflets propagating its vision of a peaceful postwar society within a pan-European framework.

The group was betrayed in September 1943, and many of its members were sentenced to death. Havemann himself sat on death row, but former colleagues used their influence to save him by arguing that he was needed for research that was essential to the war. He was given a stay of execution from month to month. A special laboratory was set up in the Brandenburg penitentiary, where Havemann carried out studies on poison gas compounds for the Army Ordnance Office. In this way Havemann survived the Nazi period, and it bespeaks his extraordinary courage and fearlessness that his lab became one of the centers of resistance in the penitentiary. The "Brandenburg experience" and living in the shadow of death for nearly two years undoubtedly shaped Havemann's subsequent life and attitudes: he is said to have told one of his sons that life after the death sentence was like a bonus.

The collapse of the Nazi dictatorship and the opening of Brandenburg prison by the Red Army was a genuine liberation for Havemann. At the same time, he later confessed in his autobiography, these events made "me into a Stalinist without my even noticing it." On the occasion of Stalin's death in the spring of 1953, he even declared publicly that he owed the ruler his life, because he had saved the German people.[2] This attitude and his unconditional loyalty toward the Soviet occupying powers, in addition to his scientific qualifications, made him the ideal "president" of the Kaiser Wilhelm Society institutes that still existed in and around Berlin. Inevitably a conflict arose with the society's main administration in Göttingen and with the American occupying authorities. The Americans had taken over the administration of the institutes, most of which were located in Berlin-Dahlem.[3] Havemann's role was progressively restricted and eventually limited solely to the Institute for Physical Chemistry. Then a scandal broke out in early 1950, at the height of the Cold War. The occasion was an article by Havemann in *Neues Deutschland,* the official party paper of the SED.[4] The article was a polemical analysis of American nuclear policy, and it led to Havemann's immediate dismissal. Thereafter he lived and worked in the GDR, "absolutely convinced that the path taken in the GDR under the leadership of the Soviet Union and our unity party of social democrats and communists was the only correct one, and that it was not the Russians and the 'East'—the communists—who had divided Germany."[5]

Havemann became the director of the venerable Physical-Chemical Institute at Humboldt University in Berlin and was appointed associate professor of colloid chemistry as early as 1946. With great commitment he devoted himself to rebuilding this institute, which had been severely damaged in the war, and he developed it into an internationally renowned center for research in physical chemistry. Here Havemann created a place where he and his colleagues could expand their earlier research on the chemistry of proteins. The institute itself, once headed by famous scientists like Hans Landolt, Walther Nernst, and Max Bodenstein, flourished once again. Havemann ran the institute between 1950 and 1964, and these years were without a doubt the most successful of its postwar history.

In addition to his earlier work on the physical chemistry of proteins, Havemann founded two new fields of research in the GDR.[6] One, magneto-chemistry, began with the "magnetic scale according to Havemann." Using this instrument, researchers pushed ahead with studies of diamagnetism and paramagnetism of hemoglobin, hemin, and other substances, and later also investigated related problems of inorganic and organic chemistry. Theoretical questions, especially the study of quantum-chemical problems, figured prominently in this research.

The second field that drew Havemann's research interests was the study of photochemical processes. In fact, one could call it the true focus of his research work in the 1950s and 1960s, since about a quarter of his publications were devoted to it. Among other things, he investigated the photochemical reactions and photoelectric phenomena in organic coloring matter; the study of photosynthesis held a particular fascination for Havemann. The committees that awarded him the National Prize in 1959 and voted him a corresponding member of the German Academy of Sciences in 1961 singled out these scientific accomplishments for special mention. When Havemann was banished from the university to the academy in 1964, Western papers reported that the Soviet Union was urging that his work be continued because of its importance to that country's space program.

From the very beginning of its existence, the GDR had problems providing scientific researchers and teachers with equipment that was up to international standards. The lack of funds to buy or even import the necessary instruments and apparatuses was chronic and not limited to the immediate postwar years. An East German scientist who wanted to compete in the international scientific arena had to compensate for these problems by building instruments and coming up with innovative homemade designs. Robert Havemann had already made a name for himself as an inventor of scientific instruments before the war. After the war he continued this side of his scientific work; shortages and hardships forced him to perfect it. Havemann's work in this field was so outstanding that it is still applied today. He devised alternatives for the items not available on the domestic market and found solutions in instrument technology that grew out of the needs of research itself. As a result, a number of specialized instruments and testing methods have gone down in the history of East German science under his name. His 1957 textbook *Introduction to Chemical Thermodynamics* came in response to very similar needs. He wrote it "to meet the urgent needs of students at our universities, who until now have had to depend largely on foreign or West German books to study chemical thermodynamics."[7]

Havemann served socialism. As a deputy of the People's Chamber, dean of students, and vice-chancellor for research at Humboldt University, as a member of the Peace Council, and in many other social functions, he was active in pivotal political areas in the young East German society. In the spring of 1953, immediately prior to Stalin's death, Havemann, then the dean of students, was involved in the expulsion of students who were members of the "Young Congregation." In this instance his commitment to "social progress" was following the logic of the Cold War. Later, between 1956 and 1962, Havemann also cooperated with the East German secret service, providing reports about his contacts with Western scientists.[8]

Havemann received a good deal of public recognition and many official honors for all his contributions to the development of socialist society in East Germany, including the Patriotic Order of Merit, the German Peace Medal, and the National Prize, the highest scientific honor in East Germany.

The National Prize in 1959 was the high point of his social recognition. The events of June 17, 1953, and especially the Twentieth Party Congress of the Communist Party of the Soviet Union in 1956, began a profound transformation in Havemann's political views: as he later confessed, "the edifice of my belief collapsed from the jolts of this earthquake."[9] During the following years he found himself increasingly in conflict with the power structures of the GDR. Little by little he began to realize the extent to which he was a victim of hopes, illusions, and deceptions. The clash between Havemann and the Party was triggered in 1956 by several articles, the first of which,[10] incidentally, had been suggested by Walter Ulbricht, the powerful general secretary of the SED.[11] In these articles Havemann lashed out against the dogmatism and scholastic character of the official party philosophy and challenged its right to sit in judgment over the insights of modern science. It is no coincidence that a prominent scientist joined Havemann in his crusade against the "Central Office of Eternal Truths"—as the biting title of one article put it.[12] Science had suffered a good deal from dogmatic patronizing by party philosophers and ideologues. Entire fields of research—from genetics to psychology to cybernetics—were denounced in the Soviet Union and in the other socialist countries as "idealistic" and "bourgeois" heresies, and their development was effectively blocked. Havemann now took a public stand against this, thereby provoking the growing wrath of the party apparatus and its submissive "lectern-philosophers." In part the authorities reacted this way because Havemann's polemic was part of a much broader debate at the time. This debate ranged from demands for sensible limits on the planned economy and state guidance to discussions in intellectual circles around W. Harich, W. Janka, and R. Just, and it very quickly exceeded the boundaries the SED had in mind.

After the violent suppression of the uprising in Hungary, the tendencies toward dogmatism and neo-Stalinism were revived in East Germany, and the so-called thaw, which had also shaken the SED's Stalinist conception of power, came to an abrupt end.[13] Members of the Third University Conference in the spring of 1958 denounced Havemann and other critical scientists, maintaining that by opposing the cult of personality and dogmatism and by pleading for an open and free clash of ideas they had misunderstood what the Twentieth Party Congress was all about. In this context Havemann was even accused of engaging in factionalism and propagating revisionist views: "In many instances," claimed an internal paper of the Central Committee in schoolmasterly fashion, "he has demonstrated the arrogance and the conduct of a petty bourgeois, individualistic intellectual."[14] A number of factors explain why Havemann was not condemned outright at that time: his impeccable antifascist past, his great contributions to building East German society, and the central role that science and technology played in the years when East Germany was rebuilding itself. At the same time, Havemann was still willing to make compromises and engage in public self-criticism. For instance, at the Third University Conference he confessed in an act of devotion: "In principal questions of our policy I have no difference of opinion with the Party . . . My experience in the political struggle has taught me that deep truth lies in the sentence: 'The party is always right.'"[15]

Havemann's disagreements with official party policy came to a head again in the early 1960s. A lecture entitled "Has philosophy helped modern natural science in solving its problems?" delivered at a meeting in Leipzig on German science in the nineteenth and twentieth centuries, led to the final break between Havemann and the SED party leadership in September 1962.[16] Havemann answered his own question with a resounding "no." With regard to the official state philosophy of dialectical materialism, he stated plainly: "Something terrible has happened: for decades dialectical materialism has been progressively discredited by its official representatives among all the scientists of the world, including the leading scientists in the Soviet Union."[17] Havemann went on to argue that there was an intimate connection between philosophical dogmatism and political ossification.

Havemann had now questioned basic philosophical and political positions, and the organizer of the Leipzig conference refused to print his paper. At the same time, the party leadership set in motion a series of political and administrative measures, eventually "resolving" its conflict with Havemann. At the center of these measures was the controversy surrounding Havemann's lecture series on scientific aspects of philosophical problems. He had been offering this series at Humboldt University since 1960–1961, and by 1963–1964 attendance had reached 1,000. Havemann discussed basic philosophical questions critically and addressed a number of forbidden topics. His lectures struck a chord and provided an intellectual outlet for many students and young scientists. Such an outlet was sorely needed. The building of the Wall had reinforced centralist and statist traits in East German politics. It had also allowed the authorities to discontinue the preferential treatment accorded to intellectuals in the face of a large exodus from East Germany. At the same time, however, the SED was also under considerable pressure to modernize East German society, and it had to react to relevant demands from the political and scientific elite.

In this atmosphere of political and ideological tensions, Havemann's lectures became another source of conflict with the authorities. The fact that this occurred in philosophy of science, a field seemingly remote from politics, reveals the SED's all-embracing claim to power and the central importance of philosophy in the political system of East German society. Philosophy, in its

Marxist-Leninist version, was part of the canon of everyone's scientific education in East Germany. It was taught to students of all fields following principles laid down by the Polituburo, and was thus an important component of the ideological mechanism of rule. It was inevitable that nonconformist and unorthodox philosophical thinking, the sort exhibited in Havemann's lectures, would draw opposition from the "main office of eternal truths." Incidentally, Havemann was not the only scientist whose lectures met with the disapproval of the Party. When the physicist Max Steenbeck raised questions very similar to Havemann's in a talk he gave in the spring of 1963 at a meeting of the philosophy section of the Academy of Sciences, alarm bells went off at the Central Committee.[18] However, the dissent was eventually resolved in an exchange of views between Steenbeck and Kurt Hager, the chief ideologist of the SED.[19]

It was not easy to resolve the Havemann conflict "quietly behind closed doors." Owing to his charismatic personality, Havemann's influence did not remain limited to a narrow circle of specialists. Quite the opposite: through his lectures he was able to inspire a broad range of students, from scientists interested in philosophy and epistemology to students who felt a general discomfort with the political situation and articulated their opposition to East German society in this indirect way. To that extent, Havemann's lectures and the strong response they met with were a seismograph for the political situation in East Germany. The party apparatus of the SED saw it the same way, particularly as Havemann's criticism of the way socialism was being constructed in East Germany had attracted attention outside of East Germany. By now the Western mass media, especially the West Berlin radio station RIAS (Radio in the American Sector), was commenting with interest on his conduct. The Party met these challenges to its monopoly of power and opinion with standard party punishment and bureaucratic restrictions. In the summer of 1963, following the tried and true (Stalinist) method, a process was set in motion that began with Havemann's removal from the party leadership of Humboldt University. The occasion for this disciplinary measure was characteristically trivial. At a meeting of the board of directors of the academy's Adlershof Research Center, the question of birthday greetings for Walter Ulbricht's upcoming seventieth birthday was discussed. Havemann was the only party member who spoke out against it, arguing that it was "meaningless to pay lip service" to Ulbricht.[20] His dissent had not only violated party discipline but also encouraged those directors who were not affiliated with the Party to vote against the proposal, and in the end it was dropped. When Paul Verner, the head of the Berlin SED, heard of this "conduct detrimental to the Party," he sent word to the party organization of Humboldt University, asking it to look into "removing comrade Havemann from the central party leadership of the university."[21] The party organization immediately and dutifully complicd with this "wish," and in the summer of 1963 Havemann was expelled from the university leadership to which he had belonged since 1958.

It was against this backdrop that Havemann, in October 1963, began the lecture series "Dialectic without dogma?" which later became famous. His talks were accompanied by increasingly sharp attacks from the Central Committee's Science Section and the SED district leadership in Berlin. At the same time, a lively discussion began—or was orchestrated—in the mass media. The weekly magazine *Humboldt-Universität* devoted a series of articles to Havemann's lectures and the basic philosophical positions he advocated. In the beginning this coverage was marked by a desire for scholarly debate.[22] However, as the leadership of the SED pushed ahead with a radical break from Havemann and "hawks" in the party apparatus gained the upper hand, a scholarly dialogue was not what the authorities wanted. In the end, the strategy of political defamation prevailed: in February 1964 Havemann was publicly branded a "philosophical crackpot" and revisionist at the Central Committee Plenary Meeting. In January 1964, Paul Verner had told the party leadership of the university in unequivocal terms: "It is now necessary to spread our Marxist ideology in the party paper [meaning the journal *Humboldt-Universität*] and to refute Havemann's erroneous ideas . . . To that end ten articles should be run dealing with all basic issues, for example, the Marxist-Leninist concept of freedom."[23] This kind of engagement with Havemann's ideas was also practiced at subsequent party meetings and *Aktivtagungen*. For example, Kurt Hager, the chief party ideologist, gave a lengthy lecture arguing against Havemann's revisionism at a *Aktivtagung* of the university, and he even used the yearly Humboldt lecture as a forum for his political disagreement with Havemann. The purpose of these public appearances was obviously not to engage in objective communication and discussion of the problems that had been raised, but to "unmask" and "denounce" Havemann.

Still, in the winter of 1963–1964, Havemann did not believe that the noose would be tightened for good or that the Stalinist dogmatists would gain the upper hand. This was not a sign of nonchalance or even political naiveté on his part. Many party members—including those in his own organization at the Chemical Institute—and even some in the party leadership at the university, among them its first secretary, Werner Tzschoppe, sympathized with his philosophical and political opinions. These reform-oriented segments of the SED were placing their hopes in the anti-Stalinist discussions that had been going on since the early 1960s, especially in the communist parties of Western Europe. In the wake of the Twenty-second Party Meeting of the Soviet

Communist Party, these discussions had already led to further de-Stalinization in the socialist countries, as well, particularly in Poland and Czechoslovakia. Events like the "Kafka Conference," widely regarded as the beginning
of the Prague Spring, attracted interest in East Germany and nourished the
hope for a broad democratization and modernization of East German society
among many intellectuals. Such hopes were fed even by official SED decrees.
For instance, in the fall of 1963 the Politburo had published the "Youth
communiqué," generously noting that it was no longer acceptable "to dismiss
inconvenient questions from young people as annoying or even provoca-
tive."[24] Critical poets—one thinks of the songwriter Wolf Biermann—and
problem-oriented literature, like Christa Wolf's *Der geteilte Himmel,* found
their audience at that time. In addition, there were public roundtable discussions that deviated from the conventional ritual of rehearsed question-and-
answer games and dealt openly with topical questions of genuine interest.
The leadership no longer ignored even to the youthful need for modern
dance music and Western fashions, and a generous concession was made to
East Germany's young people: "The kind of good manners young people
choose is their business; the important thing is that they remain tactful."[25]

It was of course easy to see Havemann's lectures and the controversy
surrounding him against the backdrop of these trends and the widely articulated demand that the state support an "intellectual-cultural life." Moreover,
those months were shaped by discussions about the introduction of the New
Economic System of Planning and Managing the People's Economy (NES).
This reform in economic policy was designed to modernize and rationalize
the East German economic system. It sought to replace the rigid, Soviet-style
planned economy with a system that had a stronger scientific foundation and
was self-regulating. Even Walter Ulbricht had pointed out that there was a
direct connection between modernizing the economy and democratizing society. At a Central Committee plenum in February 1963, he had said that
the state no longer had any use for those "who don't want to realize that in
the New Economic System of Planning and Managing the People's Economy, it is impossible to lead people with the old administrative and dogmatic
methods."[26]

Given that Ulbricht said this at the same Central Committee plenum at
which the dogmatists denounced Havemann and branded him a revisionist,
Havemann and his supporters entertained the illusory hope that the conflict
with the party leadership was still undecided and could be resolved. Havemann's statement in the summer of 1964 reveals that he did have such hopes:
"I consider comrade Ulbricht's speech at the Fifth Plenum the most important step toward overcoming the rigid economic principles of the Stalin era,
which were justified only for a quasi-war economy. It is my impression, at any

rate, that in this area [that of economic policy], our party has taken a decisive step toward overcoming dogmatic rigidity."[27] The Party had tightened discipline in the party committees at Humboldt University and dismissed the party secretary and Havemann confidant Werner Tzschoppe. Still, it was not unreasonable to think that change might be possible in cultural and scientific policies, and in the policies dealing with Havemann himself.

Whatever hopes might have been entertained were soon dashed. On March 12, 1964, Havemann was expelled from the Party and prohibited from teaching or doing research, that is, he was summarily dismissed from his professorship at Humboldt University. As a pretext for this step the Politburo used an interview that Havemann supposedly gave to a West German journalist, and which had been published the day before in the paper *Hamburger Echo*.[28] It did not help Havemann that he described the alleged interview as a fabrication and dissociated himself from it. He later wrote that it had been put together by a certain Karl-Heinz Neß, "who had visited me a few days earlier at my institute eager to talk to me about a few issues, not as a journalist but as someone who had attended my lectures with interest."[29] By selecting the West German public as his audience, Havemann had not only breached party discipline but also violated a general taboo of the German workers' movement, a taboo that was deeply rooted and only too readily cultivated by Stalinism: intraparty conflicts must not be taken outside, let alone fought out in or by means of the "bourgeois" press. In a postunification interview, Kurt Hager emphasized this point as well.[30]

Hardliners in the Central Committee used this violation of the code of conduct generally accepted in the SED at the time to resolve the Havemann case once and for all and to associate Havemann himself with counterrevolution.[31] This was easier to do now that Walter Ulbricht no longer protected Havemann. The stability of power always took absolute priority for the seasoned politician Ulbricht, and it was unquestionably more important than the "intellectual-cultural life" or the "autonomous and self-confident citizen" that Havemann and others were calling for. Ulbricht and the other members of the Politburo had unpleasant memories of the workers' uprising in Berlin on June 17, 1953, and of the events in 1956 in Hungary and Poland. Havemann's opponents deliberately exploited these fears. In their reports to the party leadership they repeatedly drew parallels to those events and used emotive phrases like "political platform," "faction building," or "Petöfi club." For example, the following statement appeared in the outline of Verner's speech for the meeting of the university party leadership on January 10, 1964, which dealt with Tzschoppe's dismissal: "Havemann's lectures are clearly a political platform. This is the platform of 1956, or one could go back even further."[32]

Against this background it is therefore not surprising that the Politburo itself carefully watched or directed the Havemann affair. Indeed, Havemann's expulsion from the Party had been prompted directly by a resolution of the Central Committee on March 12, 1964.[33] The Politburo had confirmed this resolution at its meeting on March 17, ordering that "the relevant measures be further implemented."[34] At the same time the Central Committee ordered that "the Secretariat of the Central Committee must be informed within eight days about ideas for Havemann's employment, in accordance with the decision that he will be given specific [*zweckgebunden*] research work."[35] On March 24, after a meeting of the university's disciplinary committee had dutifully confirmed the dismissal after the fact,[36] Kurt Hager was finally able to report to the members of the Politburo that the measure had been implemented.[37] The whole affair had now been steamrolled through. For some time the Politburo could indulge the illusory hope that Havemann had been isolated and that his voice would no longer be heard, at least in East Germany. At the same time, an example had been made of Havemann, one that demonstrated to the discontented, critical, and reform-oriented forces in East Germany the limits of any independent thinking.

More than a year later, in December 1965, the party leadership, at the Eleventh Central Committee Plenum, gave a much more general warning to East Germany's critical intelligentsia. Erich Honecker, at the time the Central Committee secretary for security, railed against the "latest outpourings of dissoluteness from capitalist Germany," the authorities criticized the "harmful trends" in movies, plays, and contemporary literature, and artists like Wolf Biermann, Stefan Heym, and Manfred Bieler were publicly denounced. In this way the leadership of the SED initiated a new ideological ice age in East Germany and ended a phase of liberal cultural policy that had begun after the building of the Wall. Of course, such measures not only contributed to the tightening of ideological discipline and inflicted lasting damage on the cultural life in East Germany, but also had consequences for science and technology. The denunciation of Western influences and Western thinking as "reactionary" and "revisionist" made the transfer of scientific and technological ideas more difficult or called it into question altogether. And it hardly needs to be said that creative and innovative thinking does not flourish in a climate of intellectual tutelage, indoctrination, and intolerance.

The belief of the Politburo and the party bureaucracy that Robert Havemann had been permanently isolated and silenced proved to be a serious miscalculation. Moreover, the case permanently undermined confidence in the political leadership of East Germany and damaged its reputation at home and abroad. For example, on April 21, 1964, the American chemist and Nobel laureate Linus Pauling wrote a letter to the president of the academy, Werner Hartke, calling upon him to intervene with Humboldt University on behalf of Havemann to allow him to resume his teaching and research. At that time, the academy was preparing to convert corresponding memberships into regular memberships, as called for in the statutes. Pauling expressed his hope that Havemann, like most of his colleagues, would be considered for regular membership: "I hope that the German Academy of Sciences in Berlin will not permit political considerations to prevent them from taking this step, which, in my view, is fully justified on the basis of Professor Havemann's scientific accomplishments."[38] Needless to say, the leadership of the academy did not do what Pauling had expected it to—in fact, it did exactly the opposite.

Havemann had been a corresponding member of the academy since 1961. He also headed a small section in the academy that studied photochemistry. This section had been set up in 1960 as the germ cell of a planned institute for photosynthesis and photochemistry, and Havemann had run it from the outset in addition to his other duties. After his dismissal from the university he was made full-time head of this research section with a handsome salary of 4,000 marks. However, the academy did not do this for Havemann's benefit. The Central Committee apparatus, "in agreement with the relevant institutions," had ordered this step to avoid being accused by the West, and particularly by Western communist parties, of prohibiting Havemann from pursuing his profession. Nevertheless, Havemann's days at the academy were numbered.

The brief final phase in the official silencing of Havemann in East Germany began around Christmas 1965. By then, Havemann's lectures had been published by the Hamburg Rowohlt Verlag in book form under the title *Dialectic without Dogma?* and his popularity had, if anything, increased rather than decreased. The pre-Christmas issue of the Hamburg news magazine *Der Spiegel* printed Havemann's lengthy response to the banning of the KPD in West Germany.[39] Havemann left no doubt that in his view the ban damaged the reputation of the Federal Republic, and that a democratic self-understanding made the lifting of the ban an urgent necessity. At the same time, however, he asked whether it might be possible and tactically shrewd to simply circumvent this prohibition or reduce it to absurdity by founding a new Communist Party. In fact, this was later done. But what mattered to Havemann was not merely the re-establishment of the Party, but the clear need for an inner renewal of the Party in the direction of democratic socialism: "The Communist Party that is now emerging from the shadowy existence of illegality, or will emerge in the future, must be a thoroughly new and transformed party."[40]

The party apparatus reacted immediately: on December 20, 1965, the Politburo addressed "comrade Havemann's ideas about the split of the KPD" as the second item on its agenda. It approved the draft of a statement by the

Politburo of the KPD, and the following day it was published by *Neues Deutschland,* the chief mouthpiece of the SED.[41] The statement was a sharp attack on Havemann. It accused him of pleading against the lifting of the ban against the KPD and arguing that the Communist Party must be "completely purged of all Marxist-Leninist ideas and function as a handmaid to the [West German] Office for the Protection of the Constitution [Counter Intelligence Office]."[42] The measures, however, did not stop with the language of the Cold War. The party apparatus took advantage of the opportunity to settle the score with Havemann once and for all. Havemann's ideas had already been targeted by the dogmatists at the recently concluded Eleventh Central Committee Plenum, along with those of writers like Wolf Biermann, Stephan Heym, and Peter Hacks. The Central Committee's Science Section wrote a letter to the academy on December 21, ordering "that the salary and all other allowances for Professor Havemann be stopped immediately . . . and that the executive committee of the academy immediately impose *Hausverbot* [ban on entrance] for the academy and the Research Association."[43] This docu-ment illustrates how the top party leadership pushed through measures that were deemed especially urgent and important, without granting the relevant governmental functionaries even the possibility of making their own deci-sions. The measures demanded by the Party became effective as early as December 23.

Havemann was banned from his institute but was still a corresponding member of the academy. This membership could not simply be revoked by a decree of the Politburo—such a step required a decision by the plenum of the academy. To get such a decision, party group meetings of the executive committee and of the plenum of the academy were held at the beginning of January. The statutes of the academy required that such an expulsion be discussed in the plenum and be announced four weeks ahead of time. A meeting of the managing committee of the academy on January 13, 1966, decided to proceed in exactly this manner.

Havemann, however, upset this plan by his unwillingness to give up his rights without a fight and his decision to appeal both his summary dismissal as the head of a research section and the intended expulsion. He wrote a letter to the members of the executive committee, to the secretaries of the various classes, and to all members of the academy, demanding the immediate lifting of the *Hausverbot* to allow him to exercise his right to participate in the meetings of the academy: "It appears that I am to be denied what is consid-ered self-evident by elementary legal principles, namely, that prior to such a deliberation the accused should be heard and given a chance to respond to the charges before the committee that will make the decision." He went on to say about the charges leveled against him: "Without a doubt there are considerable differences of opinion between prominent people of the GDR and myself."

A working paper of the Science Section of the Central Committee of the SED (dated February 18) reveals that all the activity at the academy had been orchestrated by the Party. Under the title "Measures for isolating Havemann politically," it made the following recommendations:

1. The measures to have Havemann terminated as a corresponding member of the German Academy of Sciences, which have already been initiated and prepared politically, should be continued and concluded with the plenary meeting of the academy on March 24, 1966 . . .
2. Articles that take issue with Havemann's philosophical and political views (without mentioning his name) should continue to appear in the press . . .
3. Havemann should be offered a suitable position as chemist in a large company, if possible outside of Berlin.[44]

This plan was approved by the Secretariat of the Central Committee on February 23, making it an obligatory directive for subordinate party organizations.[45]

By now the Havemann affair was not only attracting academic attention but also being followed and analyzed with great interest in East Germany and abroad, not least in the Western mass media. It was particularly annoying to the Central Committee apparatus that Havemann attracted interest beyond the anticommunist cap and the so-called bourgeois intellectuals; indeed, his view also found favor with many West European communists. Support came especially from the proponents of Eurocommunism, who had been leading the anti-Stalinist debate within their parties since the mid-1960s and were trying, by means of fundamental reforms, to push through national concerns and democratic organizational forms. They expressed their support in articles and acts of solidarity. This was one reason Havemann's books and other publications were later increasingly translated into Italian, Spanish, and French. Of course all this was damaging to the SED and its international reputation and was carefully noted in the Central Committee apparatus. When an article that spoke positively of Havemann's book *Dialectic without Dogma?* appeared in the Italian Communist Party paper *Unita* at the beginning of January 1966, the Science Section of the Central Committee immediately asked seven leading East German philosophers to oppose it in an open letter.[46]

While the Central Committee was thus pushing ahead with the exclusion of Havemann from the academy and seeking to discredit him within the world communist movement, Havemann, who had been informed about the confidential report of the academy leadership, sent a letter of his own to the members of the academy on March 9. In it he responded to the charges and criticized the procedure that denied him any opportunity to respond publicly and violated the most elementary legal principles. Still, the leadership of the academy declined to let him appear before the plenum; he was allowed to deliver his statement only to a meeting of the executive committee, where the intended expulsion was explained to him.

The unfolding of this carefully planned scenario, which was to take place entirely out of public sight, was seriously upset. The internal report that the president of the academy had sent to all regular members—listing all of Havemann's alleged lapses, violations of discipline, and other offenses, and charging that "by means of an affair deliberately kept alive for two years . . . [he] has done everything to drag the academy and its institutions into public conflicts with tasks, laws, and the government"—had been leaked to the West German weekly *Die Zeit.*[47] The German paper published it on March 18 together with Havemann's response. Although Havemann and his supporters were not able to bring about a postponement of the plenary meeting or even stop the planned expulsion, the article and Havemann's tenacious fight were having an effect. Prominent scientists—from the Nobel laureates Adolf Butenandt, Peter Debye, Werner Heisenberg, and Linus Pauling, to Helmut Gollwitzer, Jürgen Habermas, and Ranuccio Bianchi-Bandinelli, the Italian classical philologist and member of the Central Committee of the Italian Communist Party—expressed their objections or strong protests in letters and telegrams to the leadership of the academy.[48] As the Heidelberg astronomer Hans Kienle put it in a telegram, "the reputation of the academy is in serious jeopardy."[49] A weak opposition emerged also among the regular members, that is, members who resided in East Germany. Kurt Mothes, a biochemist in Halle and the president of the Leopoldina, and Wolfgang Steinitz, a linguist in Berlin, voiced serious misgivings about the way the affair was being handled.

Still, these expressions of concern and protest essentially had no effect. At best they influenced the "atmospherics" of the vote on March 24. There are several reasons for this. First, the basic decision to expel Havemann had been made by the Central Committee, hence any protest seemed pointless, indeed dangerous. Second, Havemann's position was highly controversial even among his academic colleagues. Because of his strong political commitments, there had already been misgivings when he was elected to the academy, and subsequent developments had not entirely eliminated them. The members were therefore only too willing to submit to the political constraints. This submissive attitude continued a tradition among German academics that granted the state—any state—the right to demand loyalty from its professors with civil servant status, and loyalty included abstaining from "excessive" political activities or from pursuing such activities in certain directions. For instance, neither the case of the Social Democratic physicist Leo Arons in the era of William II, nor that of the mathematician Emil Gumbel in the Weimar Republic, let alone the unparalleled expulsion of Jewish scholars in the Third Reich, had triggered broad protests or even acts of solidarity among colleagues.

Nevertheless, a bombshell dropped on the day of the vote: of the 100 academy members who cast votes, only 70 voted for Havemann's expulsion, while 13 voted against it and 17 abstained; some, moreover, had chosen not to attend at all. This meant that the three-quarters majority required by statute for the expulsion of a member had not been reached. As a letter from the executive committee put it, this created "a thoroughly complicated situation" for the academy.[50] However, considering itself duty-bound toward the party leadership, the executive committee felt authorized, in its meeting on March 31, 1966, to simply strike Havemann from the members roll by decree and thus terminate his membership. Incidentally, the material for this meeting of the executive committee was thoroughly prepared by deliberations of the party group the day before (March 30). The decisions of the party group in turn were probably condoned by the Politburo. The Politburo met that same day, and the minutes record the following under the item entitled "Information on the Havemann matter": "The information from comrade Hager is noted. The procedure as planned by the executive committee of the academy to expel Havemann from the academy is noted."[51]

On April 1, *Neues Deutschland*—as well as other East German papers—published a statement from the executive committee of the academy. It provided brief information about the striking of Havemann's name from the list and offered the following justification: "The corresponding member Havemann, through his own publications or those initiated by him in a press hostile to our state, has not only violated the loyalty that every scientist in the world today must have for his state; by his conduct, which is harmful to the academy and the GDR, he has also violated the duties of a citizen in a socialist state."[52]

Now began the period of Robert Havemann's complete exclusion from public life in East Germany; from here on he lived the life of a dissident. However, the hopes nourished by the Politburo and its willing accomplices that Havemann had been permanently isolated and silenced, and that his voice would no longer be heard in East Germany, proved to be illusory. The Western mass media were still willing to devote coverage to the "Havemann affair." Havemann's ideas met with a broad resonance within the Eurocommunism movement in Western Europe. Lastly, the Prague Spring consolidated Havemann's critique of individual phenomena into a principled systemic criticism of the Stalinist model of socialism.[53] The magnetism of his ideas and his charismatic personality led the authorities continually to tighten governmental repression and surveillance. In the summer of 1975, there was even talk of bringing criminal charges against Havemann for "subversive agitation." The protocol of the deliberations in the Politburo even noted that, if Havemann did not understand the stated warnings, "the relevant authorities of the GDR . . . should proceed to expel him from the GDR. Inform Havemann, within the legally permissible period of twenty-four hours, that his East German citizenship is being revoked and take him across the border."[54]

In the fall of 1976, such drastic steps were in fact taken against Wolf Biermann, probably Havemann's closest friend and comrade-in-arms. Havemann used every opportunity—including articles of protest in the Western mass media and an open letter to SED General Secretary Erich Honecker, once a fellow inmate at the prison in Brandenburg—to protest this arbitrary act, and to organize a successful solidarity movement among East German artists and intellectuals. The East German leaders reacted to this with heightened repression and surveillance, eventually placing Havemann under illegal house arrest that lasted nearly two years. In the last decades of his life—he died on April 9, 1982, not long after his seventy-second birthday—Robert Havemann thus lived under conditions that, as the Berlin political scientist Hartmut Jäckel once noted, "no Western intellectual could endure or free citizen would tolerate. They would run away if no redress was forthcoming or in the offing."[55] Instead, Havemann remained in East Germany and endured all forms of harassment. As he confessed in 1972, he was "a friend of the GDR and a committed socialist," and "the future of socialism in Germany will be decided, after all, here in the GDR."[56]

Thoughts about the future of socialism in the world and in East Germany were the dominant themes of Havemann's later activity and publications, in which he called attention to the existential importance of ecological and global problems for the development of humanity. In this way he had a very direct influence on the emergence of an independent environmental and peace movement in East Germany. He became the pioneer of the civic rights movement in East Germany and thus paved the way for the revolution in the fall of 1989.

—Translated by Thomas Dunlap

14

Kurt Gottschaldt and Psychological Research in Nazi and Socialist Germany

Mitchell G. Ash

The career of the psychologist Kurt Gottschaldt (1902–1991) exemplifies a pattern that can be found only in the twentieth-century history of German science; his career had four major permutations under four different regimes.[1] In the Weimar era he was an experimental researcher associated with the Gestalt psychologists in Berlin (1922–1928), then a clinical researcher at the Rhenish Hospital for Difficult Children near Bonn (1928–1933). Under Nazism (1935–1945) he was the head of the Department for "Hereditary Psychology" of the Kaiser Wilhelm Institute for Anthropology, Human Heredity, and Eugenics in Berlin. From 1947 to 1961, he headed the Psychological Institute at Humboldt University in East Berlin and was the leading psychologist in the German Democratic Republic. After a series of conflicts within his institute and with Socialist Unity Party and state officials, Gottschaldt left East Germany in 1962 and ended his scientific life at the University of Göttingen, in West Germany.

This chapter briefly summarizes Gottschaldt's career and research in the Nazi era, and then discusses his career in the Soviet Occupation Zone and the GDR in some detail, focusing on five related issues:

1. how Gottschaldt established a pragmatic, instrumental relationship with the Nazi state concerning his psychological research on twins;
2. how he recovered the data from his Nazi-era twin studies after 1945;
3. how he reconstructed the institutional conditions for his research in the Soviet Occupation Zone and during the early years of the GDR, while mobilizing government support from both East and West Germany;
4. how he continued or changed the research practices he had developed in the 1930s; and
5. how he elaborated and modified after 1945 the model of the interaction of heredity and environment in psychical development that he had pursued under Nazism.

Two arguments hold all of these topics together. The first argument is that, in cases like that of Kurt Gottschaldt, polemical references often made to "effortless," "natural," or "unbroken" continuity in scientific careers from the Nazi to the postwar eras do not capture the historical processes involved. What occurred, rather, are what might be called *constructed continuities*.[2] Reestablishing scientific careers after 1945 was not automatic, but required effort; it is therefore necessary to analyze more closely both the strategies that scientists employed to reestablish themselves and the government policies that made their success possible.

The second argument is that these strategies and policies can be conceptualized as the mobilization of resources of many kinds, ranging from physical objects, money, and personnel to concepts, research methods, and ideological or technocratic rhetoric. Recent work in science studies has suggested, among other things, that the alliances scientists establish with external funding agencies or with political actors function as resources that can be drawn upon in various ways to resolve scientific controversies.[3] Acknowledging that governments also mobilize scientists for purposes of their own, this chapter extends that approach both to relationships negotiated between scientists and the state, and to the changing shapes of scientific work in different political situations.[4]

Gottschaldt's Career and Research under Nazism

In April 1935, Kurt Gottschaldt became the head of the newly founded department of "Hereditary Psychology" *(Erbpsychologische Abteilung)* of the Kaiser Wilhelm Institute (KWI) for Anthropology, Human Heredity, and Eugenics.[5] He remained in that position until the end of the war, except for a brief interval of service in the psychological department of the Wehrmacht. He also directed a Polyclinic for Difficult Children and Youths at the Berlin Children's Hospital in Wedding, and was thus one of the first psychologists in Germany to combine basic research with clinical practice. In 1938 he became a nontenured associate professor in the Mathematics and Natural Science Faculty of the University of Berlin.[6] From these three positions, Gottschaldt was able to deploy and to recombine in creative ways the intellectual and practical resources he had acquired in the 1920s as a student of the Gestalt psychologists and as a clinical researcher in Bonn, in an institutional setting that had already come into close proximity with the eugenical practices of the Nazi state.

Gottschaldt's primary research achievement in the Nazi era was made possible by the organization of a set of "twin camps" *(Zwillingslager)* for the purpose of observing children. The first was set up on a trial basis in the summer of 1936 with support from the Berlin division of the National Socialist Welfare Organization (NSV).[7] Two much larger twin camps ran for eight-week periods in the summers of 1937 and 1938 at a children's summer home on the island of Norderney, in the North Sea. In a 1939 article, Gottschaldt described the central research issue, taking a term from the KWI institute head Eugen Fischer, as "phenogenetics"—the heritability of psychical characteristics in general and personality traits in particular, as well as their emergence in the course of human development.[8] The term "phenogenetics" is no longer in use, but the subject matter is no different in principle from what is now called "developmental" or "behavioral genetics."

Gottschaldt and his coworkers approached this topic by skillfully combining five methods:

1. long-term phenomenological observation *(Dauerbeobachtung)* of the behavioral styles of children;
2. experimental studies of children's problem-solving behavior;
3. social-psychological experiments on children's reactions to success and failure, and their behavior in conflict situations;
4. daily ratings of children's "basic temperaments," that is, their degree of extra- or introversion, their degree of emotional expressiveness, and their degree of activity or passivity; and
5. statistical treatment of the quantifiable results using the technique of concordance and discordance, comparing the average performances of identical and fraternal twin pairs, rather than those of individual twins. This method belonged to the internationally recognized standard repertoire of the time, but was still quite unusual in German psychology.[9]

With its creative combination of field observation and laboratory research with statistical assessment, the research of Gottschaldt and his coworkers provides additional evidence for the view already advanced for other disciplines that modern science was quite compatible with Nazism, particularly in fields perceived to be closest to its core projects—in this case the creation of a "new," dominant German race.[10]

Gottschaldt summarized the primary results of his own work in two claims, both expressed in quantitative terms:

1. The intelligent problem-solving behavior of identical twins, as measured by the method of concordance and discordance, was more similar than that of fraternal twins by a factor of more than two to one. To Gottschaldt this meant that such expressions of a higher intellectual "giftedness function" *(Funktion der Begabung)* are "largely hereditary."[11]
2. The ratings of "basic temperament" and "emotionality" were still more

similar in identical than in fraternal twins, by a factor of more than five to one. To Gottschaldt this meant that these "endothymic" functions are even more heritable than intellectual functions.[12]

The implications of these results for eugenical research and policy in Nazi Germany were deeply ambiguous. Both Eugen Fischer and Otmar von Verschuer, who succeeded Fischer in 1943, presented Gottschaldt's work as valuable basic research that could eventually lead to instruments for positive eugenical selection. Fischer had stressed the importance of psychological research for "positive race hygiene" and "the biological foundations of culture" in a report to the Kaiser Wilhelm Society as early as 1933.[13] In his book *Leitfaden der Rassenhygiene* (Guide to Race Hygiene), Verschuer, who had introduced the term "hereditary psychology" *(Erbpsychologie)* into the institute's research program in the 1920s, called Gottschaldt's "twin camp" method "a particularly valuable supplement and deepening" of the test and genealogical methods that were standard in eugenical research; he claimed that it could increase the precision with which researchers could determine hereditary capacities of all kinds, particularly in the German peoples.[14] Gottschaldt himself never joined the NSDAP and said hardly a word about Nazi ideology or eugenical practices in his published work.[15] He did place his work squarely within the eugenical research tradition begun by Francis Galton, however, and he maintained a connection with the Race Policy Office of the NSDAP and its head, Walter Gross.[16]

Although he made strong claims for the heritability of psychological characteristics, Gottschaldt consistently emphasized the interaction of heredity and environment in human behavior, the difficulty of separating the "environment" factor from the subject's reception of exterior environmental conditions, and above all the complexity of the entire topic. Intelligence, for example, was so complicated that even its hereditary aspect could not be carried by any single gene.[17] All this seemed to stand in the way of any effort to carry out a program of rapid positive eugenical selection. In an interview, Gottschaldt insisted that his emphasis on method was not a tactic, but rather a fundamental criticism of Nazi race policy, albeit on practical rather than moral grounds.[18] Such hints at the practical difficulties of realizing the Nazis' version of the eugenical project should not, however, be confused with fundamental opposition to that project.

Recovering the Data after 1945

For many German scientists during the immediate postwar period, professional survival depended on their ability to mobilize material and financial as well as conceptual, rhetorical, and ideological support. Gottschaldt was hardly alone in his efforts to reconstruct his career and his research by these means. It is fair to say, however, that in the Soviet zone and especially during the early years of the GDR, he was able to combine these resources more effectively than many other scientists, and thus to achieve a remarkable degree of status, relative autonomy, and institutional power. By the mid-1950s, he had thus become a primary exemplar of the Socialist Unity Party's effort to mobilize nonparty scientists as a resource to build socialism. But this success proved impossible to sustain for long.

Gottschaldt joined Verschuer when the institute's inventory was transferred to the Hessian countryside in February 1945, but returned to Berlin with his superior's permission in August. He wrote to Verschuer that he wanted to organize a Research Center for Psychology of the Kaiser Wilhelm Society (KWG) with the support of the Health Office of the Berlin district of Steglitz. The aim was to conduct studies on juvenile delinquency, one of the major social problems in bombed-out Berlin. The officiating president of the KWG, Max Planck, refused to allow Gottschaldt to use the KWG's name; nonetheless, Gottschaldt kept that name on his stationary and used his Dahlem office as a base of operations until the 1950s, though he received no KWG funds.[19]

In the spring of 1946, while Verschuer was trying to recover his earlier professorship in Frankfurt am Main, a public controversy broke out over his role in the Third Reich—a controversy in which Gottschaldt played a significant part that has been noted only rarely in the existing literature. Already in February, Robert Havemann, who had been named "president" of the KWG by the Berlin city government, had written to Samuel Shulits, the education and religious affairs officer in the American sector of Berlin, to demand the return of the institute's entire inventory. In that letter he denounced Verschuer's political position under Nazism and connected him with the infamous "experiments" of Joseph Mengele on inmates in Auschwitz.[20] In May 1946, Gottschaldt wrote to Shulits in an effort to recover his research data from Verschuer.[21] In the same month the issue became public with articles that Havemann placed both in the *Neue Zeitung,* published in the American sector of Berlin, and in the *Tägliche Rundschau,* published in the Soviet sector.[22]

Both Boris Rajewsky, the head of the KWI for Biophysics, and Verschuer himself clearly believed that it was Gottschaldt who began the investigation that led to the discovery of Verschuer's connection with Mengele and Auschwitz.[23] There is no doubt that he was the chief informant of both Robert Havemann and American war crimes investigators in this matter.[24] In reaction to an inquiry from Shulits, the American university officer in Frankfurt at the time, the sociologist Charles Hartshorne, had the institute's inventory brought to Frankfurt in July 1946, but also had Verschuer put under temporary house arrest by CIC (Counterintelligence Commission) officers. At Hartshorne's invitation, Gottschaldt went in August 1946 to inspect the institute's inventory. By the end of the year the twin research data he wanted to retrieve were on their way from Solz to Frankfurt; early in 1947 Gottschaldt was still petitioning for their transfer back to Berlin.[25]

Thus, the denunciation of Verschuer and the exposure of the Mengele connection were not only an act of political courage, but also a means of obtaining scientific resources. In such a situation it was not possible simply to continue working as before; continuity had to be constructed. Since he was not a geneticist, Gottschaldt did not belong to the inner circle of colleagues in that field who protected one another. Lacking allies in his own field of psychology, he was dependent on his wits and on alternative alliances, in this case with both a communist, Robert Havemann, and the American military— a combination that was possible only while the political future of Berlin was still unclear.

Constructing a New Career

While he struggled to recover the data from his twin research, Gottschaldt also established rather different alliances in order to construct a new career in Berlin. The wider context for this part of the story is a shift in de-nazification policy in the universities of the Soviet Occupation Zone, from initial, schematic purges to more selective, controlled integration of former NSDAP members and others, like Gottschaldt, who had collaborated directly or indirectly with Nazism's aims without actually joining the party.[26]

In January 1946, Gottschaldt had assumed the equivalent of his former academic position by becoming associate professor for psychology in the Mathematics and Natural Science Faculty of the newly opened university, which was located in the Soviet sector of Berlin.[27] Soon he was proposed for a full professorship. External opinions solicited by the faculty on the issue were mixed, but only one opposed him on the basis of his scientific support for positive eugenics under Nazism. The process clearly shows that the impact of collegial networks in this period went beyond the boundaries between the occupation zones. Testimonials came primarily from colleagues in the Western zones, including Gottschaldt's pre-1933 superior in Bonn, Erich Rothacker. Apparently no one noticed or cared about Rothacker's earlier active membership in the NSDAP. A highly critical opinion from Hermann Muckermann, who had been dismissed from the KWI for Anthropology in 1933 for political reasons, and the vehement opposition of Hans Nachtsheim,

Gottschaldt's former colleague and fellow department head at the KWI for Anthropology, apparently only served to delay matters for a few months.[28]

On December 1, 1947, Gottschaldt was appointed by Theodor Brugsch, the vice-president of the German Administration for Education in the Soviet Occupation Zone (DVV), without a faculty vote. Brugsch, a medical scientist, belonged to no political party at the time. He based his decision both on practical considerations—the need to establish instruction in psychology in the Mathematics and Natural Sciences Faculty—and on explicit support for Gottschaldt's natural scientific approach to psychology.[29] With such a negotiating partner, it was not necessary either to make any reference to dialectical materialism or to claim that natural scientific thinking was in itself materialistic. Such rhetorical moves came later.

In 1948 Gottschaldt received permission to establish a psychological institute in a building located at Oranienburger Strasse 18, near the university, that he had already sought out in 1946 and stoutly defended against claims from other university offices as well as from the Berlin city government.[30] Within ten years he had built the institute into one of the largest psychological research establishments in Europe.[31] Most important here is that Gottschaldt mobilized his energy, self-confidence, and organizational skills to expand his institutional power and position still further in the early 1950s, accumulating honorific and influential positions in a manner consistent with the reorganization of East German science along slightly modified Soviet lines. In 1953 he was elected associate dean of the Mathematics-Natural Science Faculty; in 1955 and 1957 he was chosen dean. Also in 1953, a plenary session of the Deutsche Akademie der Wissenschaften elected him a full member. He signaled his leading position at the discipline level in 1954, when he became the editor of the *Zeitschrift für Psychologie,* formerly the leading German journal in the field. In 1955, he expanded his control over research resources when he became the founding director of a Research Center for Experimental and Applied Psychology that was nominally attached to the academy, but was actually located in the upper floors of his institute building.[32]

Gottschaldt was elected to the Academy of Sciences in 1953 with the votes of 33 out of 36 full members, while the Marxist social scientists Leo Stern and Jürgen Kuczynski failed to be chosen at that time and another Marxist, Fred Oelsner, was elected with far fewer votes. This indicates the advantage that natural scientific standing still conveyed five years after the founding of the GDR. However, the technical chemist Heinrich Frank, who nominated Gottschaldt, apparently thought it necessary to emphasize the contribution that Gottschaldt's twin research would make "toward clarifying the *dialectical* interaction of hereditary disposition and environmental aspects," and added that "a strong link with the natural sciences and therefore the connection to dialectical materialism is not to be denied."[33]

In contrast to the strongly theoretical and basic science orientation of the Berlin Psychological Institute in the Weimar period, basic research took second place in Gottschaldt's institute after 1948 behind training for pedagogical, labor psychological, and medical-psychological practice.[34] To achieve these aims, the institute maintained its own children's clinic and day care center. Work in these places, or assisting Gottschaldt and his senior coworkers in preparing clinical diagnoses of youthful or adult suspects and criminals for the court system, was a normal part of students' training. As early as 1946 Gottschaldt gave practical experience a role in the training of psychologists, and practical psychology a role in the training of teachers, that was comparable in each sphere to the importance of hands-on experience in medical training.[35] These linkages of natural science and technocratic practice brought Gottschaldt's views into complete harmony with those of the East German state, independent of any explicit statement of ideological loyalty. The unity of theory and practice was, after all, a foundation stone both of Marxist theory and of Marxism-Leninism; of course, it is also the basis of modern technocracy in non-Marxist regimes.

Reformulating a Model

As early as 1950, in his work on juvenile delinquency, Gottschaldt offered a clear hint of the ways in which he would refashion conceptual resources from his earlier work and modify the model of heredity and environment he had articulated under Nazism. He argued on the basis of 560 case studies from the years 1945 and 1946 that the basic behavioral symptoms of delinquency had hardly changed since the 1920s. The problem had simply spread to all classes. It was therefore "obvious that all of these phenomena of delinquency are connected with the general collapse of social order" in the immediate postwar period.[36] "Thus—to reconfirm an old result—we are often dealing not with 'delinquent children,'" he wrote, "but with 'delinquent circumstances' in the widest sense of the word." The stand that Gottschaldt took on heredity and environment in this context could hardly have been clearer nor more different from what he had written in the 1930s:

> It cannot be overlooked that in recent decades precisely in the field of applications of human-heredity research much misuse has occurred. Even today we encounter chatter [*Geschwätz*] about a general "hereditary burden" in many welfare agency documents, which easily leads to an inhibiting fatalism.[37]

To continue his twin research, Gottschaldt obtained financial resources from both German states; when he presented his results, however, he subtly employed terminology from the vocabulary of Marxism and modified his position on heredity still further. Because the twins he had studied in the 1930s now lived in both German states, he applied for and received support for a follow-up study from both the Department of Higher Education (Abteilung Fach- und Hochschulwesen) of the GDR Education Ministry and the Notgemeinschaft der Deutschen Wissenschaft, soon renamed the Deutsche Forschungsgemeinschaft, in Bad Godesberg.[38] Neither of his funding applications contains a word of ideology. This was hardly necessary, since the issue of heredity and environment, or the limits of educability *(Grenzen der Erziehbarkeit)*, was of interest in both German states.

In an initial report on a follow-up study of seventy of the Norderney twin pairs, given as a plenary lecture at the Academy of Sciences and published in 1954, Gottschaldt presented a theory of personality with three dimensions: the "concrete situational field" of the person, the person's inherited "constitution," and—this was the new factor—the socioeconomic determinants of personality. Specifically, he argued that in identical twins, higher cortical functions, in comparison with "endothymic" functions like basic temperament, vitality, or emotionality, are "essentially dependent upon the [person's] biographical, educational, and child-rearing situation."[39] In 1939, he had claimed only that the probability of environmental influences on higher cortical functions was higher. Thus, a vague suggestion of a possible connection had become the assertion of a straightforward functional dependency.

Two of Gottschaldt's formulations were completely new: the claim that the relationship between hereditary and environmental dimensions of psychical development was a "dialectical interaction," and the concept of the *social superstructure (sozialer Überbau)* of the person, which he claimed began to develop in puberty. He worked through this idea most thoroughly in research on so-called persona-phenomena (which would be called body imagery today), and in studies of the role of so-called we-groups among youths and in the workplace.[40] Using the method of systematic long-term observation he had developed for his 1930s twin research in the Berlin institute's children's clinic and in other situations, he distinguished ten levels of "we-group" structures and behavioral styles. Going still further, he proposed applying this typology to studies of working groups in factories and offices, and opined that such work could contribute to improvements in productivity.[41] Though such ideas were potentially relevant to socialist pedagogy as well as to social and labor policy, and the terms "dialectical interaction" and "social superstructure" were plainly borrowed from Marxist theory, Gottschaldt never stated that relevance explicitly in his publications, leaving it to his readers to construct such connections for themselves. In this way Gottschaldt built upon his earlier work, kept in touch with current trends in Western, especially West German, personality theory and social psychology, while at the same time taking an occasional bow toward the political discourse of his own environment. He did all this, as he had in the Nazi era, in a way that assured him, at least in the short term, a considerable degree of autonomy and institutional power.

Gottschaldt achieved such prominence and generous support for his two research institutes without joining the Socialist Unity Party. But he declared his loyalty to the GDR. For example, in a newspaper article published in 1954, he said that he would vote for the National Front to express the "thanks of a scientist" for the revival and generous support of science in the GDR; in doing so he engaged in a subtle bit of historical reconstruction, emphasizing the "severe decline" of science under Nazism without mentioning the extensive support his own twin research had received.[42] Later, Gottschaldt represented studies done in his Research Center for Experimental and Applied Psychology of the Academy of Sciences on the role of psychological factors in the prevention of industrial accidents in state-owned enterprises as contributions to the development of socialism.[43] For their part, officials in the office of the State Secretary for Higher Education made much of the fact that Gottschaldt chose in 1952 to move from West to East Berlin, and had thus "consciously made a decision for our Republic."[44] This did not mean a loss of comfort, since he moved from middle-class Steglitz to Hessenwinkel, a quiet, wooded district in southeastern Berlin with fine homes, where many prominent East German academics lived. Nonetheless, it would be a mistake to dismiss such statements as rhetorical flourishes. Gottschaldt's commitment to working in the GDR could have been motivated by both careerism and by a sincere commitment to some version of socialist ideals.

Far more important is the discursive context in which such statements and actions were embedded—a context in which both psychological science and its human subjects are treated as social and economic resources. A document from 1956 is revealing in this respect. In it Gottschaldt protested because the architectural planning for the expansion of his research center into a full-fledged academy institute seemed to be making little progress. In this connection he also replied to criticism from the experimental physicist and former DVV Vice-President Robert Rompe, who was then the secretary of the Mathematical and Physical Sciences Class of the academy, and who had been a member of the Communist Party since 1932. At a meeting of the academy's presidium Rompe had remarked in his typically pointed way that "we are a poor country and cannot afford an institute for psychology." Pointing to the costs of high sickness and accident rates in socialist production plants, Gottschaldt wrote, "People are the most expensive and largest capital in the GDR. It is senseless and irresponsible to continue ignoring the human factor as we have done until now."[45]

One of Gottschaldt's senior assistants, Hans-Dieter Schmidt, clarified the concept of science on which this opinion was based in a 1959 study of work patterns of groups in socialist industrial plants. As he put it, after the end of capitalist exploitation under socialism, psychological concepts from American "human relations" research could also become "tools or instruments of the (new) dominant class," precisely because they are "socially indifferent," that is, ideologically neutral.[46]

Gottschaldt's Struggle for Autonomy

After the SED began to work more systematically and aggressively to ensure not only the material but also the ideological integration of natural scientists and the technical intelligentsia from 1958 onward, it was no longer possible to maintain such a position without contradiction. In February 1958, the Third Higher Education Conference of the SED called for a redirection of psychological research and training toward fulfilling the needs of socialism, increased knowledge of Soviet psychology, and the strengthening of SED party groups in psychological institutes. To achieve all this, it called for a change in the composition of the Advisory Council *(Wissenschaftlicher Beirat)* for psychology attached to the State Secretariat for Higher Education, which Gottschaldt still headed. In 1956 a polemic by the émigré psychologist Alfred Katzenstein had appeared against the allegedly "idealistic" standpoint of Gestalt psychology, the tradition to which Gottschaldt still adhered.[47] Shortly thereafter a series of lectures was held by the pedagogical psychologists at Humboldt University, the stated aim of which was "to overcome the last vestiges of bourgeois psychology." Here, too, Gestalt psychology came under attack, as Ulrich Ihlefeld spoke of the "reactionary core of this apparently materialistic conception."[48] All of these statements were part of a carefully orchestrated effort to create a special brand of Marxist psychology in the GDR.[49]

The extraordinarily complex system of alliances at work in the struggle that followed can only be sketched here. For a grasp of the structure of this struggle, three levels can be distinguished:

1. At the highest level were top political officials and the holders of scientific power in a triangular relationship involving the relevant officials of the Science Department of the Central Committee of the SED, their counterparts in the State Secretariat for Higher Education, and the

leadership of the Academy of Sciences and its SED party group. All three of these agencies were in regular communication with one another, but each possessed its own potential for independent action.

2. One level below this was the university, which was technically subordinate to the State Secretariat and thus had less potential for independent action. Here there were tensions and conflicts between the SED party organization (PO) and the university leadership on the one hand, and the leadership of the university's faculties (in this case the Pedagogical Faculty and the Mathematics/Natural-Science Faculty) on the other. The dual roles of important actors as both university professors and academy members considerably complicated matters at this level.

3. Last but not least, and in certain respects decisive, were the convoluted conflicts in Gottschaldt's two institutes, particularly the university's psychological institute. One line of conflict ran between SED members and non-SED members, with Hans-Dieter Schmidt, who appears to have regarded himself as a "nonparty Marxist," in an interesting middle position.[50] Other lines of conflict, not necessarily coterminous with the first, ran between Gottschaldt loyalists and those opposed to his at times rather sharply expressed authority, and between older and younger doctoral candidates, assistants, and associates.

A strictly structural account of this kind is not sufficient, however, because none of the actors in this drama kept to his or her assigned level. All involved tried to build alliances with people at other locations on the institutional map, and thus to mobilize them as resources for their side. The most active player in this particular game was, of course, Gottschaldt himself, who tried again and again to have party officials arrange "conversations" *(Aussprachen)* either with the State Secretary for Higher Education, Wilhelm Girnus, with the head of the State Secretariat's department of research, the SED Central Committee member Franz Dahlem, or with the head of the science and propaganda section of the Central Committee of the SED and later party ideological chief, Kurt Hager, and thus to frustrate the efforts of his subordinates in league with subaltern party and state officials to bring him down.[51]

To be sure, the picture is far too complicated to be reduced easily to a simple struggle of "science" versus "ideology," or to the imposition of a single rigid party line on a basically passive or politically innocent scientist. Gottschaldt himself did this only when he was backed into a corner (see below). This was understood by all players at the time to be primarily a struggle for autonomy and institutional power—and hence for control of institutional and material resources and personnel—by an institute director who was very strong in his own area, and highly respected in other fields as well.

On Gottschaldt's side stood the leadership of the academy, in particular the heads of the natural science and medical institutes; on the other stood the official responsible for psychology in the State Secretariat for Higher Education, Gretel Junge, and officials of the Science Department of the Central Committee, especially Walter Mäder, who was responsible for psychology. It would, however, be false to represent this as a struggle between "scientists" and "ideologues," for many of those involved from the academy were also members of the SED. Thus, when Gretel Junge differentiated in one memorandum between the "correct theory," which of course came from the SED leadership, and the "bourgeois" emphasis on "correct method," SED members in the academy accused her of "incorrect party work," or even of "sectarianism" *(Sektierertum)*.[52]

Further unpacking of the conflicts at the institute level indicates that here, too, a simple dichotomy of "ideology" versus "science" is insufficient. The struggle had begun with disputes within the institute between Gottschaldt and his younger staff, only a few of whom belonged to the institute's party organization. Lengthy reports by institute staff members in November and December 1958 stated that Gottschaldt had hindered SED party work, but complained mainly of his tight control of research, even that conducted by senior assistants, and his rigid insistence that the methods he had developed in the 1930s be used wherever possible.[53] Such complaints showed how little had changed in the traditional top-down, authoritarian structure of German academic research institutions under socialism. Another document in the Central Party Archives indicates that the institute heads of the academy regarded this as a generational conflict and a rebellion against legitimate authority, and thus understandably came to the side of their fellow academician. This support was not without ambivalence. As one institute director, the linguist Wolfgang Steinitz, remarked, Gottschaldt was "a dictator, but a great scientist."[54]

In the midst of this struggle, Gottschaldt tried to strengthen his position by intensifying already existing alliances with colleagues in both the Soviet Union and West Germany. His first move was to cultivate contacts with Soviet psychologists such as Alexander Luria and A. N. Leontiew, whom he had met at the International Congress for Psychology in Brussels in 1957. In a trip from June 22 to July 2, 1960, arranged by the Academy of Sciences, he attended the All-Union Congress of Psychology in Moscow, one of the events that signaled the gradual liberation of Soviet psychology from Pavlovian orthodoxy under Krushchev. As Gottschaldt reported to the academy president, East German psychologists had "no essential differences" with Soviet psychologists on this score.[55] This was a clear attempt to trump local efforts to create a "Marxist" psychology by showing that his own approach was the one accepted in the "socialist motherland."

Gottschaldt also maintained links to the West, however. In September 1959 he became the first East German psychologist elected to the governing board of the German Society for Psychology. Because he advanced his candidacy and accepted his election as an individual and not as an official representative of the GDR, this move brought him public denunciations and behind-the-scenes criticism for associating himself with the so-called Hallstein doctrine of the West German government.[56] At the society's meeting he also conferred with Johannes von Allesch, an associate of the Gestalt psychologists whom he had known since the 1920s, and who had just retired as professor in Göttingen. Academy of Sciences President Werner Hartke repeatedly emphasized the possibility that Gottschaldt could leave the GDR and succeed von Allesch in Göttingen, thus causing "lost prestige" for East German science.[57]

Ultimately, this dispute exemplified the contradictory character of science policy in the early years of the GDR. A certain respect for "objective" knowledge, or at least the power it could provide, was certainly present in the SED leadership. Along with this went a desire to make use of the prestige, the image of modernity, and the potential applications of high-level research, as well as an urgent need to compensate for the obvious scarcity of ideologically secure "cadres" in the natural sciences. Such considerations made cooperation with nonparty scientists seem necessary for some time. To make this policy work, and also to keep top scientists from accepting offers from West Germany, considerable financial concessions and other privileges were often granted.[58] In this case, Gottschaldt went still further; he demanded, and achieved, the removal of all SED members from the Psychological Institute. The last to leave, revealingly, was Friedhart Klix, who transferred to a professorship at a new psychological institute in Jena in the spring of 1960, but became Gottschaldt's successor after the latter's departure.[59]

On the other side of the equation, however, stood *die Machtfrage*—the need to maintain SED control, and with it the ideological predominance of Marxism-Leninism, in all areas of society. Such claims were advanced with particular vehemence and dogmatism in disciplines like psychology, which was located on the boundary between the natural and the social sciences, but had undoubted relevance to the ideologically charged fields of education and labor policy.

Dénouement

Ultimately, the Socialist Unity Party's effort to mobilize high-level work by nonparty scientists as a resource to build socialism while gradually increasing party influence in their institutes failed in this case. In the spring of 1960, party and state officials followed up the removal of SED party members from Gottschaldt's institute with a sharp cut in the number of students admitted to study psychology, citing, of course, the lack of staff. At the same time, three former members of Gottschaldt's staff were reassigned to a psychological institute in the Pedagogical Faculty. To Gottschaldt, it must have looked as though he had been led into a trap. He protested vehemently against the plan to reduce his student and staff contingent at a meeting of the Mathematics-Natural Sciences Faculty in June 1960, claiming that the move virtually excluded him from teaching.[60] To this the Central Committee cynically replied that Gottschaldt himself had said he was overburdened, and that no one was preventing him from engaging in teaching or research. The dispute appeared resolved with a new agreement giving Gottschaldt an increase in staff at his academy research center. But the next year his student contingent was cut still further; in February 1961, a close friend, the Humboldt University chemist Erich Thilo, wrote to Kurt Hager and Franz Dahlem that Gottschaldt was again threatening to leave for the West.[61]

On August 8, 1961, five days before the building of the Berlin Wall, Gottschaldt reported to the State Secretariat that he had received an offer from Göttingen and intended to accept it. A semester of utter confusion followed. The State Secretariat for Higher Education released Gottschaldt from his university post, but he was not allowed to leave the GDR. Instruction continued on the lower floors of the Psychological Institute, while Gottschaldt and a small staff kept working in the academy research center on the upper floors of the same building, with the explicit endorsement of the academy's SED party organization.[62] Gottschaldt left the GDR illegally on February 14, 1962. The details of this *Republikflucht* remain as yet unclear.

Conclusion

Though Gottschaldt apparently departed in great haste and left most of his research library in Berlin, documents in his rather brief Stasi file and other sources indicate that he was able to take some of his enormous collection of twin research data with him.[63] On this basis he organized a third study of the Norderney twins—that is, of those twins who lived in the West—in 1967, with renewed support from the DFG (German Research Council). The published results reflected both the life patterns of the twins, now in their forties, and Gottschaldt's own changed situation. He reported that, despite in some cases radical changes in their environments since the 1950s, the intelligent behavior patterns of identical twin pairs were roughly the same, or at least as little different from one another as they had been in earlier studies. Instead of drawing strong hereditarian conclusions from this, as he did in the 1930s, Gottschaldt now wrote that hereditary dispositions affected intelligent behavior only "to a certain degree"; and he limited their impact to "the variability that people *of a particular cultural epoch* show in school and IQ test achievement." In contrast to his emphasis in the GDR on the "social superstructure" of personality, and at the height of West German educationalists' concentration on programs for gifted children, Gottschaldt portrayed hereditary dispositions more clearly than ever before as "developmental potential, the manifestation of which is codetermined" by both environmental "and other genetic conditions."[64]

Gottschaldt had brought only part of his data to the West. For decades he tried to retrieve the remainder, but this time his efforts were fruitless; East German officials did not even reply to his letters. He nonetheless continued to work from his retirement in 1970 until his death in 1991 on a "phenogenetics of human personality" based on his twin research.[65] Only after his death were the data reunited at the Max Planck Institute for Psychological Research in Munich—a representative of the legal successor of the Kaiser Wilhelm Society. A fourth study with the surviving Norderney twins, now in their sixties, is now in progress.[66]

The persistence with which Gottschaldt pursued a single research program in multiple contexts justifies our asking whether he had four careers or only one. However, it is equally important to emphasize the enormous effort Gottschaldt put into reconstructing his scientific activity, and to a certain extent his scientific persona as well, especially after 1945. His case may have been more spectacular than others, but it is instructive for precisely that reason. Gottschaldt's story indicates both the possibilities and the limits of continuity for scientific personnel and research in the Soviet Occupation Zone and in the early years of the GDR. It also shows how much and what kinds of effort were required to achieve these constructed continuities and reconstructions in content. One ironic result was that such heavy investments in reassembling scientific resources and continuing research programs from the past may have inhibited innovation in both East and West Germany.

Scientists like Kurt Gottschaldt proved to be capable—up to a point—of adapting themselves and their practices to varied political and institutional circumstances, and thus, in a sense, of mobilizing even the East German state as a resource for their own work. But this was always a two-way street. Stories like Gottschaldt's were possible only because, within certain limits, Socialist Unity Party officials in the GDR regarded scientists as they presented themselves—as human resources whose knowledge and skills could be used to build a socialist society, regardless of what they had done under Nazism. Gottschaldt's struggle for autonomy and its outcome indicate that the limits of that policy had already been reached in this case long before the closing of the border in 1961.

Abbreviations

Notes

Biographies of Contributors

Index

Abbreviations

AAW	Archiv der Akademie der Wissenschaften Berlin
AUB	Archiv der Humboldt Universität Berlin
BAP	Bundesarchiv, Potsdam
BStU	Der Bundesbeauftragte für die Unterlagen des Staatssicherheitsdienstes der ehemaligen Deutschen Demokratischen Republik
HSA	Hessisches Staatsarchiv Marburg
SAPMO	Stiftung Archiv der Parteien und Massenorganisationen der DDR im Bundesarchiv, Zentrales Parteiarchiv, Berlin
SPK	Staatliche Plankommission

Notes

Prologue

1. In this book the terms "socialism" and "communism" have been used interchangeably by most authors. We are not using socialism in the broadest sense of "we are all socialists now" or as another name for a welfare state. The German Democratic Republic itself used the term "socialism" to mean the last stage before fully developed communism.
2. The German word *Wissenschaft* (science and educational institutions) most accurately describes what is covered in this book. Please note that it was not our intention to offer a large-scale comparison with other countries (that is, a separate chapter dealing with another East Bloc country) or times, but each contribution touches upon these comparisons. This is not a study of Science in Eastern Europe but a study of science in East Germany. For two pre-Wall anthologies on Science in Eastern Europe see György Darvas, ed., *Science and Technology in Eastern Europe* (Essex: Longman, 1988), and Craig Sinclair, ed., *The Status of Civil Science in Eastern Europe* (Dordrecht: Kluwer Academic Publisher, 1989). Both use more of a policy approach than a historical approach.
3. For a detailed discussion see Anna-Sabine Ernst, "Between 'Investigative History' and Solid Research: The Reorganization of Historical Studies about the former German Democratic Republic," *Central European History 28* (1995): 373–395.
4. *Die Enquete-Kommission 'Aufarbeitung von Geschichte und Folgen der SED-Diktatur in Deutschland' im Deutschen Bundestag* (Frankfurt: Nomos/Suhrkamp, 1997).
5. See Timothy Garton Ash's review essay "The Truth about Dictatorship," *New York Review of Books,* February 19, 1998, pp. 38–39, for a brief discussion of the Inquiry Commission report.
6. Rainer Eckert, Ilko-Sascha Kowalczuk, and Ulrike Poppe, eds., *Wer Schreibt die DDR-Geschichte?: Ein Hisktorikerstreit um Stellen, Strukturen, Finanzen und Deutungskompetenz* (Berlin: Evangelische Akademie-Brandenburg, 1995), p. 3.
7. For example, Armin Mitter and Stefan Wolle, *Untergang auf Raten: Unbekannte Kapitel der DDR-Geschichte* (Munich: Bertelsmann, 1993).

8. For example, Charles S. Maier, "Geschichtswissenschaft und Ansteckungsstaat," *Geschichte und Gesellschaft 20* (1994): 616–624. Many of the newspaper and journal articles are reprinted in Rainer Eckart, ed., *Hure oder Muse: Klio in der DDR* (Berlin, 1994).
9. Joyce Appleby, Lynn Hunt, and Margaret Jacob, *Telling the Truth about History* (New York: W. W. Norton & Company, 1994).
10. Ibid., p. 241.
11. Mary Fulbrook, *Anatomy of a Dictatorship: Inside the GDR, 1949–1989* (New York: Oxford University Press, 1995). See also the special issue of *German Studies Review* entitled "Totalitäre Herrschaft—totalitäres Erbe," Fall 1994, with contributions by Germans in German.

Introduction

1. I would like to thank Dolores Augustine, Michael Fisher, Jeffrey Herf, and Michael and Lily Macrakis for their comments on a draft Introduction.
2. Stephen Kinzer, "Talk about Theme Parks! Is Live Ammo Next?" *New York Times,* November 9, 1993, p. A4. This idea was not realized by the middle of 1997.
3. It is beyond the scope of this book to include a number of Cold War science issues as described by recent studies on science and the Cold War in the United States; these studies stress the role of the Department of Defense and the military stimulation of many innovations in America. Other Cold War–related issues—such as the impact of the division of Germany on science and the senior-junior partner relationship between the Soviet Union and the GDR—do emerge, however.
4. Of course, researchers at American universities like MIT and Harvard build their own equipment too, but East German scientists had to more inventive with limited materials.
5. For scientists' complaints about old equipment see the biographies in Guntolf Herzberg and Klaus Meier, *Karrieremuster: Wissenschaftlerporträts* (Berlin: Aufbau Taschenbuch Verlag, 1992). For figure on old equipment and physical plant see R. H. Brocke and E. Förtsch, *Forschung und Entwicklung in den neuen Bundesländern, 1989–91* (Stuttgart: Raabe 1991), p. 38.
6. For science citation index studies see George E. Vladutz and David Pendelbury, "East European, Soviet, and Western Science Compared: A Scientometric Study," in C. Sinclair, ed., *The Status of Civil Science in Eastern Europe* (Dordrecht, Boston, London: Kluwer Publishers, 1989), pp. 113–128; Brocke and Förtsch, "Determining Outputs and Excellence" chapter 6 in R. H. Brocke and E. Förtsch, *Forschung und Entwicklung in den neuen Bundesländern*. See also Raymond Bentley, *Research and Technology in the Former German Democratic Republic* (Boulder: Westview Press, 1992), especially pp. 116–128, for a comparison between the FRG and the GDR. For patent data see Hansjörg F. Buck, "Forschungs- und Technologiepolitik in der DDR—Ziele, Lenkungsinstrumente, Mobilisierungsmittel and Ergebnisse," in Gernot Guttmann, ed., *Das

Wirtschaftssystem der DDR (Stuttgart: Gustav Fischer Verlag, 1983), pp. 229–309.

7. See Bentley, *Research and Technology,* pp. 126 and 188.

8. Patent award tables are reproduced in Buck, "Forschungs- und Technologie-politik," pp. 300–301.

9. Analysis of Science Council reports.

10. See Brocke and Förtsch, "Determining Outputs and Excellence," p. 91; table taken from work by W. Meske.

11. Ibid., p. 102.

12. See p. 63.

13. The German edition also has chapters on the aviation industry, biologists, and the early history of computing: Dieter Hoffmann and Kristie Macrakis, eds., *Naturwissenschaft und Technik in der DDR* (Berlin: Akademie Verlag, 1997).

14. Ibid. See Burghard Ciesla's chapter on the aviation industry, pp. 193–211.

15. See Mary Fulbrook, *The Anatomy of a Dictatorship: Inside the GDR, 1949–1989* (New York: Oxford University Press, 1995), pp. 283–286, for a brief analysis of the distinction between comparisons made to equate or contrast.

16. See *Protokoll der Verhandlungen der 3. Parteikonferenz der sozialistischen Einheitspartei Deutschlands* (Berlin: Dietz Verlag, 1956), p. 606.

17. John Lewis Gaddis, *We Now Know: Rethinking Cold War History* (Oxford: Clarendon Press, 1997), p. 136.

18. See "Überholen ohne einzuholen"—ein wichtiger Grundsatz unserer Wissenschaftspolitik" in *Die Wirtschaft,* no. 7 (1970), pp. 8–9.

19. Paul R. Josephson, *Totalitarian Science and Technology* (New Jersey: Humanities Press, 1996).

20. Robert Merton, "Science and the Social Order," in *The Sociology of Science* (Chicago: University of Chicago Press, 1973), reprinted from a 1938 article.

21. See Chapter 1, p. 29.

22. See Chapter 5.

23. See Chapter 8.

24. See Chapter 10.

25. Sinclair, ed., *The Status of Civil Science in Eastern Europe.*

26. Ibid., especially the Conclusion.

27. It would require a separate, lengthy chapter to analyze such a question. As stated, it was not the point of this book to analyze the respective achievements of science in the West and in the East.

28. See David Joravsky, *The Lysenko Affair* (Cambridge, Mass.: Harvard University Press, 1970), for the argument that is was an agricultural crisis that allowed Lysenkoism to take hold in the Soviet Union.

29. SAPMO, Abt. Wissenschaft, 304, May 17, 1962, "Information about a Meeting for Further Developing Biology."

30. For a more detailed analysis including the transition from divided Germany to unified Germany, see my forthcoming article in *Minerva:* "Political Transitions and Science in Twentieth-Century Germany."

31. In fact, I argue this in Kristie Macrakis, *Surviving the Swastika: Scientific Research in Nazi Germany* (New York: Oxford University Press, 1993).

32. Ibid.
33. Ralph Jessen, "Professoren im Sozialismus. Aspekte des Strukturewandels der Hochschullehrschaft in der Ulbricht-Ära," in Hartmut Kaelbe, Jürgen Kocka, and Hartmut Zwahr, eds., *Sozialgeschichte der DDR* (Stuttgart: Klett-Cotta, 1994), pp. 217–253, especially pp. 241–243.
34. See Chapter 6.
35. See, for example, Abott Gleason, *Totalitarianism: the Inner History of the Cold War* (New York: Oxford University Press, 1995). Although the concept of totalitarianism was widely used in the 1950s and 1960s in the aftermath of Nazi Germany, it slowly faded from the scene and was discredited by some academics until the collapse of Eastern European communism, when it experienced a revival.
36. See, for example, the book by Marianne and Egon Erwin Müller, "... *Stürmt die Festung Wissenschaft!" Die Sowjetisierung der mitteldeutschen Universitäten seit 1945* (Berlin: Colloquium-Verlag, 1953).
37. See Chapter 1.
38. See Chapter 11, p. 237.
39. See Chapter 12.
40. See Jeffrey Koppstein, *The Politics of Economic Decline in East Germany, 1945–1989* (Chapel Hill: The University of North Carolina Press, 1997). See Chap. 2 for an excellent reappraisal of the 1960s reform period.

1. Science, Higher Education, and Technology Policy

1. J. Habermas, *Theorie des kommunikativen Handelns*, vol. 2, *Zur Kritik der funktionalistischen Vernunft* (Frankfurt a. M.: Suhrkamp, 1981), serves as a basic reference and foundation for this chapter.
2. See the contributions by B. Weiss in this volume and B. Cisela in the German edition.
3. The most urgent task of research and development in this period was seen as reestablishing the prewar state of technology; see, for example, the Main Department for Science and Technology, Bundesarchiv (BA) Coswig 4, no. 14.
4. G. Becker, "Über die Ursachen der Überlegenheit der sowjetischen Wissenschaft," *Einheit*, no. 3 (1953), 293ff.
5. G. Harig, "Der weitere Ausbau unseres Hochschulwesens," *Einheit,* no.20 (1951) 1575ff.
6. K. Hager, "Unsere Wissenschaft im Dienste des Aufbaus des Sozialismus," *Einheit* (1953), no.1: 84ff.
7. Minutes of the sessions of the Central Committee Working Group (later Department) of Research and Technical Development, SAMPO Dy 30/IV A 2607/58; Tasks of the Department of the Sciences (April 3, 1964), SAPMO Dy 30/IV A 2/904/58.
8. See D. Hoffmann, "Die Physikalische Gesellschaft (in) der DDR," in T. Meyer-Kuckuk, ed., *Festschrift 150 Jahre Deutsche Physikalische Gesellschaft* (Weinheim: 1995).

9. See the minutes of the Research Council for the session of June 10, 1959.

10. *Dokumente der SED,* vol. 5 (Berlin: 1956), pp. 337ff.

11. I.-S. Kowalczuk, "Volkserhebung ohne 'Geistesarbeiter'? Die Intelligenz in der DDR," in I.-S. Kowalczuk, A. Mitter, and S. Wolle, eds., *Der Tag X- 17. Juni 1953. Die "Innere Staatsgründung" der DDR als Ergebnis der Krise 1952/54* (Berlin: 1995), pp. 144–145.

12. K. Hager, "Für eine verantwortungsbewußte idcologische Arbeit," *Einheit,* no. 9 (1956), 847ff.

13. E. Förtsch, "Literatur als Wissenschaftskritik," in I. Spittmann, ed., *Lebensbedingungen in der DDR, Edition Deutschland Archiv* (Cologne: 1984), 157ff.

14. "Information der ZK-Abteilung Forschung und technische Entwicklung über Probleme bei der Realisierung des Staatsplanes Wissenschaft und Technik" (1973), SAPMO Dy 30/17372. The Research Council noted similar problems, making clear reference to the dumping of products (November 13, 1958).

15. Compare Chapter 7.

16. Minutes of the sessions of the Central Committee Working Group (later Department) of Research and Technical Development, SAMPO Dy 30/IV A 2607/58; Tasks of the Department of the Sciences (April 3, 1964), SAPMO Dy 30/IV A 2/904/58.

17. See Chapters 2 and 6.

18. See the Introduction and Chapters 2, 8, and 9 for more on the New Economic System and its influence.

19. K. Hager summed up the situation in an interview with K. Macrakis on August 10, 1993, as "a larger independent role" for the Academy of Sciences in the natural sciences and technology (responsibility for deciding on research projects, in consultation with the Ministry of Science and Technology and the Communist Party), "while we [*Wissenschaft* Department] had direct responsibility for the social sciences."

20. SAPMO Dy 30/38507. In a report from April 29, 1988, for the Council of Ministers entitled "Analysis of the Level, Scope, and Utilization of Research Technology in the Academy of Sciences of the GDR, with Attached Conclusions," the authors listed the age of research equipment in percentages: in 1986, 15 percent of all large installations worth more than 100,000 marks were more than ten years old, and 19 percent more than fifteen years old. 52 percent of all equipment with a value below that was more than ten or fifteen years old.

21. See the Politburo minutes, SAPMO Dy 30 J IV 2/2–2325, session of April 18, 1989, attachment 13: "Tasks of the State Plan for Science and Technology 1989, which will become more economically effective in production in honor of the fortieth anniversary of the founding of the GDR."

22. Minutes of the sessions of the Central Committee Secretariat, SAPMO Dy 30 J IV 2 3 A/4806, show how time-consuming procedures were: as a rule, four or five different Central Committee departments were involved in approval of a trip to a Western country; the Central Committee secretaries (all members of the Politburo) had to confirm approval in circulating memos.

23. R. H. Brocke and E. Förtsch, *Forschung und Entwicklung in den neuen Bun-*

desländern 1989–1991 (Stuttgart: Raabe, 1991). Chapter 6 in Brocke and Förtsch discusses output and strengths. It would be interesting to study in detail which outstanding achievements were reached as a result of research policy and which in spite of it.

24. F. Klinger, "Die Probleme der Intensivierung des wissenschaftlichen-technischen Fortschritts und seine ökonomische Nutzung," manuscript (Berlin and Erlangen, 1989).

25. E. Förtsch, "Die bedrohliche Produktivkraft," in G.-J. Glaeßer, ed., *Die DDR in der Ära Honecker. Politik-Kultur-Gesellschaft* (Opladen: Westdeutscher Verlag, 1988), pp. 563ff.

2. The Reform Package of the 1960s

1. W. Gruhn and G. Lauterbach, "Die Organization der Forschung in der DDR," in *Das Wissenschaftssystem in der DDR*. Published by the Institut für Gesellschaft und Wissenschaft (Erlangen: Verlag Deutsche Gesellschaft für zeitgeschichtliche Fragen e.V., 1977), 114.

2. Ibid., p. 116.

3. For example, Peter-Christian Ludz wrote in 1968: "The utopian dreams connected with the GDR as well as the interpretation of the GDR as a Stalinist holdover are today rejected nearly unanimously by East German studies." Ludz, "Aktuelle oder strukturelle Schwächen der DDR-Forschung?" *Deutschland Archiv* 1, vol. 4 (1968), p. 255.

4. *Vergleich von Bildung und Erziehung in der Bundesrepublik Deutschland und in der Deutschen Demokratischen Republik. Materialien zur Lage der Nation.* Published by the Bundesministerium für innerdeutsche Beziehungen (Cologne: Verlag Wissenschaft und Politik, 1990).

5. Ibid., p. 3.

6. Ibid.

7. The terms "socialism" and "socialist" as used in this Chapter refer solely to the way in which the relevant societies characterized themselves and carry no value-judgments of any kind.

8. Hans Freyer, Jindřich Filipec, and Lothar Bossle, *Die Industriegesellschaft in Ost und West. Konvergenzen und Divergenzen.* Schriftenreihe des Instituts für Staatsbürgerliche Bildung in Rheinland-Pfalz 5 (Ingelheim/Rhein: Pädagogische Arbeitsstelle für Ostfragen, 1966).

9. *Man, Science, Technology: A Marxist Analysis of the Scientific-Technological Revolution* (Moscow and Prague: Academia Prague, 1973).

10. H. Weber, *DDR. Grundriß der Geschichte. Vollständig überarb. und erg. Neuauflage* (Hannover: Fackelträger Verlag, 1991), 129.

11. R. Karlsch, *Alles bezahlt? Die Reparationsleistungen der SBZ/DDR 1945–53* (Berlin: Links Verlag, 1993).

12. *Protokoll der Verhandlungen des V. Parteitages der SED. 10.–16. Juli 1958,* vol. 1 (Berlin: Dietz Verlag, 1959), 70.

13. D. J. de Solla Price, *Little Science, Big science: Von der Studierstube zur Großforschung* (Frankfurt am Main: Suhrkamp Verlag, 1974).

14. W. Ulbricht, "'Überholen ohne einzuholen'—ein wichtiger Grundsatz unserer Wissenschaftspolitik," *Die Wirtschaft*, Feb. 26, 1970, p. 8.

15. E. Förtsch, "Institutionen und Prozesse der forschungspolitischen Lenkung und Planung," in *Das Wissenschaftssystem in der DDR*, p. 97.

16. Weber, *DDR*, pp. 126–127.

17. Förtsch, "Institutionen und Prozesse," p. 60.

18. G. Lauterbach, *Wissenschaftspolitik und Ökonomie. Wandel der Konzeptionen im Rahmen der Wirtschaftsreformen (1963–1971)*. Published by the Institut für Gesellschaft und Wissenschaft (IGW) at the University of Erlangen-Nürnberg (Erlangen: Verlag Deutsche Gesellschaft für zeitgeschichtliche Fragen e.V., 1980), pp. 12–13.

19. P. K. Hensel, "Der Zwang zum wirtschaftspolitischen Experiment in zentral gelenkten Wirtschaften," in his *Systemvergleich als Aufgabe* (Stuttgart and New York: G. Fischer Verlag, 1977), p. 176.

20. E. Richert, *Die DDR-Elite oder Unsere Partner von morgen?* (Reinbek b. Hamburg: Rowohlt Verlag, 1968).

21. "Richtlinie für das neue ökonomische System der Planung und Leitung der Volkswirtschaft," *Gesetzblatt für die DDR*, 1963, Part II, no. 64.

22. Ibid., p. 482.

23. Ibid.

24. Ibid., p. 488.

25. Ibid., p. 485.

26. Lauterbach, *Wissenschaftspolitik und Ökonomie*, p. 79.

27. *Wissenschaftszentrum für Sozialforschung—FG Wissenschaftsstatistik. Forschungsbericht (Abschlußbericht): Struktur und Funktionsweise der industrieorientierten Forschung an der ADW unter den forschungspolitischen Bedingungen der DDR* (Berlin, 1993), pp. 19–20.

28. Ibid., p. 28.

29. Lauterbach, *Wissenschaftspolitik und Ökonomie*, p. 77.

30. Gruhn and Lauterbach, "Die Organisation der Forschung," p. 128.

31. A. Bauerkämper, B. Ciesla, and J. Roesler, "Wirklich wollen und nicht richtig können. Das Verhältnis von Innovation und Beharrung in der DDR-Wirtschaft," in J. Kocka and M. Sabrow, eds., *Die DDR als Geschichte. Fragen—Hypothesen—Perspektiven* (Berlin: Akademie Verlag, 1994), p. 117.

32. J. Roesler, "Einholen wollen und Aufholen müssen. Zum Innovationsverlauf bei numerischen Steuerungen im Werkzeugmaschinenbau der DDR vor dem Hintergrund der bundesrepublikanischen Entwicklung," in J. Kocka, ed., *Historische DDR-Forschung. Aufsätze und Studien* (Berlin: Akademie Verlag, 1993), pp. 263–285.

33. "Beschluß über die Grundsatzregelung für komplexe Maßnahmen zur weiteren Gestaltung des ökonomischen Systems des Sozialismus in der Planung und Wirtschaftsführung für die Jahre 1969 und 1970," *Gesetzblatt für die DDR*, 1968, Part II, no. 66, pp. 433 ff.

34. *Gesetzblatt für die DDR* 1968, Part II, No. 110.

35. *Vergleich von Bildung und Erziehung*, p. 15.

36. The Ten-Year School was introduced beginning with the academic year 1950–

51. Implementing it took some time; in 1967–68, 77 percent of students continued their schooling after the eighth grade.

37. "Beschluß über die Grundsätze der weiteren Systematisierung des polytechnischen Unterrichts, der schrittweisen Einführung der beruflichen Grundausbildung und der Entwicklung von Spezialschulen und -klassen," *Gesetzblatt für die DDR*, 1963, Part II, no. 65, p. 504.

38. "Beschluß zur Verbesserung und weiteren Entwicklung des Mathematikunterrichts in den allgemeinbildenden polytechnischen Oberschulen der DDR vom 17.12.1962," *Gesetzblatt für die DDR*, 1962, Part II, no. 100, p. 853.

39. L. Froese, R. Haas, and O. Anweiler, eds., *Bildungswettlauf zwischen Ost und West* (Freiburg, Basel, and Vienna: Herder Verlag, 1961).

40. A. Hearnden, *Bildungspolitik in BRD und DDR* (Düsseldorf: Schwann Verlag, 1973), p. 9.

41. Ibid., p. 266.

42. G. Picht, *Die deutsche Bildungskatastrophe* (Olten/Freiburg im Breisgau: Walter Verlag, 1964).

43. A. Hearnden, *Bildungspolitik in BRD und DDR*, p. 269.

44. Ibid., p. 246.

45. "Beschluß über die Grundsätze," p. 505.

46. W. Bergsdorf and U. Göbel, *Bildungs- und Wissenschaftspolitik im geteilten Deutschland* (Munich and Vienna: Olzog Verlag, 1980), p. 111.

47. "Prinzipien zur weiteren Entwicklung der Lehre und Forschung an den Hochschulen der DDR," *Das Hochschulwesen*, 14, no. 2 (1966), supplement.

48. *Protokoll der 4. Hochschulkonferenz* (Berlin: Dietz Verlag, 1967).

49. *Die Weiterführung der 3. Hochschulreform und die Entwicklung des Hochschulwesens bis 1975. Materialien der 16. Sitzung des Staatsrats der DDR*. Schriftenreihe des Staatsrates der DDR, no. 8 (Berlin: Staatsverlag der DDR, 1969).

50. *Das Hochschulwesen* 17, no. 4 (1969), p. 223.

51. G. Lauterbach, *Wissenschaftspolitik und Ökonomie*, p. 196.

52. W. Buchow, "Aktuelle Aspekte und Tendenzen der Hochschulreform in der DDR," *Deutschland Archiv*, 1, no. 3 (1968), p. 240.

53. F. Bolck, "Die Hochschulreform an der FSU Jena," *Das Hochschulwesen*, 16, no. 10 (1968), p. 652.

54. H. Koß, *Ökonomische Probleme der Zusammenarbeit zwischen der sozialistischen Großindustrie und der Hochschule* (Berlin: Verlag der Wirtschaft, 1970).

55. See Chapter 6.

56. R. Landrock, *Die Deutsche Akademie der Wissenschaften zu Berlin 1945 bis 1971—ihre Umwandlung zur sozialistischen Forschungsakademie. Eine Studie zur Wissenschaftspolitik der DDR*, vol. 2. Published by the Institut für Gesellschaft und Wissenschaft (IGW) at the University of Erlangen-Nuremberg (Erlangen: Verlag Deutsche Gesellschaft für zeitgeschichtliche Fragen e.V., 1977), Chapter 5, "Phase 1968–1971," pp. 267–406.

57. *Akademiegedanke und Forschungsorganisation im 20. Jahrhundert. Materialien des Wissenschaftlichen Kolloquiums zum Leibniz-Tag 1994.* Sitzungsberichte der Leibniz-Sozietät, vol. 3, no. 3 (1995).

58. J. Roesler, *Zwischen Plan und Markt: Die Wirtschaftsreform in der DDR zwischen 1963 und 1970* (Berlin: Haufe Verlag, 1991).
59. J. Roesler, "Das Neue Ökonomische System—Dekorations- oder Paradigmenwechsel?" *Forscher- und Diskussionskreis DDR-Geschichte: Zur ddr-geschichte*, 3 (1993), p. 21.
60. Ibid., p. 19.
61. N. Podewin, ". . . Der Bitte des Genossen Walter Ulbricht zu entsprechen.' Hintergründe und Modalitäten eines Führungswechsels," *Forscher- und Diskussionskreis DDR-Geschichte: hefte zur ddr-geschichte*, 33 (1996), pp. 21, 34–44.
62. G. Lauterbach, "Die Organisation der Forschung," p. 1.

3. The Shadow of National Socialism

1. H. Mommsen, "Der lange Schatten der untergehenden Republik. Zur Kontinuität politischer Denkhaltungen von der späten Weimarer zur frühen Bundesrepublik," in K. D. Bracher, M. Funke, and H. A. Jacobsen, eds., *Die Weimarer Republik, 1918–1933* (Bonn: Bundeszentrale für politische Bildung, 1988), pp. 552–586.
2. H. G. Hockerts, "Zeitgeschichte in Deutschland. Begriff, Methoden, Themenfelder," from *Aus Politik und Zeitgeschichte* (supplement to *Das Parlament*), B 29–30/93, July 16, 1993, pp. 3–19.
3. J. Danyel, O. Groehler, and M. Kessler, "Antifaschismus und Verdrängung. Zum Umgang mit der NS-Vergangenheit in der DDR," in J. Kocka and M. Sabrow, eds., *Die DDR als Geschichte* (Berlin: Akademieverlag, 1994), pp. 148–152.
4. R. Siegmund-Schultze, "Dealing with the Political Past of East German mathematics," *Mathematical Intelligencer*, 15, no. 4 (1993): pp. 27–36.
5. Ch. Klessman, "Verflechtung und Abgrenzung. Aspekte der geteilten und zusammengehörigen deutschen Nachkriegsgeschichte," from *Aus Politik und Zeitgeschichte* (supplement to *Das Parlament*), B 29–30/93, July 16, 1993, pp. 30–41.
6. R. Siegmund-Schultze, "Probleme der Wissenschaftspolitik und Wissenschaftsorganisation im NS-Staat," in G. Grau and P. Schneck, eds., *Akademische Karrieren im "Dritten Reich"* (Berlin, 1993), pp. 89–102.
7. AUB, Math. Nat. Fak., Dekanat 1, fol. 136.
8. AUB, Math. Nat. Fak., Dekanat 13, fol. 233.
9. W. J. Mommsen, "Die DDR in der deutschen Geschichte," from "Aus Politik und Zeitgeschichte" (supplement to *Das Parlament*), B 29–30/93, July 16, 1993, pp. 20–29.
10. AUB, personal file of H. Grell (Ministerialakte), fol. 17.
11. H. Mommsen, "Der lange Schatten der untergehenden Republik," p. 575.
12. S. Segal, "Helmut Hasse in 1934," *Historia Mathematica*, 7 (1980), pp. 46–56.
13. HSA, Bestand 307 d., acc. 1967/11, no. 368.
14. AUB: UK H 134.
15. Ibid.

16. Ibid.

17. Ibid.

18. W. Blaschke, *Reden und Reisen eines Geometers* (Berlin: Deutscher Verlag der Wissenschaften, 1957), p. 115.

19. AAW, Akademieleitung 709, Teil 7, fol. 2. *Zeitschrift für Angewandte Mathematik und Mechanik.*

20. Archive of East German Organizations and Parties, located in the Federal Party Archives, Berlin. (Stiftung Archiv der Parteien und Massenorganisationen der DDR im Bundesarchiv, Zentrales Parteiarchiv, Berlin) SAPMO IV 2/904/285, fol. 61.

21. SAPMO IV 2/904/280, fol. 51.

22. Schröter to K. H. Schulmeister, April 3, 1957, AUB, Institut für Mathematische Logik, Allgemeiner Schriftverkehr 1956/57, nr. 212, Laufender Bestand.

23. SAPMO IV 2/904/280, fol. 51.

24. SAPMO IV 2/904/280, fol. 36. See also *Unsere Regierung fördert die Intelligenz. Eine Zusammenstellung der Gesetze und Verordnungen zur Förderung der Angehörigen der Intelligenz,* v. 4.2.49–11.10.52 (Berlin: Kulturbund, 1953).

25. H. Mehrtens, "Angewandte Mathematik und Anwendungen der Mathematik im nationalsozialistischen Deutschland," *Geschichte und Gesellschaft,* 12 (1986), pp. 317–347; R. Siegmund-Schultze, "Zur Sozialgeschichte der Mathematik an der Berliner Universität im Faschismus," *NTM- Schriftenreihe,* 26 nr. 1 (1989) pp. 49–68.

26. H. Mehrtens, "Ludwig Bieberbach und 'Deutsche Mathematik,'" in E. R. Phillips, ed., *Studies in the History of Mathematics* (Washington, D.C.: American Mathematical Society, 1987), pp. 195–241.

27. T. Gnedenko and L. Kaloujnine, "Über den Kampf zwischen Materialismus and Idealismus in der Mathematik," *Wissenschaftliche Zeitschrift der Technischen Hochschule Dresden,* 3, nr. 5 (1953/54), pp. 631–638.

28. G. Schenk, "Zur Logikentwicklung in der DDR," *Modern Logic,* 5 (1995), pp. 248–269.

29. R. Siegmund-Schultze, "Mathematics and Ideology in Fascist Germany," in W. R. Woodward and R. S. Cohen, eds., *World Views and Scientific Discipline Formation* (Dordrecht: Kluwer, 1991), pp. 89–95.

30. SAPMO IV 2/904/285, fol. 68.

31. *Geschichte der Universität Rostock,* vol. 2 (Berlin: Deutscher Verlag der Wissenschaften, 1969), p. 117.

32. Ibid.

33. For more information on the role of the SED's cadre policy in the buildup of the socialist universities, see also the chapter by J. Connelly.

34. SAPMO IV 2/904/285, fol. 62.

35. AUB, Math.-Nat. Dekanat 1, fol. 91/92.

36. "Beschluß des Politbüros des ZK der SED und des Ministerrats der DDR vom 17. Dezember 1962: Zur Verbesserung und weiteren Entwicklung des Mathematikunterrichts in den allgemeinbildenden polytechnischen Oberschulen der DDR," *Mathematik und Physik in der Schule,* 10, no. 2 (1963), pp. 141–150.

37. AUB, UK Sch 831, fol. 80.

38. "Beschluß des Politbüros des ZK der SED und des Ministerrats der DDR vom December 17, 1962," p. 141.
39. Hockerts, "Zeitgeschichte in Deutschland, Begriff, Methoden, Themenfelder," p. 13.
40. Siegmund-Schultze, "Dealing with the Political Past of East German Mathematics," p. 35.
41. AAW, Nachlaß Kurt Schröder, no. 256.
42. AAW, Nachlaß Heinrich Grell.
43. AAW, Akademieleitung 159, Nachtrag IV, fol. 39.
44. R. Siegmund-Schultze, "Zu den ost-westdeutschen mathematischen Beziehungen bis zur Gründung der Mathematischen Gesellschaft der DDR 1962," *Hochschule Ost* 5, nr. 3, 1996, pp. 55–63.
45. SAPMO IV 2/904/285, fol. 99–104.
46. SAPMO IV 2/904/280, fol. 164.
47. Siegmund-Schultze, "Zur Sozialgeschichte der Mathematik an der Berliner Universität im Faschismus."
48. J. Petzold, "Vergleichen, nicht gleichsetzen," in J. Kocka and M. Sabrow, eds., *Die DDR als Geschichte* (Berlin: Akademieverlag, 1994), pp. 101–103.
49. R. Siegmund-Schultze, "The Problem of Anti-Fascist Resistance of 'Apolitical' German Scholars," in M. Renneberg and M. Walker, eds., *Science, Technology, and National Socialism* (Cambridge, England: Cambridge University Press, 1994), pp. 312–323, 408–411.
50. R. Kühnau, "Zur Situation der Mathematik und der Mathematiker in der ehemaligen DDR," *Mitteilungen der Deutschen Mathematiker-Vereinigung*, no. 2 (1992), pp. 57–63, quote p. 59.

4. Espionage and Technology Transfer

1. MfS branch office in Leipzig, Workbook, nr. 4106.

   I would like to thank Bernhard Priesemuth for his help in finding former agents, officers, and other sources of information, and for the many animated discussions on espionage in the GDR. I would also like to thank the staff at the Ministry for State Security Archives (officially *Der Bundesbeauftragter für die Unterlalgen des Staatssicherheitsdienstes der ehemaligen Deutschen Demokratischen Republik* (BStU), especially Heidemarie Beidokat for her tireless help and interest in the project. Finally, I am grateful to former officers for their confidential information.
2. Interview with former SWT officer, December 1994.
3. See, for example, BStU, ZAIG, nr. 3627, "Information from experts in several industry branches in Karl-Marx-City on scientific-technical solutions achieved with the support of the MfS and the achieved result of increased effectivity in combines and companies." For a preliminary article on industrial espionage see Jörg Roesler, "Industrieinnovation und Industriespionage in der DDR: Der Staatssicherheit in der Innovationsgeschichte der DDR," *Deutschland Archiv*, October 1994, pp. 1026–1040.
4. Before the fall of the Wall and the opening of the MfS archives, the most

comprehensive study of the MfS was Karl Wilhelm Fricke, *Die DDR-Staats-sicherheit: Entwicklung, Strukturen, Aktionsfelder* (Cologne, 1982). One of the first major books to have appeared after the fall of the Wall, concentrating on the structure of the MfS, is David Gill and Ulrich Schroeter, *Das Ministerium für Staatssicherheit: Anatomie des Mielke-Imperiums* (Reinbek: Rowohlt, 1991).

5. Markus Wolf interview, July 1995, on polyurethane. Werner Stiller and other officers have also described the polyurethane project as an "MfS project."

6. BStU, BKK 1587, "Information on the Federal Office for the Protection of the Constitution [*Verfassungsschutz*—West Germany's Counter-Intelligence office] Work on Special Problems of the GDR's Economy," p. 2. Information on Schalck's salary is from printed salary lists published by *Die Andere*, 12/1991. For a detailed investigation into the activities of Schalck and KoKo see "Beschlußempfehlung und Bericht des 1. Untersuchungsausschusses nach Artikel 44 des Grundgesetzes: Beschlußempfehlung," *Deutscher Bundestag 12. Wahlperiode. Drucksache 12/7600,* Bonn.

7. Figure of 90 percent reported in *Der Spiegel*, 20/1991, p. 35. Amount given to SWT reported by a confidential source.

8. BStU, Leipzig branch, Workbook nr. 4106, p. 31; Workbook nr. 4133, p. 10.

9. BStU, MfS, HA, XVIII, nr. 563, pp. 137–149. See also MfS, JHS, 20089, research results on the topic "Work with People in and toward the Operational Aera through Line XVIII," Gert Grund and Wolfgang Meinel, Ministry for State Security University, Potsdam, December 20, 1985. The Soviet liaison officers included Budachin, Lubimov, and Gubkin.

10. Rainer O. M. Engberding, *Spionageziel Wirtschaft: Technologie zum Nulltarif* (Düsseldorf, 1993), pp. 98–100 on Schalck's defection. Ingrid Köppe, a member of the Schalck committee, wrote a 163-page report deviating from the official published investigation. It details activities such as the BND's knowledge of illegal activities of Western firms. This so-called deviating report, which was made secret, was described in more than fifty newspaper articles in May and June 1994. See, for example, Wolfgang Hoffmann, "Strohmänner, Hintermänner, V-Männer: Schalck-Ausschuß: Der Western hat jahrzehntelang im KoKo-Imperium mitgemischt," *Die Zeit* 3 (June 1994).

11. "Beschlußempfehlung und Bericht des 1. Untersuchungsausschusses nach Artikel 44 des Grundgesetzes," Deutscher Bundestag, 12, Wahlperiode, Drucksache, 12/7600, 1994. (Hereafter Untersuchungsausschuß.)

12. Ibid., 126.

13. This information has been gleaned from hundreds of Department XVIII files. As this book was going to press, a useful brochure on the structure, organization, and history of Department XVIII was produced by the Gauck Office as part of the MfS handbook: Maria Haendcke-Hoppe-Arndt, "Die Hauptabteilung XVIII: Volkswirtschaft," *MfS-Handbuch*, Teil III/10, 1997. See p. 4 for total IM estimate; see Untersuchungsausschuß, p. 146, for figure of 15.

14. For the Thailand route see "Beschlußempfehlung und ergänzender Bericht des 1. Untersuchungsausschusses nach Artikel 44 des Grundgesetzes," Deutscher Bundestag, 12, Wahlperiode, Drucksache 12/8595, pp. 18–19 (a supplementary report to the first investigative committee).

15. Untersuchungsausschuß, pp. 112–113, 253.
16. This discussion is based on the few HVA, SWT documents that survive, e.g., BStU, MfS, HVA, nr. 678.
17. Ibid. For lists see BStU, HVA, SWT, nr. 678, for sample applications.
18. "Die Milliarden-Geschäfte der SED," *Frankfurter Allgemeine Zeitung*, June 22, 1994.
19. Untersuchungsausschuß, pp. 133 134, 261.
20. Ibid.
21. See Peter Richter and Klaus Rösler, *Wolfs West-Spione: Ein Insider-Report* (Berlin: Elephanten Press, 1992), pp. 52–53, where Weiberg is described as an engineer.
22. BStU cadre card on Heinrich Weiberg.
23. Rainer O. M. Engberding, *Spionageziel Wirtschaft: Technologie zum Nulltarif* (Düsseldorf: VDI Verlag, 1993), p. 94; "Soviet Acquisition of Militarily Significant Western Technology." White Paper, 1982, p. 16.
24. KGB to Nikita Krushchev, "Report for 1960," February 14, 1961, in CC CPSU Secretariat's "special dossier," cited in Vladislav M. Zubok, "Spy vs. Spy: The KGB vs. the CIA, 1960–1962," *Cold War International History Project Bulletin* (Fall 1994), pp. 22–33, here p. 24.
25. Ibid.
26. Philip Hanson, "Soviet Industrial Espionage: Some New Information" (London: The Royal Institute of International Affairs, 1987), Discussion Paper no. 1, pp. 1–39, here p. 8. VPK stands for *Voenno-promyshlennaya komissiya*.
27. Ibid., pp. 29 and 33.
28. Werner Stiller, interview, September 3, 1993. Stiller claims that Rompe was an IM and that he saw his file; another former officer says Rompe was a "contact person" and that someone of his stature did not need to be an IM. See also Werner Stiller, *Im Zentrum der Spionage* (Mainz: Hase und Koehler Verlag, 1986), translated as *Beyond the Wall: Memoirs of an East and West German Spy* (Washington: Brassey's, 1992).
29. SAPMO, personnel file on Robert Rompe in the Central Party Commission.
30. Werner Stiller, interview at author's workshop in Berlin on science in the German Democratic Republic, September 3, 1993. See also Stiller, *Beyond the Wall*.
31. BStU, policy documents, commands.
32. BStU, policy documents, "Instructions on the Installation of Directors for Cadre and Security Questions in the Office for Technology and Atomic Research and Technology . . ." February 28, 1956.
33. BStU, policy documents, commands, April 2, 1962, order nr. 172/62.
34. BStU, order nr. 23/69, 24, June 1969, on the coordination of measures for the procurement of prototypes of important military technology, signed by Erich Mielke.
35. BStU policy documents, commands, order nr. 23/69, June 24, 1969, and order nr. 29/69, August 9, 1969.
36. BStU, policy documents, commands, Council of Ministers, MfS, Office of the Director, February 12 and July 9, 1971. Other information was pieced together from the cadre cards of SWT staff.

37. Since I cannot assume that I have a complete card file, I am hesitant to pinpoint precisely how many SWT officers there were by the end of the GDR. My data includes officers who had been with the MfS since the 1950s and had died or retired by 1989, but most of the data are on officers whose last appointment was in an SWT department.

38. Council of Ministers meeting, April 21, 1976, Paul Bilke relieved of duties as head of Department V in order to be freed for "other important tasks" (he became an officer on a special mission). Harry Herrmann was appointed director in July 1976.

39. I have received 384 cadre cards from the Gauck-Behörde.

40. Interview with Werner Stiller, March 20, 1994, pp. 62–63.

41. BStU, cadre card on Horst Vogel and interview, July 1995. See Stiller interview (March 20, 1994) for defector assessment.

42. Interview with Stiller, March 1994, during which he described the cooperation with the BND and the real time scale (from April 1978 to January 1979, the time of his defection). See also my review of the English translation of the book: "Spy Secrets," *Science,* December 1994. This review was written before I heard the full story of the real cooperation with the BND.

43. BStU, Stiller files, XV/2277/79.

44. Interview with intelligence officer, March 1994 (HV); BStU, policy documents, Command nr. 2/87, March 12, 1987, "On the Coordination of Tasks and Measures in order to Acquire Embargoed Goods from Nonsocialist Countries and West Berlin."

45. Interview with former officer, December 1994.

46. *Protokoll der Verhandlungen der 3. Parteikonferenz der sozialistischen Einheitspartei Deutschlands, 24. März bis 30. März 1956* (Berlin: Dietz Verlag, 1956).

47. BStU, minister's orders, instruction nr. 14/56 on the creation of a working group for scientific-technical evaluation, June 8, 1956.

48. BStU, "Guidelines for the Working Group for Scientific-Technical Evaluation," December 19, 1956, signed by Major Last; interview with officer in evaluation unit, December 1994.

49. Ibid. p. 3.

50. BStU, MfS, archive nr. 8810/84, fol. 207, "Confirmation" from H VA/SWT/Abt. V about a scientist who could be used as an evaluator.

51. Ibid., pp. 3, 4.

52. Ibid., p. 6. For the figure of 150,000 marks see Stiller, *Beyond the Wall,* p. 126. For the figure of 32 percent see Workbook nr. 4106, Gerhard Idaszek.

53. Peter Siebenmorgen, *"Staatssicherheit" der DDR: der Westen im Fadenkreuz der Stasi* (Bonn: Bouvier Verlag, 1993), p. 190.

54. Linda Melvern, Nick Anning, David Hebditch, *Techno-Bandits,* (Boston: Houghton-Mifflin Company, 1984), pp. 142–143.

55. BStU, ZAIG nr. 3701, Uhde-Hoechst, Komplex Anlage/Polyester.

56. Hans Pretterebner, *Der Fall Lucona: Ost-Spionage, Korruption und Mord im Dunstkreis der Regierunsspitze* (Vienna: Knaur, 1989).

57. Ibid., p. 104.

58. Ibid., p. 113.

59. Ibid.
60. Ibid.
61. Stiller, *Beyond the Wall*, p. 165.
62. See, for example, MfS, HA XVIII, nrs. 638, 4729, 4715. The Politburo Resolution was passed on May 24, 1983.
63. MfS, HA XVIII, nr. 7827; order nr. 9/84, April 13, 1984, "Acquisition of Information and Prototypes of New Military Technology from Nonsocialist Countries"; order nr. 11/84, May 30, 1984.
64. See MfS, HA XVIII, Nrs. 5590, 3007.
65. For information on the MfS's reaction to the Politburo Resolution see BStU, MfS, HA XVIII, Nrs. 638 and 4729. It should be noted that the MfS was just one of eleven GDR ministries involved in implementing this program. For minutes of Honecker's visit to Carl Zeiss Jena and a list of technology provided to CZ Jena from the MfS, see BStU, MfS, XVIII, nr. 9505.
66. See BStU, MfS, HVA, SWT, nr. 686 (1977–1982), nr. 687 (1982–1986), and nr. 688 (1986–1990) for lists of money spent and orders.
67. For an application for a laser system, see BStU, MfS, HVA, nr. 678, fol. 142–143 (applications for acquiring embargoed goods).
68. MfS, HA XVIII, nr. 9505, fols. 3–5, 7–8.
69. BStU, Leipzig branch office, Workbook nr. 4106, p. 1.
70. BStU, cadre file, Werner Stiller, p. 55.
71. Newspaper articles appeared in *Frankfurter Allgemeine Zeitung, Die Welt, Frankfurter Rundschau, Stuttgarter Zeitung,* and even the sensationalist *Bild am Sonntag* and *Abendpost,* among others. Werner Stiller reported on Gottfried's 1,500-page file in his book *Im Zentrum Der Spionage* (see pp. 79–93 of the German edition and pp. 48–55 of *Beyond the Wall*). See also author's interview with Harold Gottfried, July 1995.
72. Ibid. see the newspaper articles, especially "Spionageprozeß gegen Diplomingenieur," *Frankfurter Allgemeine Zeitung,* October 17, 1969; also author's interview with Harold Gottfried, July 1995.
73. Confidential source. Köhler's conviction was also reported in the newspaper.
74. Confidential source.
75. Confidential source.
76. BStU, Department XIII. For more information on extant documents from Stiller's Department XIII see the German edition of *Science under Socialism*.
77. See, for example, Jay Tuck, *High-Tech Espionage* (New York: St. Martin's Press, 1986), who uses the newspaper articles as sources.
78. Ibid.
79. Ibid.
80. Ibid., pp. 139–140.
81. Stiller, *Beyond the Wall,* and interview, March 1994.
82. Confidential source and newspaper article on sentence.
83. Tuck, *High-Tech Espionage.*
84. SED archive, J IV 2/2, various files from 1964 to 1989 in the Central Committee's minutes. All references to computers have been culled from the daily minutes.

85. Gauck-Behörde, ZAIG, nr. 2666.

86. Gauck-Behörde, MfS, XVIII/8, nr. 4705, fol. 53.

87. This is Wolfgang Biermann's figure. See Wolfgang Biermann, "Mikroelektronik in der Volkswirtschaft der DDR." *Die Einheit* (1989), pp. 27–32, here p. 28.

88. Gauck-Behörde, MfS, XVIII/8, nr. 4705, February 20, 1986, Supplementary Information, p. 28.

89. For characterization of the billion-mark grave, see "Hohe Gewinne, kleines Risiko." *Der Spiegel*, 2/1996, pp. 74–82.

90. Gauck-Behörde, MfS, Dept. XVIII/8, nr. 4705, from a report of February 21, 1986, on the preconditions necessary for the GDR to produce a 1-megabit chip (p. 19 of file, p. 2 of report).

91. Ibid. Discussion of Biermann's letter, fol. 43–45, "Information about Letter of the Workers of the Combine Carl Zeiss Jena to the General Secretary of the Central Committee of the SED," "Jeder an seinem Platz das Beste für unseren sozialistischen Friedenstaat," *Neues Deutschland*, January 7, 1986, vol. 41, front page.

92. Ibid., fols. 18–20, "Information on the Preconditions Necessary in the GDR for the Development and Production of the 1-Megabit Chip," February 21, 1986; fols. 30–32, same title, different text, dated February 20, 1986; fols. 21–29, "Point of View on Accelerating the Technological Level for the 1-Megabit Chip," report signed by Tautenhahn, Meier, Biermann, and Nendel.

93. Ibid., pp. 82–93 of Gauck-Behörde pagination; Schalck to Nendel, December 10, 1985: "Suggestions for Further Action to Import Know-How and Equipment for a Factory for 256-Kilobit Chips from Nonsocialist Economies."

94. Ibid., pp. 3–5 of archive pagination. "Information on the State of and Suggestions for the Further Development of the Project 256-Kilobit Chip," March 6, 1986; "Hohe Gewinne, kleines Risiko." *Der Spiegel*, 2/1996, pp. 74–82.

95. Ronneberger to Schalck and Nendel, February 10, 1988. At the end of this letter Ronneberger noted to Schalck that Toshiba/Mitsui had returned the 7.8 million dollars. Letter reprinted in Engberding, *Spionageziel Wirtschaft*, pp. 101–102.

96. "The Toshiba Scandal Has Exporters Running for Cover," *Business Week*, July 20, 1987, pp. 86–87. This is just one of many articles on the topic. See also "Legislation to Prohibit the Importation of Products Made by Toshiba Corp. and Kongsberg Vaapenfabrik Co.," *Hearing before the Subcommittee on Ways and Means, House of Representatives* (Washington: U.S. Government Printing Office, July 14, 1987).

97. "Republicans Say Toshiba Broke COCOM Rules Often," *MDN* [?], November 12, 1987. I found this clipping in the Stasi file BKK 1172, "Pentagon: Illegale Geschäfte Toshiba—DDR nicht bewiesen," *ADN-Information* in BKK 1172, p. 13.

98. Gauck-Behörde, BKK 1172, "Note on the meeting with Toshiba on November 26, 1987," November 26, 1987.

99. Ronneberger to Schalck and Nendel, February 10, 1988.

100. Ibid.

101. Werner Jarowinski provided the number of 256-kilobit chips produced, SED

archive, IV 2/1/709, quoted in Charles Maier, "Vom Plan zur Pleite. Der Verfall des Sozialismus in Deutschland," in Jürgen Kocka and Martin Sabrow, *Die DDR als Geschichte: Fragen—Hypotheses—Perspektiven* (Berlin: Akademie Verlag, 1994), p. 113.

102. "Toshiba beginnt mit Megabit-Chips," *Frankfurter Allgemeine Zeitung,* January 7, 1986, and "Hier kommt der 1-Megabit DRAM: Toshiba bereitet Massenproduktuion des ersten 1-Megabit DRAM vor," *Toshiba,* October 1985. Both newspaper clippings were found in MfS XVIII/8, nr. 4705.

103. Gauck-Behörde, MfS, XVIII/8, nr. 4705, February 20, 1986, Supplementary Information, p. 28.

104. Biermann to Honecker, September 12, 1988, party archive, J IV 2/2A/3155.

105. Party archives, Vorl. SED 41668, Büro Honecker, notes on the meeting between Honecker and Gorbachev in Moscow, September 28–29, 1988.

106. Interview with former MfS officer in SWT, summer 1995.

107. "Volker Kempe," in Jochen Cerny, ed., *Wer war wer-DDR: Ein biographisches Lexikon* (Berlin: Ch. Links Verlag, 1992), p. 225.

108. Gauck-Behörde, MfS, HA XVIII/8, nr. 4706, pp. 21–23, information on meeting between Kempe and Mittag.

109. Ibid., pp. 38–40. The Leader's Group for Key Technologies was made up of the department head, Central Committee Comrade Tautenhahn; the state secretary for foreign trade, Schalck; the secretary of the State Plan Commission, Wenzel; and the state secretary for the Ministry of Electronics, Nendel.

110. Ibid.

111. I will address this issue in a chapter in the book I am writing on science and the MfS.

112. Compare John Harris, *Industrial Espionage and Technology Transfer: Britain and France in the Eighteenth Century* (Hampshire: Ashgate, 1998), and John J. Fialka, *War by Other Means: Economic Espionage in America* (New York. W. W. Norton & Co, 1997).

113. BStU, Außenstelle Gera, Jena, ZMA, 003003, fol. 120, p. 4 of a report signed by "source," 1982.

114. See BStU, HVA, nr. 678–688, application and finances, 1986–1990.

115. See, for example, BStU, HVA, Auftragsbücher, where scientists ask for disks.

116. BStU, MfS, HA XVIII, nr. 9505, 7.

5. Foundations of Diversity

1. I would like to thank the Fulbright Commission, the Krupp Foundation at the Center for European Studies at Harvard University, and the Sheldon Foundation at Harvard University for assistance in the research for this article. Thanks go also to the International Research and Exchanges Board (IREX), which provided assistance with funds from the NEH, the USIA, and the State Department, which administers the Title VIII Program. Finally, I would like to express my deep gratitude to Fiona Grigg for her helpful criticism.

L. P. Morris, *Eastern Europe since 1945* (London: Heinemann Educational Books, 1984), p. 51. Jürgen Kocka has coined the word "durchherrscht" ("ruled pervasively") to describe East German society. See Jürgen Kocka, "Eine durchherrschte Gesellschaft," in Hartmut Kaelble, Jürgen Kocka, and Hartmut Zwahr, eds., *Sozialgeschichte der DDR* (Stuttgart: Klett-Cotta, 1994), pp. 547–553. See also George Schöpflin, *Politics in Eastern Europe, 1945–1992* (Oxford: Blackwell, 1993), esp. p. 199.

2. Anke Huschner, "Der 17. Juni 1953 an Universitäten und Hochschulen der DDR," *Beiträge zur Geschichte der Arbeiterbewegung* 5 (1991), pp. 681–692; Malte Sieber and Ronald Freytag, *Kinder des Systems. DDR-Studenten vor, im und nach dem Herbst '89* (Berlin: Morgenbuch, 1993).

3. East German sociologists, economists, and historians were renowned even in the late 1980s for their adherence to hardline Marxist-Leninist positions at international gatherings, far surpassing their East European colleagues. Within East Germany, only the police and the military were more implicated in the work of the Stasi than were universities. Rainer Eckert, "Die Humboldt-Universität im Netz des MfS," in Dieter Voigt and Lothar Mertens, eds., *DDR-Wissenschaft im Zwiespalt zwischen Forschung und Staatssicherheit* (Berlin: Duncker u. Humblot, 1995), pp. 169–186. See also Chapter 4 in this book.

4. A leading expert on Leninist politics has characterized pre-1989 East European governments as "geographically contiguous replica regimes." See Ken Jowitt, *New World Disorder: The Leninist Extinction* (Berkeley: University of California Press, 1993), p. 176.

5. For the view that foundations for the higher education regimes of mature socialism were laid in the Stalinist period, see the leading Polish expert Piotr Hübner's *Polityka naukowa w Polsce w latach 1944–1953 geneza systemu* (Wrocław: Zakład Narodowy im. Ossolińskich, 1992).

6. For a handy summary of the components of a Soviet-style system of higher education, see Andrej P. Nikitin, "Die sowjetische Militäradministration und die Sowjetisierung des Bildungssystems in Ostdeutschland, 1945–1949," *Bildung und Erziehung* 45:4 (1992), pp. 406–407. See also Marianne and Egon Erwin Müller, *". . . stürmt die Festung Wissenschaft!" Die Sowjetisierung der mitteldeutschen Universitäten seit 1945* (Berlin: Colloquium Verlag, 1953), esp. Part III.

7. "Stenographische Niederschrift des Referats des Genossen Anton Ackermann auf der Arbeitstagung über die Frage der Auswahl und Zulassung zum Hochschulstudium; Freitag den 6. Mai 1949," p. 15, in SAPMO-BA, IV 2/9, 04/464 (unnumbered).

8. "Vládní nařízení ze dne 27. června 1950 o některých změnech v orgánizaci vysokých škol," *Sbírka Zákonů republiky Československé, C,"* 81/1950, p. 201. It was understood that the closing of the legal faculty had freed resources for technical education. Letter from Central Committee to regional KSC secretariats of June 24, 1950, in Archiv ÚV KSČ, f. 19/7, a.j. 104/55.

9. "Zpráva pro Pana Předsedu vlády o osnově vládního nařízení o některých změnech v orgánizaci vysokých škol," June 30, 1950; "Důvodová zpráva," June 22, 1950, in Státní Ústřední Archiv (Prague), ÚPV 833/812/30/10.

10. "Vládní nařžení ze dne 2. října 1951 o organizačních změnách na vysokých školách," *Sbírka Zákonů republiky Československé,* 80/1951, pp. 285–286; "Vládní nařízení ze dne 8. července 1952 o některých změnech v orgánizaci

vysokých škol," Ibid., 30/1952, pp. 193–194; "Vládní nařízení ze dne 19. srpna 1952 o dalších změnech v orgánizaci vysokých škol," Ibid., 40/1952, pp. 211–212.

11. The initial plan had been to set up this school in Berlin, but Ulbricht objected owing to the proximity of Western secret services. Letter to Paul Wandel, November 1, 1948, BAP, R2/1478/252.

12. BAP, E1/17038/113.

13. BAP, E1/9107/7.

14. Bolesław Krasiewicz, *Odbudowa szkolnictwa wyższego w Polsce Ludowej w latach, 1944–1948* (Wrocław: Zakład Narodowy im. Ossolińskich, 1976), pp. 338–339.

15. Hübner, *Polityka naukowa*, pp. 600–603.

16. On the Soviets' perception that they had to "write the first study plans" for the East German universities, see Norman M. Naimark, *The Russians in Germany: A History of the Soviet Zone of Occupation, 1945–1949* (Cambridge, Mass.: Harvard University Press, 1995), p. 444.

17. Müller, *". . . stürmt die Festung,"* pp. 231–238; quote p. 231.

18. The ten-month year of study had been announced by Walter Ulbricht at the SED's Third Party Congress in July 1950.

19. Ernst Richert, *"Sozialistische Universität": Die Hochschulpolitik der SED* (Berlin: Colloquium Verlag, 1967), pp. 84–86.

20. Rudolf Urban, *Die Organisation der Wissenschaft in der Tschechoslowakei* (Marburg/Lahn, 1958), pp. 198, 206.

21. Müller, *". . . stürmt die Festung,"* pp. 299–301.

22. The natural and technical sciences were far more successful in protecting scholarly standards than were the social sciences and humanities. From the early postwar years, differing standards applied in graduate admissions. On July 30, 1947, a DVV staff member made the following note: "The candidates for medical and mathematical-technical subjects must in every case have sufficient academic [*fachlich*] preparation. Besides this we need to demand that they have a positive political outlook [that is, that they support the state]. For candidates in social scientific faculties, political dependability is to be the primary condition. In these fields candidates can be accepted who have not had the prescribed course of development, as long as they can prove that they have established the prerequisites for successful training in private studies. BAP, R2/1447/196–197. During the 1950s, the humanities and social sciences in Leipzig maintained high standards, however. See the discussion in Walter Markov, *Zwiesprache mit dem Jahrhundert* (Cologne: Volksblatt, 1990), pp. 179–218. Emigration and death gradually eroded this milieu in the following decades.

23. Státní Ústřední Archiv (Prague), ÚPV K2474 12/3/38.5; ÚV KSČ, f. 19/7 a.j. 284/11.

24. For a record of appointments in the late 1940s, see Státní Ústřední Archiv (Prague), ÚPV 514–516.

25. See Naimark, *The Russians*, pp. 440–452; Nikitin, "Die sowjetische Militäradministration," pp. 405–416; Alexandr Haritonow, *Sowjetische Hochschulpolitik in Sachsen (1945–49)* (Cologne: Böhlau, 1995).

26. BAP, R2/936/38; 934/137.

27. For the practice of antifascism in early postwar East Germany, see John Connelly, "East German Higher Education Policies and Student Resistance, 1945–1948," *Central European History* 28:3 (1995).

28. For records of visits of Soviet delegations in the early 1950s, see SAPMO-BA, J IV 2/3/160, IV 2/2/134; Nachlass 182 (Ulbricht), 934/109; Urban, *Die Organisation*, p. 195; Státní Ústřední Archiv (Prague), ÚV KSČ, f. 19/7, a.j. 275/51–60, 275, 279, 283; AAN KCPZPR 237/XVI/10/77–78, 237/XVI/184/129–130, 237/XVI/190/76–79; MSW 24/219; communications by rector of January 31, 1951, June 27, 1950, and June 28, 1950, in Archiwum Uniwersytetu Jagiellońskiego (AUJ), WP III 202.

29. Státní Ústřední Archiv (Prague), ÚV KSČ, f. 19/7, a.j. 272/109–110. In February 1947 Ferdinand Sauerbruch, the renowned surgeon, wrote to Paul Wandel after the dismissal of the dean of the medical faculty Else Knake: "I have on several occasions spoken with leading Russian professors, and reached the conclusion that they do not expect us to share their opinion in every issue . . ." BAP, R2/1142/315–316.

30. For a discussion of the *rabfaky*, see Sheila Fitzpatrick, *Education and Social Mobility in the Soviet Union, 1921–1934* (Cambridge: Cambridge University Press, 1979).

31. James C. McClelland, "Proletarianizing the Student Body: The Soviet Experience during the New Economic Policy," *Past and Present* 80 (August 1978).

32. In 1935–36, 9.9 percent of freshmen at Polish universities had worker backgrounds, and 11.7 percent had peasant backgrounds. "Skład społeczny studentów I roku szkoł wyższych w/g wydziałów w porównaniu ze stanem przedwojennym 1935/36," AAN MO/2879 (unnumbered). Only 7 percent of Czech University students before 1948 were from the working class. In 1948, manual laborers constituted 40 percent of society; in 1947, 3 percent of Brno's students were of the working class, and 11 percent of agricultural background. *Dějiny University v Brně* (Brno: Universita J. E. Purkyně, 1969), p. 269. In 1932, 3 percent of Germany's students were from the working class, and 2.2 percent of peasant background. "Zehn-Jahresstatistik des Hochschulbesuchs 1943." Zentralarchiv des FDGB, Bundesvorstand, 11–785, cited in Hans-Hendrik Kasper, "Der Kampf der SED um die Heranbildung einer Intelligenz aus der Arbeiterklasse und der werktätigen Bauernschaft über die Vorstudienanstalten an den Universitäten und Hochschulen der sowjetischen Besatzungszone Deutschlands (1945/46 bis 1949)," unpublished Ph.D. dissertation (Freiberg, 1979), p. 269.

33. "Grundlegende Hinweise über die Zulassung zum Studium an Universitäten und Hochschulen vom 8. Dezember 1945," in Herbert Stallmann, *Hochschulzugang in der SBZ/DDR: 1945–1959* (Sankt Augustin: Richarz, 1980), pp. 426–431.

34. See the remarks of Walter Markov in "Vierte Tagung des zentralen Hochschulausschusses der SED am 7. und 8. Februar 1948," p. 250; SAPMO-BA, ZPA, IV 2/9.04/6 (unnumbered).

35. See the comments of Party General Secretary Władysław Gomułka at the first Polish Workers' Party Congress of late 1945. Władysław Gomułka, *Ku nowej*

Polsce sprawozdani: polityczne i przemówienia wygłoszone na I Zjeździe PPR (Łódź: Spółdzielnia Wydawnicza "Książka," 1946), p. 139.

36. In the early 1950s entering classes of ABF students constituted about 40 percent of the college freshman populations, before they decreased to one-quarter in 1954. In Czechoslovakia they varied between 4 and 6 percent. *Statistická ročenka Republiky Československé 1957* (Prague: Státní nakl. technické literatury, 1957), p. 214; *Statistisches Jahrbuch der Deutschen Demokratischen Republik 1960/61* (Berlin: VEB Deutscher Zentralverlag, 1961), pp. 132–133.

37. The peak of 57.73 percent was reached in 1958 in East Germany. Thereafter, the figure stabilized at about 38 percent of the student body. Stallmann, *Hochschulzugang*, pp. 305–307; Marianne Usko, *Hochschulen in der DDR* (Berlin: Holzapfel, 1974), p. 32. In January 1956, 30 percent of Czech students came from worker background, and 12 percent from "worker-peasant" backgrounds. The remainder, the "overwhelming majority," came from the "working intelligentsia." Státní Ústřední Archiv (Prague), ÚV KSČ, f. 19/7, a.j. 284/5. In 1951–52, worker-peasant students constituted 64 percent of the freshman class in Poland (AAN MSW 17/91). Yet by 1955, that figure had decreased to 56 percent. Thereafter the number of worker-peasant students in Polish higher education continued to drop. Stanisław Ciesielski, "Rekrutacja na studia dzienne w Uniwersytecie Wrocławskim w latach 1945–1985," *Studia i materiały* z dziejów Uniwersytetu Wrocławskiego, vol. 2 (Wrocław: Wydawnictwo Uniwersytetu Wrocławskiego, 1993), p. 168.

38. Telegram from Robert Rompe of February 3, 1946, Universitätsarchiv Rostock, Rektorat/III/A/1/1/1a.

39. BAP, R2/1060/21.

40. Ralph Jessen, "Professoren im Sozialismus. Aspekte des Strukturwandels der Hochschullehrerschaft in der Ulbricht-Ära," in Kaelble, Kocka, and Zwahr, eds., *Sozialgeschichte der DDR*, p. 226.

41. Helmut Klein, ed., *Humboldt-Universität zu Berlin, Überblick 1810–1985* (Berlin: VEB Deutscher Verlag der Wissenschaften, 1985), pp. 94, 96.

42. The percentages of NSDAP members stood in inverse relationship to the percentages of SED members: 14.5 percent of medical professors belonged to the SED, and 73.8 percent of economics, law, and Marxism-Leninism professors were SED members. Jessen, "Professoren," p. 241.

43. Robert Havemann, *Frage-Antworten-Fragen: Aus der Biographie eines deutschen Marxisten* (Munich: R. Piper, 1970), p. 85. In the early 1960s the rector, the state secretary of higher education, and the president of the Academy of Sciences who had ostracized Havemann were themselves all former NSDAP members.

44. Piotr Hübner, *Nauka polska po II wojnie światowej—idee i instytucje* (Warsaw: Centralny ośrodek metodyczny studiów nauk politycznych, 1987), p. 174.

45. See the correspondence in SAPMO-BA, ZPA NL 90/559/142–168.

46. The sociologist Zygmunt Bauman maintains that one cannot understand the political culture of East European socialist regimes without understanding the new elite's "sense of achievement and personal indebtedness to the victorious

revolution." Zygmunt Bauman, "Social Dissent in the East European Political System," in Bernard L. Faber, ed., *The Social Structure of Eastern Europe* (New York: Praeger, 1976), p. 129.

47. Ralph Jessen, "Zur Sozialgeschichte der ostdeutschen Gelehrtenschaft (1945–1970)," in Martin Sabrow and Peter T. Walter, eds., *Historische Forschung und sozialistische Diktatur. Beiträge zur Geschichtswissenschaft der DDR* (Leipzig: Leipziger Universitätsverlag, 1995), pp. 121–143.

48. The sociologist Walter D. Connor writes that "in the socialist world, a greater bureaucratization operates in combination with the abolition of concentrations of private, inheritable wealth to give education an even greater significance in the process of status attainment." *Socialism, Politics, and Equality: Hierarchy and Change in Eastern Europe and the USSR* (New York: Columbia University Press, 1979), p. 133.

49. Jeffrey Kopstein, *The Politics of Economic Decline in East Germany, 1945–1989* (Chapel Hill: The University of North Carolina Press, 1997), esp. chapters 2–3.

6. From German Academy of Sciences to Socialist Research Academy

1. The research for this chapter was supported by funds from the Stifterverband für die Deutsche Wissenschaft. I would like to thank Kristie Macrakis and Dieter Hoffmann for their intensive work on this chapter.

2. See P. Nötzoldt, "Wissenschaft in Berlin—Anmerkungen zum ersten Nach-kriegsjahr 1945/46," *Potsdamer Bulletin für Zeithistorische Studien* 5 (1995).

3. C. Grau, W. Schlicker, and L. Zeil, *Die Berliner Akademie der Wissenschaften in der Zeit des Imperialismus,* parts 2 and 3 (Berlin: Akademieverlag, 1975 and 1979); *Jahrbuch der PAW* (1939): 119ff.; "Denkschrift von 1929," *AAW* II: Ia, vol. 12, 85.

4. Johannes Stroux, who joined in 1937, was president from 1945–1951. The academy leadership included also the two class secretaries, Ludwig Diels for the class for mathematics and natural sciences, and Fritz Hartung for the class of philosophy and history. Helmut Scheel, appointed director of the PAS in 1939, remained in his position for the time being.

5. BAP, R-2 1388, Bl. 7.

6. AAW, P 1/0, p. 14 f. and p. 48ff., as well as Bestand Akademieleitung, no. 660, p. 43.

7. AAW, Bestand Akademieleitung, no. 661, p. 153.

8. Hans Kienle (member since 1946), *Deutsche Akademie der Wissenschaften zu Berlin 1946–1956* (Berlin: Akadmieverlag, 1956), p. 27.

9. AAW, P 1/0, p. 32ff., and Bestand AL, no. 660, p. 3.

10. See Wolfgang Leonhard, *Die Revolution entläßt ihre Kinder* (Leipzig, 1990), p. 405, and Otto Winzer, "Protokoll der Sitzung des Sekretariats des ZK der KPD vom 28.9.1945," SAPMO I 2/540, p. 46.

11. That is the unanimous opinion Paul Wandel and Robert Rompe expressed to the author in June 1990. See also the posthumous papers of Anton Ackermann, SAPMO NL 109/55, p. 300.

12. Appointment on July 5, 1945, by the municipal authorities. Archiv zur Geschichte der MPG, II. Abt. I A 9, p. 87ff.
13. BAP, R-2 1428, pp. 19–22. See Kristie Macrakis, *Surviving the Swastika: Scientific Research in Nazi Germany* (New York: Oxford University Press, 1993), p. 189ff.
14. AAW, Bestand Akademieleitung, no. 660, p. 156ff.
15. Information given to the author by P. Nikitin on September 16, 1992. Nikitin worked in the Soviet Military Administration's Section for People's Education, and he prepared the reopening of the academy together with the German Central Administration for People's Education.
16. Order no. 187 by the head of the Soviet Military Administration and the commander-in-chief of the Soviet occupying forces in Germany to reopen the academy (dated July 1, 1946).
17. Budgeted funds in East German marks were 905,000 (1946), 4,104,00 (1947), 6,476,000 (1948), and 8,416,000 (1949); in addition, from 1946 to 1950, investment funds totaled 14,298,000. Figures are from *Deutsche Akademie der Wissenschaften zu Berlin 1946–1956*, p. 65, and *Jahrbuch der DAW 1946–1949*, p. 53.
18. Iwan Bejdin (former staff member in the SMAG), Minutes of Colloquium for University and Science Policy of the SMAG, 31.8.–5.9.1992 in Gosen, 5th day, p. 63.
19. AAW, Bestand Akademieleitung, no. 547, and *Deutsche Akademie der Wissenschaften zu Berlin 1946–1956*, p. 59ff.
20. Discussion between the executive committee of the academy and the president of the Central Administration on March 12, 1948, AAW, P2/1, p. 33.
21. Ibid.
22. Josef Naas on March 8, 1949, AAW, Bestand Akademieleitung, no. 662.
23. Draft for the plenary session of the German Economic Commission on March 30–31, 1949, and *Zentralverordnungsblatt 1949*, part 1, no. 28.
24. Those duties included, for example, "providing scientific help to the state-owned industry in the Eastern Zone." Josef Naas, "Wer wird in die Akademie aufgenommen?," June 15, 1949, AAW, Bestand Akademieleitung, no. 662.
25. See BBAP, R-2 1895, p. 5ff.
26. Walter Friedrich (member since 1949), medical physics, president 1951–1955.
27. See the consultation on November 28, 1952, between the minister president and the academy leadership and representatives of the Central Office for Research and Technology. In M. Heinemann, *Dokumentenband der Konferenz 'Hochschul- und Wissenschaftspolitk der SMAD'* (1992), pp. 201–210.
28. Deutsche Bauakademie, December 1950, Deutsche Akademie der Landwirtschaftswissenschaften, January 1951.
29. Gesetzblatt der DDR. nr. 28, 23.3.1950.
30. H. Roloff, "Aufgaben und Ziele," in *Nacht-Express*, July 10, 1950.
31. Resolution of the Ministry for People's Education, February 27, 1959, JB, 1950–51, p. 71.
32. Minutes of plenum meeting, April 19, 1951.

33. Decision of seventh Meeting of the Central Committee of the SED, October 18–20, 1951, in *Neues Deutschland* 15, November 1951.
34. SAPMO IV 2/9.04/372, p. 41. Resolution of the Politburo on Measures to Improve the Work of the Academy, January 1953.
35. AAW, W. Steinitz Papers, nr. 71, SAPMO IV 2/9.04/377, folio page 2 (on the fight).
36. SAPMO IV 2/9.04/369, folio page 41ff., minutes of meeting.
37. SAPMO IV 2/9.04/372, pp. 37–48.
38. Resolution of the meeting of the executive committee of the Academy on November 27, 1952, item 3.
39. See the protocol of the special meeting, AAW library, as well as the protocol of the prior consultation, SAPMO IV 2/9.04/369, p. 41ff.
40. Josef Naas, IV 2/9.04/372, p. 19.
41. Plenary session on November 13, 1952, AAW, P1/2, p. 78.
42. See SAPMO IV 2/9.04/372, p. 52.
43. Ibid., p. 51.
44. See the report of the Science Section of the Central Committee, SAPMO IV 2/9.04/372, p. 146ff.
45. The first party organizer was Manfred Naumann (regular member since 1978, literary studies). See M. Naumann in *Erinnerungen und Erlebnisse verdienstvoller Mitglieder der SED* (Berlin, 1987), pp. 21–29. Statistical data from SAPMO IV/2/9.04/380, p. 334.
46. SAPMO IV 2/9.04/373, p. 97 f.
47. Max Volmer (regular member since 1934, though not confirmed by the Nazi regime), physical chemistry, president 1955–1958. Peter Adolf Thiessen (regular member since 1939, expelled 1945, readmitted 1956), physical chemistry, chairman of the Research Council of the GDR 1957–1965. SAPMO IV 2/9.04/370, pp. 42 and 51ff.
48. See W. Freund, Report on a discussion with P. A. Thiessen on January 17, 1957, to the Central Committee's *Wissenschaft* department, Party Organization, German Academy of Sciences and H. Wittbrodt, SAPMO IV 2/9.04/412, folio 51ff.
49. SAPMO IV 2/9.04/372, p. 117.
50. Ibid., p. 119ff.
51. In the academy, the institutes were subordinated to the faculties—and thus also to the West German members.
52. On the formation of cliques see SAPMO IV 2/9.04/380, pp. 32–38; personnel proposals: 380, pp. 134 and 412; AAW, meeting of the executive committee on May 2 and 9, 1957.
53. At first more than fifty research institutes were grouped together, including some from outside the academy.
54. The Founding Resolution of the Council of Ministers and the list of Research Council members were published in *New Paths for Scientific-technical Research and Progress* (East Berlin, 1967), pp. 7 and 60ff. See also the minutes of the conference on September 19, 1957 in AAW, NL, R. Rompe, nr. 1: R. Model,

"On the Position of and on Open Questions in the German Academy of Sciences, including the Research Association, December 27, 1957, SAPMO IV 2/9.04/372, folio 158.

55. SAPMO IV 2/0.04/372, p. 141ff.

56. Ibid., p. 158.

57. See SAPMO IV 2/9.04/367 and 370, pp. 13–16.

58. Hartke was the president of the academy from 1958 to 1968.

59. Hartke received only 36 of the 67 ballots that were cast. Because of the restrictions there "was not an election but a vote." See the report on the meeting of the plenum of the academy on October 23, 1958, in SAPMO IV 2/9.04/370, p. 15ff.

60. SAPMO IV 2/9.04/372, p. 158.

61. Draft for the Politburo of the Central Committee of the SED, SAPMO IV 2/9.04/372, pp. 416—483.

62. The loss amounted to 1.6 percent (1955), 1.4 percent (1956), 1.5 percent (1957), 3 percent (1958), and 2–3 percent (1959 and 1960). The exodus affected especially medicine and the natural sciences. See SAPMO IV 2/9.04/370, pp. 52 and 383, pp. 111, 210, 247.

63. There were ten resignations (seven in 1950); in three cases the reason given was pressure from employers in West Berlin.

64. Ernst Lemmer, Federal Minister for All-German Affairs (1957–1962), in *Süddeutsche Zeitung*, February 20, 1958, p. 3.

65. Dieter Hoffmann, "Ein Jubiläum wird gefeiert—die Max-Planck-Ehrung(en) in Berlin 1958," lecture at the Fifth Meeting of the History of Physics, Mainz, 1993.

66. For more details on the issues and sources discussed in this paragraph, see the German version of this chapter in Hoffmann and Macrakis, eds., *Naturwissenschaften und Technik in der DDR* (Berlin: Akademie Verlag, 1997).

67. Werner Hartke, "Bericht über die Entwicklung der Akademic," *Jahrbuch der DAW* (1961): 10.

68. Fourteenth meeting of the Central Committee of the SED, in November 1961.

69. Program of the SED. "Die Rolle der Wissenschaft bei der umfassenden Verwirklichung des Sozialismus," *Einheit* 1 (Berlin, 1963): 44.

70. Kurt Hager, consultation with W. Hartke, June 22, 1962, SAPMO IV 2/9.04/372, folio 395ff.

71. Material for the Politburo of the SED and Kurt Hager, consultation with W. Hartke on June 22, 1962, SAPMO IV 2/9.04/372, folios 416–483 and 319–405.

72. Only the academy president and four other SED representatives of the academy (none of them members) appear in the documents. SAPMO IV 2/9.04/372, pp. 405 and 411.

73. "Rolle, Aufgaben und weitere Entwicklung der Deutschen Akademie der Wissenschaften zu Berlin als eine sozialistische Akademie," SAPMO IV 2/9.04/372, p. 416ff.

74. Kurt Hager emphasized explicitly in regard to the relationship between the

Academy and the Research Council: "The Research Council is the superordinated organ." SAPMO IV 2/9.04/372, p. 398.

75. The Research Council of the DDR was established on June 6, 1957. Its first chairman was Peter Adolf Thiessen. The Council of Ministers' resolution (June 6, 1957) establishing the Research Council and a list of its members are in *Wissenschaft und Fortschritt* (Berlin, 1967), pp. 7 and 60ff. See also Chapter 2 of this book.

76. Kurt Hager, June 6, 1962, SAPMO IV 2/9.04/372, p. 404.

77. Ibid.

78. SAPMO IV 2/9.04/372, pp. 141 and 373ff.

79. See academy party leadership, December 31, 1963, in AAW, Nachlaß Rompe, no. 55, and "Beschluß über die kaderpolitische Sicherung der weiteren Entwicklung der DAW und ihrer Forschungseinrichtungen," SAPMO IV 2/9.04/383, p. 299ff.

80. SAPMO IV 2/9.04/372, p. 415.

81. As late as 1961 the belief was that "the Soviet Academy is not interested in genuine cooperation," SAPMO IV 2/9.04/370, p. 83ff.

82. Consultation of the Central Committee section with the comrades from the academy on February 15, 1962, SAPMO IV 2/9.04/370, p. 135ff.

83. Werner Hartke at the main meeting of the academy, 1967, in *Jahrbuch der DAW 1967*, p. 121.

84. Hermann Klare, "Zu Fragen der Akademiereform," talk given on August 31, 1968.

85. Hermann Klare, regular member since 1961, president 1968–1979. Klare was a chemist who came from industry. He was not a member of the SED, and from 1961 to 1968 he headed the Research Society. His appointment as the successor of Werner Hartke, a scholar in the humanities, was certainly in line with the regular rotation of the presidency, but it was also symbolic of the dominant direction of the Research Academy.

86. "Aufgaben, Profil und Struktur der DAW nach dem VII. Parteitag der SED. Akademieparteileitung 1967," p. 11, in AAW, Nachlaß R. Rompe.

87. See note 84.

88. SAPMO 2/9.04/380, p. 353, and 383, pp. 369–373, and member rosters of the academy's party leadership in Nachlaß R. Rompe, nos. 55, 66, and 123.

89. Cadre nomenclature of the party leadership of the academy, February 22, 1967. High-level positions also required the approval of the Central Committee apparatus, in Nachlaß R. Rompe, no. 66.

90. *Jahrbücher der DAW*, 1963–1968.

7. The Unity of Science vs. the Division of Germany

1. I would like to thank Cathy Carson for detailed comments on this chapter. I also received helpful feedback from the participants of the conference for this book.

2. Usually what SED policy makers meant by the term "socialist institution" was an institution with a high number of party members who adhered to state

policies and who shared the SED's values. In addition, such institutions were transformed to incorporate Marxist-Leninist ideology. Policy for these socialist institutions was usually formed at the level of the SED. See Chapters 5 and 6 for examples of "socialist institutions."

3. See, for example, Steven Dickman, "Rich Uncle or Big Brother," *Nature,* April 12, 1990, pp. 604–606, here p. 605.

4. *Nova acta Leopoldina,* bd. 36, nr. 198, 1970, p. 27. For the most recent literature on the history of the Leopoldina by an academy president, see Benno Parthier, *Die Leopoldina: Bestand und Wandel der ältesten deutschen Akademie* (Halle: Deutsche Akademie der Naturforscher Leopoldina, 1994).

5. "Glückwunschschreiben Otto Grotowohls zur 300-Jahr-Feier der Leopoldina," *Neues Deutschland,* February 16, 1952. Numerous newspaper articles, undated, are available in R-3, Ministry for Universities, BAP.

6. *Deutsche Akademie der Naturforscher, Leopoldina: Geschichte, Struktur, Aufgaben* (Halle, 1993), p. 12.

7. "Heinz Bethge," *in Wer war Wer—DDR: Ein biographisches Lexikon* (Berlin: Ch. Links Verlag, 1992), p. 40.

8. SAPMO IV 2/9.04/382, February 10, 1956, notes on a discussion between Kurt Hager and Kurt Mothes at the Central Committee Guest House.

9. SAPMO, Party Archive, IV 2/9.04/8, minutes of the University Conference of the Central Committee's Department of Science (Wissenschaft) and Universities, October 31–November 1, 1953, –Leipzig.

10. Ibid, section on the education of the new intelligentsia.

11. SAPMO IV A 2/9.04/363, F. Dahlem to Gießmann re Leopoldina, July 8, 1969, emphasis added.

12. SAPMO IV A 2/9.04/363, March 31, 1959, Leopoldina: Structure and Membership.

13. Interview, Macrakis with Hager, 10 August 1993.

14. Invitation to yearly meeting, June 1961. This copy seen in the party files.

15. Mothes to all members of the academy, end of August 1961, Seen in academy archives, Academy of Sciences, Berlin, file 431.

16. SAPMO, Walter Ulbricht papers, NL 182, vol. 936, shortened minutes of discussion between Alexander Abusch and Kurt Mothes, November 30, 1961.

17. Ibid.

18. Ibid., pp. 4, 5.

19. BAP, R-3, Ministry for Universities, 4547, minutes of meeting, Abusch and Mothes, January 24, 1962.

20. Ibid.

21. BAP, R-3, Ministry for Universities, file 4547.

22. Ibid.

23. Ibid., Goßens (deputy to state secretary) to Otto Schlüter, April 3, 1952; Schlüter to MHF, April 17, 1952.

24. Ibid., Otto Schlüter to MHF, February 12, 1953.

25. BAP, R-3, MHF, November 21, 1955, memo signed by Dahlem.

26. Ibid., Müller's memorandum, April 28, 1956.

27. Ibid., November 21, 1955, memo signed by Dahlem.

28. Ibid.
29. Dr. G. Mehnert, "Leopoldina: Tradition und Verpflichtung," November 11, 1955, no newspaper title, in BAP, R-3, 4547.
30. BAP, R-3, 4547.
31. Invitation to members of the academy from the Leopoldina, November 15, 1958. This copy seen in the party archives, but available elsewhere.
32. Ibid.
33. BAP, R-3, 4547, October 5, 1967, conception for a conversation with Mothes.
34. BAP, R-3, 1277/1, February 14, 1971, draft, conception for the tasks of the MHF in the further development of the German Academy of Natural Researchers.
35. Ibid.
36. Ibid., p. 3.
37. Ibid., pp. 10–12.
38. Ibid.
39. BAP, R-3, 1998, October 26, 1972, tips for the conversation with the comrade minister in Halle.
40. Ibid.
41. Ibid., February 12, 1974, "Leopoldina," to Minister Professor Böhme.
42. Ibid., file note on January 1, 1975, meeting with Bethge.
43. BAP, R-3, 1998, discussion with Comrade Hörnig on October 18, 1979, and October 15, 1979.
44. BAP, R-3, files 1998 and 2006, cover the 1980s.
45. BAP, R-3, 1998, first tips for the comrade minister's conversation with the president of the Leopoldina, November 11, 1986.
46. Ministry for State Security Archive, BStU, Operational Group File nr. 36/58, Archive 3557/69, April 1, 1959, first summary of material, "strictly confidential." The professors were Kurt Mothes, Erwin Reichenbach (vice-president), Bernd Lueken, Franz Runge, Robert Mark, Theodor Grüneberg, Horst Hanson, Wilhelm Messerschmidt, Rudolf Zaunick, Rudolf Käubler, and sometimes Günther Mönch.
47. Ibid., vol 1, order, December 5, 1958, p. 167 (archive pagination).
48. Ibid., "Summary of Material on the 'Halle Center' of the Academy of Nature Researchers 'Leopoldina,'" April 1, 1959, pp. 12 and 13 of text, Gauck-Behörde folio nos. 33–34, paragraph numbers of GDR basic law, 13, 19, 20, and 23.
49. Ibid., vol. 1, p. 164 (archive pagination).
50. Ibid., p. 165.
51. Ibid., vol. 3, pp. 89, 91, and 95. Fink's assignment was given on December 8, 1958, Sternheim's on January 5, 1959, and Egon's on January 8, 1959.
52. Ibid., p. 165.
53. Ibid., vol. 5, "Preliminary summary of former Nazis in the Leopoldina," p. 34; vol. 18, Closing Report, June 26, 1969; vol. 3, "Summary of material on Leopoldina, April 1, 1959, p. 99.
54. Ibid., vol. 6, "Operational Plan to Smash the Negative Influence of the Reactionary 'Halle Center,'" pp. 135–145.

55. Ibid., vol. 3, "Summary of material . . .," April 1, 1959. pp. 120–126.

56. Ibid., vol. 3, appendix to Summary, pp. 166–168.

57. Ibid., vol. 6, "Operational plan to smash the negative influence of the reactionary 'Halle Center' of the Leopoldina," pp. 135–145.

58. Ibid., and "Note on consultation with the secretary of the district office of the SED Halle, Comrade Franz Bruck."

59. Ibid., vol. 8, "Planned and partly realized measures for the operational file 'Komet,'" November 29, 1961.

60. Ibid., vol. 10, January 18, 1962, "Report on the enemy actions of the so-called Leopoldina Center with special attention to the enemy activity since August 13, 1961." A handwritten note on the text reads, "Original version made for Comrades Koenen and Bruck." A second version is entitled "Report on operational file 'Komet' . . . with special attention to enemy activity after August 13, 1961.

61. Ibid., vol. 11, March 30, 1962, "Chief measures on the further work on the operational file 'Komet.'"

62. Ibid., vol. 13, N to district office of state security, August 19, 1962.

63. Ibid., vol. 18, Closing Report, June 26, 1969, directive to close file September 4, 1969.

64. SED Party Archive, IV A 2/9.04/259: 304, "Assessment of Cadre-Political Situation Relating to August 13, 1961."

65. Karl Robert Mandelkow, *Goethe in Deutschland. Rezeptionsgeschichte eines Klassikers*, Band II, 1919–1982 (Munich: Verlag C. H. Beck, 1989). On the Goethe Society in Weimar as an all-German institution, see pp. 164–170.

66. Ibid., p. 166.

67. Ibid.

8. Frustrated Technocrats

1. Research for this chapter was supported in part by a grant from the German Exchange Service (DAAD), which is not responsible for its content.

2. See Konrad Jarausch, *The Unfree Professions: German Lawyers, Teachers, and Engineers, 1900–1950* (New York and Oxford: Oxford University Press, 1990), esp. pp. 7, 22–23.

3. On the historical manifestations of professions and professionalization, see Hannes Siegrist, "Bürgerliche Berufe. Die Professionen und das Bürgertum," in Hannes Siegrist, ed., *Bürgerliche Berufe* (Göttingen: Vandenhoeck & Ruprecht, 1988).

4. For example, engineers were unable to push through standardized requirements for entry into their profession (such as university degrees or state examinations). See Jarausch, *Unfree Professions;* Kees Gispen, *New Profession, Old Order: Engineers and German Society, 1815–1914* (Cambridge, England: Cambridge University Press, 1989); Karl-Heinz Ludwig, *Technik und Ingenieure im Dritten Reich* (Düsseldorf, 1974); Jeffrey Herf, *Reactionary Modernism: Technology, Culture and Politics in Weimar and the Third Reich* (Cambridge, England: Cambridge University Press, 1984).

5. This included all who were involved in research and teaching in technical fields, as well as technicians, chemists, and sometimes even highly skilled workers, along with engineers.

6. In terms of education, skills, and responsibility, these professionals were half way between engineers and foremen, or *Meister*. *Techniker* were trained at an engineering college (Ingenieur-Fachschule) or at a factory academy (Betriebsakademie). See Günter Erbe, *Arbeiterklasse und Intelligenz in der DDR* (Opladen: Westdeutscher Verlag, 1982), p. 76.

7. Hartmut Zimmermann, "Intelligenz," *DDR Handbuch* (Cologne: Verlag Wissenschaft und Politik, 1985), vol. 1, p. 658.

8. My calculations (no totals were given in the original source) are based on Statistisches Bundesamt (SB), Außenstelle Berlin-Mitte "Volks- und Berufszählung in der DDR am 311. August 1950," pp. 200–209 and unnumbered pages for Berlin.

9. This is my estimate, based on SAPMO IV 2/2.029/114, reports from the Department for State and Legal Questions dated December 12, 1958, January 23, 1959, and February 2, 1959.

10. BAP F-4 461, memorandum dated October 5, 1950.

11. SAPMO IV 2/2.029/114, "Analyse über die Republikfluchten aus der chemischen Industrie," undated (sent to Erich Apel on June 8, 1959), p. 3.

12. SAPMO IV 2/2.029/114, reports from the State Planning Commission (Cadre Department) dated December 12, 1958, July 30, 1960, and April 9, 1960.

13. SAPMO IV 2/2.029/114, report from the Department of Internal Affairs, no date, sent to Erich Apel's office on June 8, 1959.

14. Figures do not include education majors. Source: SAPMO IV 2/9.04/478, table entitled "Studierende an sämtlichen Ing.- und Fachschulen nach sozialer Herkunft." Cf. *Statistisches Jahrbuch der Deutschen Demokratischen Republic 1955* (Berlin: VEB Deutscher Zentralverlag, 1956), p. 65. The number of graduates rose from 15,759 in 1952 to 25,544 in 1956, an increase of 62 percent.

15. Figures include night and correspondence students. SAPMO IV A2/9.04/379, table entitled "Ing.- und Fachschulen/DDR/sämtl. Ausb.-Ziele."

16. BAP F-4, nr. SFT 47, "Vorlage über die Planung, Ausbildung, Erziehung und Verteilung der Hochschulabsolventen auf dem Gebiet der Elektrotechnik . . .," 1962, p. 3.

17. SAPMO IV 2/9.04/477 (party section on academia), report entitled "Warum können nicht alle Abiturienten studieren?" dated January 6, 1958.

18. See Siegfried Baske, ed., *Bildungspolitik in der DDR 1963–1976* (Wiesbaden: Harrassowitz, 1979), p. 469. See also *Statisches Jahrbuch der Deutschen Demokratischen Republik 1971* (Berlin: Staatsverlag der DDR, 1971).

19. In 1964, there were 159,921 people employed as engineers in the GDR; the 1971 census does not tell us how many people were actually employed as engineers. My calculations are based on SB, *Ergebnisse der Volks- und Berufszählung am 31. Dez. 1964* (Berlin [GDR], 1967), vol. 1, pp. 276–278, 291–294; *Volks-, Berufs-, Wohnraum- und Gebäudezählung am 1. Jan. 1971*, vol. 5:

Wirtschaftlich tätige und nichtwirtschaftlichtätige Wohnbevölkerung, Berlin (GDR) 1972, pp. 114–119.

20. SAPMO IV A2/9.04/422, report entitled "Information an die Abt. Wissenschaften beim ZK der SED über die Schwierigkeiten beim Einsatz von Hoch- und Fachschulabsolventen," no date.

21. Ibid. A survey conducted in the 1950s shows that half of low-level administrators and engineers had completed only an elementary school education. See Thomas A. Baylis, *The Technical Intelligentsia and the East German Elite* (Berkeley: University of California, 1974), p. 28.

22. SAPMO IV 2/9.04/607, "Einschätzung des Berufseinsatzes der Hochschulabsolventen 1957," esp. p. 3.

23. Data provided by the Kammer der Technik (R. Höntzsch), no call number. Data based on surveys and on the *Statische Jahrbücher der DDR* (no exact source cited).

24. This is noted, for example, by Franz Dahlem, Undersecretary for the Ministry of Higher Education, in SAPMO IV 2/2.029/157: "Diskussionsbeitrag des Gen. Dahlem auf der Wirtschaftskonferenz am 10./11.10.61," p. 3.

25. SAPMO 2/9.04/477, report entitled "Information über den Stand der diesjähringen Immatrikulation . . .," dated April 28, 1958, p. 1; IV 2/9.04/478, "Vorlage f. das Sekretariat des ZKs der SED über Maßnahmen zur Auswahl und Zulassung Direkt-, Fern-, und Abendstudium . . .," dated January 1, 1959, p. 2.

26. SAPMO IV 2/2.029/157, transcript of Vice-State Secretary for University Affairs Franz Dahlem at an "economic conference" on October 10–11, 1961.

27. SAPMO IV 2/5/1177, memorandum dated December 7, 1951.

28. SAPMO IV 2/9.04/618, "Probleme der Qualifizierung und Förderung von Frauen," pp. 39–40.

29. Ibid., "Zu Fragen der Beteiligung der Frauen am Hoch- und Fachschulstudium ohne Unterbrechung der Berufsarbeit."

30. SAPMO IV A2/9.04/238, "Bericht des SS für Hoch- und Fachschulwesen über Maßnahmen zur Förderung der Frauen . . .," no date (1966).

31. SB, "Volks- und Berufszählung in der DDR am 31. August 1950," pp. 200–209 and unnumbered pages for Berlin; *Ergebnisse der Volks- und Berufszählung am 31. Dez. 1964* (Berlin [GDR], 1967), vol. 1, pp. 233–235; SB, *Volks-, Berufs-, Wohnraum- und Gebäudezählung am 1. Jan. 1971,* vol. 5: *Wirtschaftlich tätige und nichtwirtschaftlichtätige Wohnbevölkerung* (Berlin [GDR], 1972), pp. 114–119.

32. SAPMO IV 2/9.04/618, "Probleme der Qualifizierung und Förderung von Frauen," pp. 13–14; SAPMO IV 2/5/998, "Information über die Durchführung des Beschlusses des Politbüros vom 5.12.1961 . . .," dated January 24, 1962.

33. SAPMO FDGB-BuVo A3383, table 3. These data include basic pay, overtime, bonuses, and other extra pay.

34. SAPMO IV 2/5/1196, "Bericht der Instrukteurgruppe des ZK der Landesleitung Berlin und der Kreisleitung Pankow im VEB Bergmann-Borsig," dated September 13, 1950, p. 6.

35. SAPMO IV 2/5/1177, "Bericht der Instrukteurgruppe des Parteivorstandes über die Tätigkeit im VEM Transformatoren- und Röntgenwerk Dresden in der Zeit vom 6.–8. Dez. 1949," p. 6.

36. SAPMO IV 2/9.04/607, "Einschätzung des Berufseinsatzes der Hochschulabsolventen 1957," p. 3; SAPMO FDGB-BuVo A3782, transcript of meeting of the chair of the national executive committe of the *FDGB* (the party-run union) with the intelligentsia on February 16, 1961, pp. 8–9, 34; SAPMO IV 2/5/1308, "Instrukteur- und Brigadeberichte" for Berlin, report dated August 5, 1955, p. 6; SAPMO IV 2/5/1349/1, report of a brigade of the Central Committee on the Chemical Works in Buna, dated June 4, 1959, p. 14.

37. These issues are discussed in BAP F-4 461, memorandum dated May 7, 1951; SAPMO IV 2/2.029/114, "Bericht über die Republikabgänge, Rückkehrer und Zuziehenden im Bereich des Maschinenbaus," dated July 30, 1960, p. 9; SAPMO FDGB-BuVo A3385, undated FDGB report (1970), pp. 10–11; SAPMO FDGB-BuVo A3384, "Probleme der materiellen und ideellen Stimulierung in den produktionsvorbereitenden Bereichen," dated December 28, 1967, appendix, p. 4.

38. See Jarausch, *Unfree Professions*, p. 19.

39. SAPMO FDGB-BuVo A3385, undated FDGB report (1970), pp. 10–11; SAPMO FDGB-BuVo A3383, "Analyse über einige Probleme der volkswirtschaftlichen und arbeitökonomischen Entwicklung im I. Halbjahr 1967."

40. A decree of January 1, 1962, ordered factory administrators to ensure that recent engineering graduates were given positions that corresponded to their qualifications. An example of how this was put into practice can be found in Landesa Archiv Berlin (LAB) Rep. 411, nr. 625: Arbeit mit der jungen Intelligenz 1960–63. An example on work and research units can be found in SAPMO IV 2/5/1349/1, report of a brigade of the Central Committee on the Chemical Works in Buna, dated June 4, 1959, p. 14.

41. SB, "Volks- und Berufszählung in der DDR am 31. August 1950," pp. 200–209 and unnumbered pages for Berlin.

42. SB, *Ergebnisse der Volks- und Berufszählung am 31. Dez. 1964* (Berlin [GDR], 1967), pp. 233–235. Only 1,829 out of the 412,490 GDR citizens under forty who had a university or college *(Fachschule)* degree (all majors) had completed their degrees by 1945. Of the 91,436 GDR citizens aged forty to forty-nine who had completed their advanced education, 13,258 (14.5 percent) finished their studies by 1945. My calculations are based on SB, *Ergebnisse*, p. 275. There is no reason to think that the age distribution of engineering graduates would be any different. However, it is impossible to estimate how many engineers without a university or college degree entered their profession by 1945.

43. This is merely an educated guess, based on the fact that 59.9 percent of the GDR college- or university-educated population aged fifty or older had completed their degres by 1945. My calculations are based on SB, *Ergebnisse*, p. 275.

44. This estimate may be somewhat low given that the percentage of engineers without advanced degrees was most likely higher in the older age cohorts.

45. This figure, calculated on the basis of 1971 census data, actually refers to employed persons with engineering degrees. This tremendous shift was primarily

caused by the educational boom, as well as by the demographic "hole" caused by the war. There may also be a distortion here because the 1964 figure is for practicing engineers, whereas the 1971 figure is for persons with engineering degrees. (There were probably more older engineers who had not completed their advanced education.) SB, *Volks-, Berufs-, Wohnraum- und Gebäudezählung am 1. Jan. 1971*, vol. 5: *Wirtschaftlich tätige und nichtwirtschaftlichtätige Wohnbevölkerung* (Berlin [GDR], 1972), pp. 114–119.

46. SAPMO IV 2/5/1196, "Bericht über die Tätigkeit der Parteiorganisation der Parteigruppen und die Lage im Transformatorenwerk 'Karl-Liebknecht' Oberschöneweide Kreis Köpenick, vom 18.-26.9.1951."

47. Ibid., "Bericht von der Kreisdelegiertenkonferenz Berlin-Lichtenberg am 24. und 25. Juni 1950." Of 368 delegates, only 4 technicians and engineers were present.

48. SAPMO IV 2/5/1177, "Bericht über den Instrukteureinsatz im Edelstahlwerk Döhlen, Dresden-Freital, vom 8.-10.3.1951," p. 1.

49. SAPMO IV 2/5/1196, "Protokoll 12. und 13.11.49 - Kreisdelegierten-Konferenz Köpenick," p. 10a.

50. Ibid., p. 4.

51. Ibid., p. 26.

52. Ibid., p. 3.

53. Examples can be found in SAPMO FDGB-BuVo A6040, resolution of the executive committee of the *FDGB* from March 9, 1956, number P7/56; SAPMO FDGB-BuVo A4131, "Stellungnahme des Sekretariats des Bundesvorstands des FDGB zur Verbesserung der Arbeit der Gewerkschaften mit der Intelligenz," dated December 14, 1951, pp. 2–5.

54. SAPMO IV 2/2.029/114, "Analyse der Republikfluchten der Intelligenz aus der Elektroindustrie," dated July 10, 1959, p. 7.

55. SAPMO IV 2/5/1308, "Material für die Bezirksdelegierten-Konferenz Berlin," February 25, 1956, p. 13.

56. My calculations and estimates are based on SAPMO IV 2/5/1367, Analysen über die Mitgliederbewegung in der Gesamtpartei, "Stand der Organisation am 31. Juli 1946"; "Stand der Organisation am 30. September 1947"; SAPMO IV 2/5/1368, "Stand der Organisation am 31. Oktober 1948"; SAPMO IV 2/5/1369, "Organisationsstatistik für den Monat Juli 1950"; SAPMO IV 2/5/1370, "Berichtsbogen zur Organisationsstatistik nach dem Stand vom Dezember 1952"; SAPMO IV 2/5/1371, "Statistischer Jahresbericht über die Zusammensetzung der Parteiorganisation Gesamtpartei zum 1. January 1957." Data on total work force and white-collar workers from *Statistisches Jahrbuch der Deutschen Demokratischen Republik 1956* (Berlin: VEB Deutscher Zentralverlag, 1956), pp. 154, 166. Unfortunately, data on party membership in the 1960s were not available.

57. Data provided by the *Kammer der Technik* (R. Höntzsch), no call number. Data based on internal records.

58. BAP, Rep. F-4, nr. 464, "Maßnahmen zur Verbesserung der Arbeits- und Lebensbedingungen der technischen Intelligenz 1950–53," draft dated August 31, 1953, p. 2.

59. SAPMO IV 2/5/1196, "Protokoll 12. u. 13.11.49—Kreisdelegierten-Konferenz Köpenick," p. 22a.

60. BAP F-4 634, "Bericht, betr.: Beobachtungen während der Dienstreise nach Jena am 27.11.50," p. 1.

61. In 1950, 60 percent of all GDR engineers worked at the "technical preparation of production" (research and development, design, planning, production technology, standarization); in 1970, the percentage had declined to 40 percent. The percentage of engineers in production rose from 5 percent to 10 percent between 1950 and 1970. Engineers in sales and material management made up 6 percent of all engineers in 1950 and 8 percent in 1970. Three percent of all engineers were in planning, computing or accounting in 1950, but only 9 percent were in 1970. In 1950, 18 percent of all engineers held administrative positions; in 1970, 23 percent held such posts. (Other specializations made up 8 and 10 percent of the totals, respectively.) See Erbe, *Arbeiterklasse*, p. 142. Source: W. Draeger, "Wachsender Bestand an Ingenieuren erfordert neue Maßstäbe für ihren effektiven Einsatz," *Sozialistische Arbeitswissenschaft*, vol. 20 (1976), p. 32.

62. Walther Ulbricht, *Die Durchführung der ökonomischen Politik im Planjahr 1964 unter besonderer Berücksichtigung der chemischen Industrie* (Berlin: Dietz Verlag, 1964), p. 53, quoted according to Bentley, *Technological Change*, p. 5.

63. Bentley, *Technological Change*, pp. 27ff. Bentley estimates that the institutes had an average of 250 employees in 1964, and enterprise R & D units an average of 40.

64. LAB Rep. 411, nr. 1, "Betriebs-Chronik," dated 1949 and "Entwicklung Transformatorenwerk 'Karl Liebknecht,'" probably 1966.

65. BAP F-4, nr. Sekretariat Forschung und Technik (SFT) 8, "Zusammenfegaßte Auswertung der Jahresberichte der Forschungs- und Entwicklungsstellen 1961 unter Berücksichtigung weiterer Quellen," dated 1962; BAP F-4, nr. 21184, "Information über Ergebnisse und Probleme bei der Durchführung des Staatsplanes Wissenschaft und Technik und des Planes der Naturwissenschaftliche Forschung im Jahre 1966."

66. See also Chapter 9.

67. LAB Rep. 411, nr. 654, "Jahresbericht über die Kaderarbeit im Jahre 1957," p. 3; LAB Rep. 411, nr. 804, "Analyse der Entwicklung der Effektivität der Forschungs- und Entwicklungsabteilungen des VEB TRO," dated July 25, 1968, and "Die Lage in den Entwicklungsbereichen des VEB Transformatorenwerk 'Karl Liebknecht,'" no date (1966); LAB Rep. 411, nr. 1196, "Vorlage für die Produktionskommitee: Untersuchung der Möglichkeiten und der einzuleitenden Maßnahmen zur Verkürzung der Entwicklungszeiten . . ."

68. BAP F-4, nr. SFT 8, "Zusammengefaßte Auswertung der Jahresberichte der Forschungs- und Entwicklungsstellen 1961 unter Berücksichtigung weiterer Quellen," dated 1962, pp. 23, 65, 75; BAP F-4, nr. 21184, "Information über Ergebnisse und Probleme bei der Durchführung des Staatsplanes Wissenschaft und Technik und des Planes der Naturwissenschaftlichen Forschung im Jahre 1966," pp. 19–22. See also Bentley, *Technological Change*, pp. 35–41.

69. LAB Rep. 411, nr. 780, "Jahresbericht der Forschungs- und Entwicklungsstelle

1955"; Rep. 411, nr. 800, "Jahresbericht 1956 der Forschungs- und Entwicklungsstelle."

70. See Bentley, *Technological Change*, p. 41. Original sources for GDR data: H. Kusicka and W. Leupold, *Industrieforschung und Ökonomie* (Berlin: Dietz Verlag, 1966), p. 44; Autorenkollektiv, *Das Forschungspotential im Sozialismus* (Berlin, Akademie Verlag, 1977), p. 113.

71. LAB Rep. 411, nr. 1196, "Vorlage für die Produktionskommittee: Untersuchung der Möglichkeiten und der einzuleitenden Maßnahmen zur Verkürzung der Entwicklungszeiten . . ."

72. LAB Rep. 411, nr. 1076, "Kaderunterlagen," memorandum on Joachim Jerratsch, August 8, 1955.

73. BAP F-4, nr. SFT 8, "Zusammengefaßte Auswertung der Jahresberichte der Forschungs- und Entwicklungsstellen 1961 unter Berücksichtigung weiterer Quellen," dated 1962, pp. 6–13, 68. See Bentley, *Technological Change*, p. 34.

74. BAP F-4, nr. 21184, "Information über Ergebnisse und Probleme bei der Durchführung des Staatsplanes Wissenschaft und Technik und des Planes der Naturwissenschaftlichen Forschung im Jahre 1966," pp. 10ff., 22; BAP F-4, nr. SFT 8, "Zusammengefaßte Auswertung der Jahresberichte der Forschungs- und Entwicklungsstellen 1961 unter Berücksichtigung weiterer Quellen," dated 1962, pp. 6–13, 68.

75. LAB Rep. 411, nr. 804, "Die Lage in den Entwicklungsbereichen . . . und Schlußfolgerungen zur Beschleunigung des wissenschaftlich-technischen Fortschritts" (1965, 1966), pp. 1–3; Rep. 411, nr. 800, "Jahresbericht 1958 der Forschungs- und Entwicklungsstelle"; "Jahresbericht 1962 der Forschungs- und Entwicklungsstelle."

76. SAPMO-BA IV 2/2.029/114, "Bericht über die Republikabgänge aus der chemischen Industrie," April 29, 1959, p. 10; BAP F-4, nr. SFT 8, "Zusammengefaßte Auswertung der Jahresberichte der Forschungs- und Entwicklungsstellen 1961 unter Berücksichtigung weiterer Quellen," dated 1962, pp. 24ff., 35–41, 54ff.

77. An example is the materials division of the Institute for Lightweight Construction: BAP F-4, nr. SFT 1720, memorandum on the visit of Comrades Kraaß and Richter at the Institute for *Leichtbau*, dated February 6, 1965, p. 3. See also Bentley, *Technological Change*, pp. 61–67.

78. BAP F-4, nr. SFT 8, "Zusammengefaßte Auswertung der Jahresberichte der Forschungs- und Entwicklungsstellen 1961 unter Berücksichtigung weiterer Quellen," dated 1962, pp. 18–23. Also BAP F-4, nr. SFT 1720, memorandum on the visit of Comrades Kraaß and Richter at the Institute for *Leichtbau*, dated February 6, 1965, pp. 34ff., 45ff.; BAP F-4, nr. 21184, "Information über Ergebnisse und Probleme bei der Durchführung des Staatsplanes Wissenschaft und Technik und des Planes der Naturwissenschaftlichen Forschung im Jahre 1966," pp. 22ff. See also Bentley, *Technological Change*, pp. 67, 78ff., 97ff.

79. LAB Rep. 411, nr. 804, "Analyse der Entwicklung der Effektivität der Forschungs-und Entwicklungsabteilungen des VEB TRO," dated July 25, 1968, esp. pp. 3–4. According to this report, R & D expenditures in Western corpora-

tions were about 5–11 percent of sales or production. In the GDR, expenditures for testing, pilot projects, materials, and start-up costs were included in R & D expenditures (unlike in the West). If these were subtracted, then Karl Liebknecht spent somewhat under 5 percent on R & D. LAB Rep. 411, nr. 944, Geschäftsbericht, December 31, 1953, p. 2; Rep. 411, nr. 654, transcript of a meeting of the works director with division Z on November 13, 1962.

80. LAB Rep. 411, nr. 800, "Jahresbericht 1959 der Forschungs- und Entwicklungsstelle," "Jahresbericht 1961 der Forschungs- und Entwicklungsstelle," "Jahresbericht 1962 der Forschungs- und Entwicklungsstelle," "Jahresbericht 1963 der Forschungs- und Entwicklungsstelle"; LAB Rep. 411, nr. 804, "Analyse der Entwicklung der Effektivität der Forschungs- und Entwicklungsabteilungen des VEB TRO," dated July 25, 1968, and "Die Lage in den Entwicklungsbereichen des VEB Transformatorenwerk 'Karl Liebknecht,'" no date (1966).

81. See Bentley, *Technological Change,* chapter 2, pp. 97ff.

82. LAB Rep. 411, nr. 362, letter dated October 25, 1961, to the council of the Club of the Young Intelligentsia.

83. SAPMO IV 2/5/1308, Instrukteur- und Brigadeberichte Berlin, "Material für die Bezirksdelegierten-Konferenz Berlin, February 25, 1956," p. 13.

84. On the "social de-differentiation" ("soziale Entdifferenzierung") and the lack of independent sub-systems in the GDR, see Sigrid Meuschel, *Legitimation und Parteiherrschaft in der DDR* (Frankfurt am Main: Suhrkamp, 1992).

9. Chemistry and the Chemical Industry under Socialism

1. Financial support for the research and writing of this chapter came from the National Science Foundation (HPST Grant number DIR-90–23462) and the German Marshall Fund of the United States. Thanks also to conference participants for comments and suggestions.

2. I deal with aspects of the German technological tradition in West German industry in "Technology and the West German *Wirtschaftswunder,*" *Technology and Culture* 32 (1991): 1–22; in *Opting for Oil: The Political Economy of Technological Change in the West German Chemical Industry, 1945–1961* (New York: Cambridge University Press, 1994); and in "In Search of the Socialist Artefact: Technology and Ideology in East Germany, 1945–1962," *German History* 15 (1997): 221–239. On the "German path" in technology more generally, see Joachim Radkau, *Technik in Deutschland. Vom 18. Jahrhundert bis zur Gegenwart* (Frankfurt: Suhrkamp, 1989).

3. VEB Leuna-Werke "Walter Ulbricht" [hereafter Leuna-Werke], Büro des Generaldirektors, "Die Bedeutung der chemischen Industrie für die Volkswirtschaft in der DDR 1945–1956," n.d. [ca. late 1956], Leuna-Werke A. G. Werksarchiv, Merseburg, [hereafter LWA] 3225.

4. *Dokumentation zur Chronik der wissenschaftlich-technischen Entwicklung der Leuna-Werke von 1945 bis 1961* (Leuna, 1982), [*Zahlen und Fakten zur Betriebsgeschichte,* Heft 16], pp. 4, 7–9. Seen in LWA.

5. Quote from *Dokumentation zur Chronik der wissenschaftlich-technischen*

Entwicklung der Leuna-Werke von 1945 bis 1961, p. 8. Information on ammonia synthesis from *Geschichte der Leuna-Werke 'Walter Ulbricht': 1945 bis 1981*, p. 32. More generally, see Martin McCauley, *The German Democratic Republic since 1945* (NY: St. Martin's, 1983), p. 25.

6. My point here is to imply not that there was such a thing as a "Nazi" chemical reactor, but rather that the reactors built during the NS period were designed in the social, political, and economic context of the period and represented assumptions that derived from that context (this is particularly true of the way they were put together with other chemical production apparatus into technological systems). Among these assumptions were that German coal was the preferred starting material for chemical production; that production for war needs took priority over production for civilian requirements; and that a key goal in design and implementation of production was the attainment of German domestic self-sufficiency. A particularly good example of this was the Wolfen factory, formerly part of I. G. Farben, where there was a "forced marriage" of film and artificial fiber production under the National Socialists, something that GDR authorities had to deal with throughout the 1950s and 1960s. For the latter example, see Rainer Karlsch, "Forschung und Entwicklung in der chemischen Industrie der DDR. Ein Überblick (1945–1970)," unpublished manuscript (1994), p. 35. My thanks to Dr. Karlsch for providing me with a copy of the paper.

7. On this notion, see Langdon Winner, "Do Artifacts Have Politics?" in Winner, *The Whale and the Reactor* (Chicago: University of Chicago Press, 1986); and Wiebe Bijker, Thomas Hughes, and Trevor Pinch, eds., *The Social Construction of Technological Systems* (Cambridge, Mass.: MIT Press, 1987).

8. Ernst Homburg, "The Emergence of Research Laboratories in the Dyestuffs Industry, 1870–1900," *British Journal of the History of Science* 25 (1992): 91–111.

9. See also Raymond G. Stokes, "Autarky, Ideology, and Technological Lag: The Case of the East German Chemical Industry, 1945–1964," *Central European History* 28 (1995): (29–46). The point is also made in Karlsch, "Forschung und Entwicklung," esp. with regard to cold rubber development, pp. 29–30. The practice of setting technological targets based on developments in the West was not restricted to the chemical industry, but was characteristic of GDR technological culture generally. On machine tools, see Jörg Roesler, "Einholen wollen und Aufholen müssen. Zum Innovationsverlauf bei numerischen Steuerungen im Werkzeugmaschinenbau der DDR vor dem Hintergrund der bundesrepublikanischen Entwicklung," pp. 263–285 of Jürgen Kocka, ed., *Historische DDR-Forschung* (Berlin: Akademie-Verlag, 1992); more generally, see Eckart Förtsch and Clemens Burrichter, "Technik und Staat in der Deutschen Demokratischen Republik (1949–1989/90)," pp. 205–228 of Armin Hermann and Hans-Peter Sang, eds., *Technik und Staat* (Düsseldorf: VDI Verlag, 1992).

10. Peter Hayes, "Carl Bosch and Carl Krauch: Chemistry and the Political Economy of Germany, 1925–1945," *Journal of Economic History* 47 (1987): 353–363.

11. Stokes, *Opting for Oil*, esp. pp. 49–50.

12. A. Lawrence Waddams, *Chemicals from Petroleum: An Introductory Survey*, 3rd ed. (New York: John Wiley, 1973), pp. 7–13.

13. Ministerium für chemische Industrie, Arbeitsgruppe Entwicklung der chemischen Industrie, "Zentraler Plan Forschung und Technik 1958–1960," October 1, 1957, p. 39, LWA 15621.

14. Leuna-Werke, Org. Abt. Gr. 1 [Dr. Geiseler], "Überblick über die bisherigen Versuchsarbeiten und Ergebnisse zur Herstellung von Polyäthylen," October 10, 1958, LWA 9109.

15. Leuna-Werke, Org. Abt. Gr. 1, Schmierölfabrik [Dr. Geiseler], "Stand der Versuchsarbeiten zur Herstellung von Polyäthylen nach dem Höchstdruck- und dem Niederdruck-Verfahren," August 21, 1957, pp. 1–2, LWA 9109.

16. E.g., Leuna-Werke, Org. Abt. Gr. 1 [Dr. Geiseler], "Aktenvermerk: Erfahrungsaustausch und Zusammenarbeit zwischen der DDR und der UdSSR bezüglich der Herstellung von Polyäthylen," December 15, 1955, LWA 9109; Org. Abt. Gr. 1 [Dr. Geiseler], "Bericht über den Erfahrungsaustausch hinsichtlich der Herstellung von Polyolefinen mit Vertretern der sozialistischen Länder in Berlin," June 3, 1958, LWA 9109; Leuschner to W. Nowikow, Stellv. des Vors. des Ministerrates der UdSSR, November 17, 1962, Anlage, p. 3, in SAPMO IV 2/607/16.

    For additional examples of Soviet technical assistance to the GDR, see Wolfgang Mühlfriedel and Klaus Wießner, *Die Geschichte der Industrie der DDR bis 1965* (Berlin: Akademie-Verlag, 1989), pp. 242, 292, 294ff.

17. On the impact of the changing market for GDR chemicals on R & D, see Karlsch, "Forschung und Entwicklung," p. 9.

18. Leuna-Werke, Org. Abt. Gr. 1 [Dr. Geiseler], "Aktenvermerk: Erfahrungsaustauch und Zusammenarbeit zwischen der DDR und der UdSSR bezüglich der Herstellung von Polyäthylen," December 15, 1955, LWA 9109.

19. For similar problems plaguing various GDR industries, see Mühlfriedel and Wießner, *Die Geschichte der Industrie der DDR bis 1965;* and André Steiner, "Technikgenese in der DDR am Beispiel der Entwicklung der numerischen Steuerung von Werkzeugmaschinen," *Technikgeschichte* 60 (1993): 307–319.

20. On talks with the French, see SPK, Abt. Chemie, Fachgebiet internationale ökonomische Beziehungen, "Aktennotiz über eine am 13.1.1959 durchgeführte Besprechung," January 14, 1959, BAP DE1/15277. On negotiations with the British on the transfer of chemical technology, see "Unterlagen des Außenhandels für Gespräche mit englischen Messebesuchern anläßlich der Leipziger Frühjahrsmesse 1961," February 21, 1961, pp. 1–2, 6, SAPMO IV 2/607/20; Selbmann to Prof. Winkler, January 9, 1961, BAP DE1/14195.

21. Leuna-Werke, *Jahresbericht 1958*, p. 86, LWA 16086.

22. BASF, *Bericht über das Geschäftsjahr 1956*, p. 33.

23. Leuna-Werke, *Zahlen und Fakten zur Betriebsgeschichte* Heft 54: "Zur Entwicklung von 'Leuna II' (1958–1986)" (1986), pp. 5, 9–10.

24. Karlsch provides case studies of other chemical production technologies in "Forschung und Entwicklung," pp. 22–41. In every case, regardless of the reasons, technological development started in a timely fashion in the GDR, but fell

irretrievably behind the West by the mid- to late-1960s. A similar story in development of numerically controlled machine tools is in Steiner, "Technikgenese in der DDR," esp. p. 317.

25. Sima Liebermann, *The Growth of European Mixed Economies, 1945–1970* (New York: Schenkman, 1977), pp. 276, 279–280, and 284–285. For more details on planning for GDR science and technology, see Raymond Bentley, *Research and Technology in the Former German Democratic Republic* (Boulder, Colo.: Westview, 1992).

26. Leuna-Werke, *Zahlen und Fakten zur Betriebsgeschichte* Heft 54: "Zur Entwicklung von 'Leuna II' (1958–1986)" (1986), p. 4.

27. *Chemie gibt Brot—Wohlstand—Schönheit. Chemiekonferenz des Zentralkomittes der SED und der Staatlichen Plankommission in Leuna am 3. und 4. November 1958* (published by the Central Committee of the SED, Sections on Agitation and Propaganda and Mining, Coal, Energy, and Chemistry, n.d.), quote, p. 22, seen in LWA.

28. Ibid., pp. 29–30, 35, quote, p. 35, emphasis in original.

29. *Chemie gibt Brot—Wohlstand—Schönheit*, p. 28, emphasis in original. Most scholars who have examined this program tend to mention only the development of petrochemicals in their discussion, ignoring or downplaying its backward-looking aspects. See, for example, Mühlfriedel and Stießner, *Die Geschichte der Industrie der DDR bis 1965*, p. 152.

30. For the chemical industry specifically, see Ministerium für Chemische Industrie, Arbeitsgruppe Entwicklung der Chemischen Industrie, "Zentraler Plan Forschung und Technik 1958 bis 1960," October 1, 1957, p. 24, LWA 15621. More generally, see Burrichter and Förtsch, "Technik und Staat in der DDR," quote from p. 218.

31. See Stokes, "Technology and the West German *Wirtschaftswunder*"; Joachim Radkau, "'Wirtschaftswunder' ohne technologische Innovation? Technische Modernität in den 50er Jahren," pp. 129–154 in Arnold Sywottek and Axel Schildt, eds., *Modernisierung im Wiederaufbau: Die westdeutsche Gesellschaft der 50er Jahre* (Bonn: Verl. J. H. W. Dietz Nachf., 1993).

32. Stokes, *Opting for Oil*, pp. 233–243.

33. SPK, Abteilung Chemie, "Bilanzbetrachtung zur Entwicklung der Petrolchemie 1966–1975," November 29, 1965, quotes from pp. 1–2, LWA 13132.

34. See Stokes, *Opting for Oil*, esp. chapter 10.

35. Bentley, *Research and Technology in the Former German Democratic Republic*, p. 54.

36. Karlsch, "Forschung und Entwicklung," pp. 18–20.

37. Bentley, *Research and Technology in the Former German Democratic Republic*, pp. 53–54, 76–79, 142; quote p. 53.

38. See Stokes, *Opting for Oil*, esp. chapter 4. An examination of rapid change in another research-intensive industry in the 1950s and beyond is in Paul Erker, "The Challenge of a New Technology: Transistor Research and Business Strategy at Siemens and Philips," *History and Technology* 11 (1994): 131–143.

39. Such is the contention of Thane Gustafson, *Selling the Russians the Rope?* (Santa Monica, Calif.: RAND, 1981), and Nigel Swain, *Hungary: The Rise and Fall of*

Feasible Socialism (London: Verson, 1992). My thanks to Paul Josephson and John Connelly for advice on the literature in this area.

40. Jörg Roesler, in *Zwischen Plan und Markt. Die Wirtschaftsreform 1963–1970 in der DDR* (Berlin: Haufe, 1991), contends that the NES could have changed the system.

10. Nuclear Research and Technology in Comparative Perspective

1. Joachim Radkau, *Aufstieg und Krise der deutschen Atomwirtschaft, 1945–1975. Verdrängte Alternativen in der Kerntechnik und der Ursprung der nuklearen Kontroverse* (Reinbek bei Hamburg: Rowohlt, 1983); Paul R. Josephson, *Physics and Politics in Russia* (Berkeley: University of California Press, 1991).

2. Joachim Kahlert, *Die Kernenergiepolitik in der DDR. Zur Geschichte uneingelöster Fortschrittshoffnungen* (Cologne: Verlag Wissenschaft und Politik, 1988).

3. Wolfgang D. Müller, "Zfk Rossendorf, ein Nachruf," *Atomwirtschaft* (March 1992): 136–139; S. Ulbig, "Keine Großforschung in Rossendorf," *Physikalische Blätter* 50 (1994): 814.

4. "Beschluß des Plenums der AdW vom 16.11.1950"; "Beschluß des Präsidiums der AdW vom 8.12.1951"; "Schreiben der AdW, Otterbein, an die Staatlichen Plankommision, Zentralamt für Forschung und Technik, vom 7.7.1952" (AAW, Bestand Akademieleitung, no. 29, unpaginated folder). I am grateful to Peter Nötzoldt for calling my attention to this material.

5. "Die Situation der kernphysikalischen Forschung in der DDR und die Möglichkeiten ihrer Erweiterung," signed Zöllner, July 7, 1954 (SAPMO, DY 30, IV 2/9.04/288).

6. Ibid. Participants were Zöllner, Rompe, Wittbrodt, and Lanius. See also Central Committee of the SED, Science and Propaganda Section, "Vorlage an das Politbüro des Zentralkomites der SED," 13.1.1955 (SAPMO, DY 30, IV 2/9.04/288).

7. Ulrich Albert, Andreas Heinemann-Grüder, and Arend Wellmann, *Die Spezialisten. Deutsche Naturwissenschaftler und Techniker in der Sowjetunion nach 1945* (Berlin: Dietz, 1992).

8. Richard G. Helewett and Jack M. Holl, *Atoms for Peace and War, 1953–1961: Eisenhower and the Atomic Energy Commission* (Berkeley: University of California Press, 1989).

9. Bertram Winde and Lotar Ziert, *Organisation der Kernforschung und Kerntechnik in der Deutschen Demokratischen Republik* (Leipzig: VEB Deutscher Verlag für Grundstoffindustrie, 1961), pp. 7–13.

10. On the history of the West German centers, see Gerhard A. Ritter, *Großforschung und Staat in Deutschland. Ein historischer Überblick* (Munich: Beck, 1992).

11. Resolution of the Ministerial Council of the GDR entitled "Measures for the application of atomic energy for peaceful purposes," November 10, 1955.

12. See Chapter 6.

13. Heinz Barwich, J. Schintlmeister, and F. Thümmler, *Das Zentralinstitut für Kernphysik am Beginn seiner Arbeit. Aus Anlaß der Inbetriebnahme des ersten Forschungsreaktors der Deutschen Demokratischen Republik gehaltenen Vorträge* (East Berlin: Akademie-Verlag, 1958).

14. Michael Eckert, "Neutrons and Politics. Maier-Leibnitz and the Emergence of Pile Neutron Research in the FRG," *Historical Studies in the Physical Sciences* 19 (1988): 81–113.

15. Heinz Barwich, "Über den Forschungsreaktor der DDR und seine Ausnutzungsmöglichkeiten," in Heinz Barwich et al., *Das Zentralinstitut,* pp. 7–23, here p. 7.

16. Burghard Weiss, *Großforschung in Berlin. Geschichte des Hahn-Meitner-Instituts 1955–1980* (Frankfurt: Campus, 1994), p. 123.

17. "Abkommen über die Gewährung technischer Hilfe seitens der Union der Sozialistischen Sowjetrepubliken für die Deutsche Demokratische Republik beim Bau eines Atomkraftwerkes," drawn up in Moscow on July 17, 1956, copy dated July 21, 1956 (SAPMO DY 30, NL 90/471).

18. "Thesen der Parteiorganisation des Zentralinstituts für Kernphysik zur Vorbereitung der Tagung des Zentralkommitees der Sozialistischen Einheitspartei Deutschlands," containing "Aufbau und Perspektive des Zentralinstituts für Kernphysik," ca. 1959 (SAPMO DY 30, IV 2/9.04/290).

19. Radkau, *Aufstieg und Krise der deutschen Atomwirtschaft,* p. 137.

20. Michael Lemke, "Kampagnen gegen Bonn. Die Systemkrise der DDR and die West-Propaganda der SED 1960–1963," *Vierteljahrshefte für Zeitgeschichte* 41 (1993): 153–174. See also documents in SAPMO DY 30, J IV 2/202/28–29.

21. Wolfgang Behr, *Bundesrepublik Deutschland—Deutsche Demokratische Republik. Systemvergleich Politik—Wirtschaft—Gesellschaft,* 2nd ed. (Stuttgart: Kohlhammer, 1985), p. 109.

22. In July 1960, for instance, the SED had reached the following conclusion: "The departure of more than two million people for West Germany during the last ten years is attributable primarily to the fact that West Germany surpasses us in the economic sphere. The flight from our Republic intensifies to the same degree that our difficulties are mounting. This year the exodus is once again particularly high. Among those leaving are many specialized workers, technicians, engineers, doctors, and teachers. This question of emigration is one of the most difficult problems for us. In the final analysis, it concerns a critical foundation of our production. However, we can change this situation fundamentally only by strengthening the economy of the GDR." Draft of the Politburo to the Central Committee of the SED, July 1960. The text, including the passage cited here, was incorporated into a letter from Ulbricht to Khrushchev, dated October 19, 1960 (SAPMO DY 30, J IV 2/202/29).

23. Max Steenbeck, Energy Commission of the Research Council, to Winde, Commissarial Head of the Office for Nuclear Research and Technology, November 11, 1961 (SAPMO DY 30, IV 2/6.07/67). See additional documents in ibid.

24. "Bemerkungen und Schlußfolgerungen zur Vermeidung jetzt nicht notwendigen Aufwandes bei der Erzeugung von Elektroenergie aus Kernergie," signed

Fabian (member of the Central Committee), March 8, 1962 (SAPMO DY, IV 2/6.07/67). For a discussion of coal and oil see Chapter 9.

25. The state-owned "WIB Vakutronic Dresden," "Vakutronik Pockau," and "Construction and Project Planning of Nuclear Installation" in Dresden were placed under the National Economic Council; the "Institute for Applied Physics of Super Pure Materials" in Dresden and the "Institute for Molecular Electronics" were incorporated into the academy. Only the state-owned "Development and Project Planning of Nuclear Installations" in Berlin and the Central Institute for Nuclear Research in Rossendorf remained within the sphere of the Office for Nuclear Research and Technology for the time being (until 1963, when the office was dissolved).

26. See "Beschlußvorlage über die Organisation der Arbeit auf dem Gebiet der Kernforschung und Kerntechnik" and additional material in SAPMO DY 30, IV 2/6.07/26.

27. Central Committee of the SED, Research and Technological Development section, file note concerning "Conduct of Gen. Prof. Dr. Fuchs . . .," January 5, 1963 (SAPMO DY, IV A 2/6.07/181).

28. "Interner Bericht über die 3. Internationale Konferenz für die friedliche Anwendung der Atomenergie vom 31.8. bis 9.9.1964 in Genf," pp. 11, 23 (SAPMO DY 30, IV A 2/6.07/56).

29. Collected in SAPMO DY 30, IV A 2/6.07/181.

30. Werner, first secretary of the institute party leadership, to Günter Müller, Central Committee of the SED, Research and Technology section, August 31, 1964 (SAPMO DY 30, IV A 2/6.07/109).

31. See Chapter 8.

32. See Chapter 2.

33. On the Soviet program see A. M. Petrosjanz, *Ot naitschnogo poiska k atomnoi promyschlennosti,* 2nd ed. (Moscow: Atomisdat, 1972); German translation of the second edition, *Das Atom. Forschung und Nutzung* (East Berlin: Akademie-Verlag, 1973).

34. Petrosjanz, *Das Atom,* pp. 90–109.

35. Ibid., p. 118.

36. Ibid., pp. 79–84, 114–115.

37. Though this was also true of the United States, the U.S. faced other competitive pressures from its Western partners on account of the larger Western market for nuclear power plants.

38. On this see Kahlert, *Die Kernenergiepolitik in der DDR,* pp. 36 and 45.

39. Klaus Fuchs, "Zum Problem der Entwicklung der Kernenergie in der DDR," February 10, 1964 (SAPMO DY 30, NL 182/978).

40. "Protokoll über die Beratung . . . der Kernenergiekommission, zu der Vertreter der Partieorganisation des VEB AKW . . . und des ZfK Rossendorf eingeladen sind," SFT, March 30, 1966 (SAPMO DY 30, IV A 2/6.07/181).

41. On this see Werner Stiller, *Im Zentrum der Spionage* (Hamburg: Hase & Koehler, 1986), pp. 81–93, and 272–276; as well as my interview with Karl Rambusch and Robert Rompe on September 23, 1993, in Berlin.

42. "Über die Zusammenarbeit mit der Sowjetunion auf dem Gebiet der Kernforschung und Kerntechnik," signed H. Barwich, October 8, 1958, copy (SAPMO DY 30, NL 182/978). See the summary of this document in Heinz and Elfi Barwich, *Das rote Atom* (Munich: Scherz, 1967), pp. 187–188.

43. Barwich, *Das rote Atom,* pp. 187–188.

44. Vice-chairman of the SPK to Ulbricht, August 5, 1959 (SAPMO DY 30, J IV 2/202/29). For a further discussion of the GDR's computer industry see the contribution by Naumann in the German edition of *Science under Socialism* and Chapter 11 of this book.

45. "Bericht über die 1. Tagung der Arbeitsgruppe für Reaktorwissenschaft und -technik sowie Kernenergetik der ständigen Kommission des RGW für die friedliche Ausnutzung der Kernenergie," Dubna, April 20–22, 1965 (SAPMO DY 30, IV A 2/6.07/32).

46. See Petrosjanz, *Das Atom,* p. 342.

47. "Bericht über die 1. Tagung der Arbeitsgruppe für Reaktorwissenschaft und -technik sowie Kernenergetik der ständigen Kommission des RGW für die friedliche Ausnutzung der Kernenergie," Dubna, April 20–22, 1965 (SAPMO DY 30, IV A 2/6.07/32).

48. "Bericht über die Beratung zur Verbesserung der Organisationsstruktur im Vereinigten Institut für Kernforschung," Dubna, April 13–15, 1965, signed Dr. Winde (SAPMO DY 30, IV A 2/6.07/108).

49. Ulbricht to Khrushchev, February 11, 1956; Khrushchev to Ulbricht, March 13, 1956 (SAPMO DY 30, J IV 2/202/28).

50. Kahlert, *Die Kernenergiepolitik in der DDR,* p. 26.

51. Ibid., p. 53.

11. Politics and Computers in the Honecker Era

1. Bruce Parrott, *Politics and Technology in the Soviet Union* (Cambridge, Mass.: The MIT Press, 1983), p. 4.

2. Henry Nau, "National Policies for High Technology Development and Trade," in Francis W. Rushing and Carole Ganz Brown, eds., *National Policies for Developing High Technology Industries* (Boulder, Colo.: Westview, 1986), p. 9.

3. Erich Honecker, "Bericht des ZK der SED an den X. Parteitag der SED," *Neues Deutschland* (April 12, 1981), p. 6.

4. Kurt Hager, *Neues Deutschland* (October 29–30, 1988). See also Kurt Hager, "Wissenschaft und Gesellschaft," Speech to the Bergakademie Freiburg (June 23, 1981), in *Wissenschaft und Wissenschaftspolitik im Sozialismus* (Berlin: Dietz, 1987), pp. 122–125, in which he offers an analysis of scientific-technical development in the past century that could have fit within mainstream philosophies of science in Western Europe or the United States.

5. Helmut Koziolek, "Wissenschaftlich-technischer Fortschritt und oekonomische Kreislaufe," *Spectrum* 12 (March 1981), p. ii.

6. Otfried Steger, "Wissenschaft und Technik setzen uns neue Maßstäbe," *Neues Deutschland* (December 11–12, 1976), p. 7.

7. Erich Honecker, *Die sozialistische Revolution in der DDR und ihre Perspektiven* (Berlin: Dietz Verlag, 1977), p. 26, as cited in Gerhard Merkel, "Mikroelektronik und wissenschaftlich-technischer Fortschritt," *Einheit* 32 (December 1977), p. 1357.

8. "Direktive des XI. Parteitages der SED zum Fünfjahrplan für die Entwicklung der Volkswirtschaft der DDR in den Jahren 1986 bis 1990," *Neues Deutschland* (April 23, 1986), p. 3.

9. Hans-Jürgen Wunderlich, "Schlüsseltechnologien und internationale Wirtschaftsbeziehungen," in Werner Sydow, ed., *Technologien im Umbruch* (Berlin: Die Wirtschaft, 1988), p. 91.

10. Irene Fischer and Karl Hartmann, eds., *Schlüsseltechnologien—warum und für wen?* (Berlin: Dietz, 1987), p. 209.

11. Parrott, *Politics and Technology in the Soviet Union*, p. 9.

12. This dilemma is raised by Erik P. Hoffmann, "Technology, Values, and Political Power in the Soviet Union: Do Computers Matter?" in Frederic J. Fleron, ed., *Technology and Communist Culture* (New York: Praeger, 1977), p. 423.

13. Christopher Freeman, *The Economics of Industrial Innovation* (London: Frances Pinter, 1982), p. 185.

14. Günter Mittag, *Um jeden Preis: Im Spannungsfeld zweier Systeme* (Berlin: Aufbau, 1991), p. 123.

15. See Chapter 4.

16. For comprehensive reviews of the two programs see N. C. Davis and S. E. Goodman, "The Soviet Bloc's Unified System of Computers," *Computing Surveys* 10 (June 1978), pp. 93–122; C. Hammer et al., "Soviet Computer Science Research," *FASAC-TAR-2020* (July 31, 1984); and S. E. Goodman and W. K. McHenry, "Computing in the USSR: Recent Progress and Policies," *Soviet Economy*, no. 4 (1986), pp. 327–354.

17. For a more detailed review of national strategies in the other East European countries, and a review of intra-CMEA trade, see Gary L. Geipel, A. Tomasz Jarmoszko, and Seymour E. Goodman, "The Information Technologies and East European Societies," *East European Politics and Societies* (Fall 1991), pp. 394–438.

18. See Siegfried G. Schoppe, "Die intrasystemaren und die intersystemaren Technologietransfers der DDR," in Gernot Gutmann, ed., *Das Wirtschaftssystem der DDR* (Stuttgart: Gustav Fischer Verlag, 1983), pp. 345–362.

19. "Elektronische Datenverarbeitung," *Analysen und Berichte aus Gesellschaft und Wissenschaft*, no. 2–3 (1979), p. 70, and numerous interviews with former GDR government and industry officials.

20. Wolfgang Marschall, "Der Osten Deutschlands—ein Standort europäischer Elektronik-Industrie," unpublished paper (October 1990), p. 61.

21. See, for example, Jan Monkiewicz, "Implementation of the Complex Programme for Scientific and Technological Progress of the CMEA Countries: The Experience of the Polish Participants," in Jan Monkiewicz and Roland Scharff, eds., *Reform, Innovational Performance and Technical Progress: The Polish Case* (Erlangen, FRG: Institut für Gesellschaft und Wissenschaft, 1989), p. 189.

22. Otto Reinhold, *Deutsche Lehrerzeitung* (January 26, 1989), p. 13.

23. Schoppe, "Die intrasystemaren und die intersystemaren Technologietransfers der DDR," p. 348.

24. Ibid., p. 357, provides a useful analysis of this dilemma. Strengthening exports to the Soviet Union and reducing raw-materials use had been stated goals of the GDR's IT drive since its inception in 1977. See Otfried Steger, "Die Durchführung der Beschlüße des IX. Partcitages der SED auf dem Gebiet der Elektrotechnik und Elektronik," *Neues Deutschland* (June 25–26, 1977), pp. 3–5.

25. "DDR-Wirtschaft unter steigendem Innovationsdruck," *Neue Zürcher Zeitung* (December 25, 1986). The stated goal for the 1986–1990 plan period was for GDR trade with the USSR to increase by 26 percent to 380 bio valuta marks (40 percent of total GDR foreign trade).

26. Thomas A. Baylis, *The Technical Intelligentsia and the East German Elite* (Berkeley: University of California Press, 1974), p. 15.

27. Fred Klinger, "Die Krise des Fortschritts in der DDR," *Aus Politik und Zeitgeschichte* (January 17, 1987), pp. 18, 19.

28. One East German source compared the GDR's pursuit of high-profile IT projects to the country's promotion of sports; both made for good press and improved the country's image. Personal interview.

29. Erich Honecker, "Exakte ökonomische Ziele," *Neues Deutschland* (March 30, 1983).

30. Günter Mittag, "Oekonomische Strategie der Partei—klares Konzept für weiteres Wachstum," *Neues Deutschland* (September 30, 1983), p. 4.

31. "Party Ideologue Charts Course for the 1990s," *JPRS-EER-87-041* April 16, 1989, p. 12.

32. See, for example, Karl Nendel, "Volkswirtschaftliche Anforderungen an die Entwicklung und Anwendung der Informatik in der DDR," *GI-Mitteilungen* 3, no. 2 (1988), p. 27; or Hans-Jürgen Wunderlich, "Schlüsseltechnologien und internationale Wirtschaftsbeziehungen," in Werner Sydow, ed., *Technologien im Umbruch* (Berlin: Die Wirtschaft, 1988), p. 91.

33. Harald Wessel, "Wissenschaftlich-technische Revolution für dem Menschen," *Neues Deutschland* (July 10, 1985). For examples of similar formulations see Hager, "Wissenschaft und Gesellschaft," pp. 118, 125; and Otto Reinhold, "Wissenschafts- und Wirtschaftsplanung in der DDR im Zeichen wissenschaftlicher Revolution," speech to the Friedrich-Ebert-Stiftung (May 24, 1984), p. 3.

34. Christian Steinauer, "Zur notwendigkeit der internationalen sozialistischen Arbeitsteilung in der Mikroelektronik," in *Zum Stand und zur weiteren Entwicklung der internationalen sozialistischen Arbeitsteilung zwischen den Mitgliedsländern des RGW* (Berlin: Akademie fuer Gesellschaftswissenschaften beim ZK der SED, 1989), no. A-75, p. 37.

35. "Die Durchführung der Beschlüsse des IX. Parteitages der SED auf dem Gebiet der Elektrotechnik und Elektronik," *Neues Deutschland* (June 25 to 26, 1977), p. 4.

36. Erich Honecker, "Bericht des Zentralkomitees der Sozialistischen Einheitspartei

Deutschlands an den X. Parteitag der SED," *Neues Deutschland* (April 12, 1981), p. 7.

37. Erich Honecker, "Aus dem Bericht des Politbüros," *Neues Deutschland* (November 23, 1984), p. 5.

38. See, for example, Heinz Wedler, "Die erreichten Ergebnisse beim Aufbau der mikroelektronischen Industrie der DDR seit der 6. Tagung des ZK der SED vom Juli 1977 und die weiteren Arbeitsrichtungen bis 1990," unpublished paper, VEB Kombinat Mikroelektronik Erfurt, Office of the General Director (October 30, 1985), and Karl Nendel, "Volkswirtschaftliche Anforderungen an die Entwicklung und Anwendung der Informatik in der DDR," *GI-Mitteilungen* 3, no. 2 (1988), p. 27.

39. Joachim Abicht, "Ausrüstungen und Leistungen für die Mikroelektronik—ein bedeutendes Exportpotential," *Jenaer Rundschau,* no. 1 (1988), pp. 6, 7. CZJ could not have developed this capability without illegal acquisitions of similar technology from the West, which permitted extensive reverse engineering. Cf. Chapter 4.

40. See, for example, Wolfgang Salecker, "Arbeitsplatzcomputer und die Freude des Entdeckens" [Workplace Computers and the Joy of Discovery], *Neues Deutschland* (July 26, 1986). One outside observer coined the phrase "CAD/CAM euphoria" to describe the GDR's excessive promotion of the technology in almost every media outlet and public forum during the late 1980s.

41. Heike Belitz, Ulrich Köhler, and Mathias Weber, *Der Ostdeutsche Markt für Personalcomputer* (Kronberg: IDC Deutschland, 1990), p. 13.

42. Schoppe, "Die intrasystemaren und die intersystemaren Technologietransfers der DDR," p. 360.

43. The difficulties of making hard-currency purchases of computers or other research equipment were described to me in several interviews with GDR scientists and enterprise officials.

44. *Szamitastechnikai Statisztikai zsebkonyv* (Budapest: Central Statistical Office, 1988), p. 14.

45. See, for example, Jan S. Prybyla, "The GDR in COMECON: Does the GDR Economy Demonstrate That Orthodox Central Planning Is Viable and has a Future?" *Comparative Strategy* 7, no. 1 (1988), pp. 39–50.

46. Parrott, *Politics and Technology in the Soviet Union,* p. 9.

47. CoCom's members consisted of all the NATO countries (except Iceland) and Japan. CoCom was formed primarily to prevent diversions of militarily relevant technology to the Soviet Union and its allies during the Cold War.

48. "Elektronische Datenverarbeitung," pp. 66, 67.

49. "Die DDR als Kooperationspartner," *DIW Wochenbericht* (November 17, 1988), p. 617.

50. "Abkommen zwischen der Regierung der Bundesrepublik Deutschland und der Regierung der Deutschen Demokratischen Republik ueber die Zusammenarbeit auf den Gebieten der Wissenschaft und Technik," *Bulletin* (September 10, 1987), pp. 714–717.

51. See, for example, Christine Kulke-Fiedler and Paul Freiberg, "Joint Ventures in

der Ost-West-Wirtschaftszusammenarbeit," *IPW Berichte,* no. 6 (1988), pp. 44–47.

52. In the late 1980s, the GDR did permit at least two West German software firms to form partnerships with East German university teams doing decision-support research. Perhaps fearful of losing some of their best scientists to Western industry, however, the East German authorities limited the amount of time the GDR scientists could devote to the cooperative projects and maintained central control over the financial proceeds of the relationships. This information was gathered in conversations at trade fairs with the principals of the two projects.

53. See Geipel et al., "The Information Technologies and East European Societies," pp. 425–429.

54. In addition, Steven Popper has suggested that differences in the willingness of East European countries to import high technology from the West were related to (1) different domestic economic cycles; (2) the status of their international trade and payment balances; (3) the broader quality of their relations with the West; and (4) "fundamental political choices," especially regarding the desirability of partnership with the United States and the USSR. See Steven W. Popper, *East European Reliance on Technology Imports from the West,* RAND/R-3632-USDP (Santa Monica, Calif.: Rand, 1988), p. 42.

55. See, for example, "Ergebnisse und Perspektiven in den 80er Jahre," in *Wissenschaftlich-technische Zusammenarbeit DDR/UdSSR* (Berlin: Staatsverlag der Deutschen Demokratischen Republik, 1986), pp. 186–202; Politburo report by Horst Dohlus, *Neues Deutschland* (June 19, 1987); and Willi Kunz, "GDR Seeks to Preserve High-Tech Role through CEMA Coordination," JPRS-EER-88–105 (December 7, 1988), pp. 33–42, translating *Wirtschaftswissenschaft* 36 (September 1988), pp. 1281–1296.

56. In 1977, there was the GDR-USSR "Agreement on Cooperation in the Field of Microelectronics"; in 1979 the GDR-USSR "Program on Specialization and Cooperation to 1990"; in 1984 the GDR-USSR "Program for the Development of Cooperation in the Fields of Science, Technology and Production to the Year 2000"; and in 1985 the CMEA's "Complex Program to the Year 2000." See Klaus Krakat, "Schluesseltechnologien in der DDR: Anwendungsschwerpunkte und Durchsetzungsprobleme," *FS-Analysen,* nr. 5 (1986), p. 140.

57. *Neues Deutschland* (March 14, 1989), pp. 1, 3.

58. Robotron, press conference attended by the author, Leipzig Trade Fair (March 15, 1989).

59. Official East German statistics valued GDR-Soviet trade at 66.5 billion marks in 1988; *Statistisches Jahrbuch der Deutschen Demokratischen Republik* (East Berlin: Staatsverlag, 1989), p. 242. Included in Robotron's exports to the Soviet Union were numerous non-IT products as well, but most embodied a fairly high level of sophistication in electronics.

60. The 5:1 ratio was claimed by Gerhard Montag, the former deputy minister for science and technology, in an interview with the author. Another official said that in the thinking of East German economic leaders, for example, 100 small mainframe computers were worth 10–15,000 Skoda cars from Czechoslovakia.

61. Robotron, press conference attended by the author, Leipzig Trade Fair (March 15, 1989), and numerous interviews.

62. Alfred Schüller, "Zunehmende Internationalisierung der Wirtschaftsprozesse: Die DDR unter Anpassungsdruck," unpublished paper presented to the Symposium der Forschungsstelle für gesamtdeutsche wirtschaftliche und soziale Fragen, West Berlin (November 17, 1988), pp. 1, 2.

63. Steven W. Popper, "Eastern Europe as a Source of High-Technology Imports for Soviet Economic Modernization," RAND/R-3902-USDP (1991), p. 28.

64. Fears of being passed over by Moscow dated to the Ulbricht regime, when the GDR appeared concerned that if it did not prove itself to be a worthy economic partner of the Soviet Union, the Soviets would be tempted to move closer to West Germany, thus isolating and weakening the GDR. East Germany's 1966–1970 trade accord with the USSR tied more than one-half of East German foreign trade to its socialist big brother in an effort to avoid that scenario, creating a bias that was to persist throughout the GDR's history. For a detailed account of how the accord was reached, see Baylis, *The Technical Intelligentsia and the East German Elite,* esp. pp. 247, 248.

65. "'Computertelegramm' aus meinem Kollektiv," *Neues Deutschland* (April 18, 1986).

66. Klaus Krakat, "DDR-Industrie setzt auf Modernisierung," *Computerwoche* (October 28, 1988), pp. 42–45.

67. Dieter Brückner and Jochen Mämecke, "Erfurter Mikroelektroniker übergaben Muster von 32-bit-Mikroprozessoren," *Neues Deutschland* (August 15, 1989), p. 1.

68. This was confirmed independently by two highly placed interviewees, who requested anonymity.

69. This view was expressed by the officials in interviews with me conducted in late 1990.

70. Erich Honecker, *Mit dem Blick auf den XII Parteitag die Aufgaben der Gegenwart lösen* (Berlin: Dietz, 1988), p. 29. That quote was repeated, for emphasis, in numerous other sources.

71. "Verpflichtung wurde eingelöst: Kombinat Carl Zeiss Jena übergab 1-Megabit-Speicherschaltkreis," *Neues Deutschland* (September 13, 1988), p. 1.

72. Brückner and Mämecke, "Erfurter Mikroelektroniker übergaben Muster von 32-bit-Mikroprozessoren," p. 1.

73. Personal interview; anonymity requested.

74. Klaus Krakat, "Mikroelektronik in der DDR unter Wirtschaftlichkeitsaspekten," *FS-Analysen,* 1990, p. 60.

75. Frank Rößler, "Woraus wir die Kraft gewannen," *Neues Deutschland* (August 15, 1989), p. 3.

76. Klaus Maier, "DDR-Wissenschaft in der Krise," *Forum Wissenschaft* (January 1990), p. 4.

77. See, for example, "Ökonomische Strategie des XI. Parteitages der SED wird erfolgreich entwickelt," *Neues Deutschland* (October 3, 1987); Wolfgang Biermann, "Mikroelektronik in der Volkswirtschaft der DDR," *Einheit* (January

1989), p. 27; and Brückner and Mämecke, "Erfurter Mikroelektroniker übergaben Muster von 32-bit-Mikroprozessoren," p. 1.

78. Erich Honecker, "Wer den Sozialismus stärkt, handelt zum Wohl des Volkes," *Neues Deutschland* (August 15, 1989), p. 3.

79. "Freundschaftliches Treffen Erich Honeckers und Michail Gorbatschows im ZK der KPdSU," *Neues Deutschland* (September 22, 1988), p. 1.

80. Those policies are discussed in detail in Gary L. Geipel, "Politics and Technology in the German Democratic Republic, 1977–1990," Ph.D. diss., Columbia University, 1993.

81. That approach was offered as an alternative by Wolfgang Marschall, "DDR-Elektronik an der Schwelle der 90er Jahre—Strategie der weiteren Entwicklung," unpublished paper, GDR Academy of Sciences (February 15, 1990), p. 10.

82. Most Hungarian leaders recognized that "modern technology based on microelectronics does not only gain ground as a result of central decisions and campaigns of investment. It is especially in the utilization of electronic products that private initiatives, i.e. the citizens' own activities assume a great importance." See Ildikó Kováts and János Tölgyesi, "Problems in Adaptation of New Communication Technology: Cases from Hungary," unpublished paper, Hungarian Institute of Public Opinion Research (July 1988), p. 7.

83. Wolfgang Biermann, "Was wir am Sozialismus ändern müssen? Zuerst den eigenen Beitrag dafür!" *Neues Deutschland* (November 2, 1989), p. 4.

84. Mittag, *Um jeden Preis,* p. 222.

85. This conclusion was expressed in separate personal interviews with numerous former technology policy-makers after the fall of the GDR.

86. Rolf Hillig, "Dissertation (A)," unpublished, Fakultät für Wirtschafts- und Rechtswissenschaften der Karl-Marx-Universität Leipzig (January 10, 1987), p. 17. For a variation on the same theme, see Heinz Wedler, "Höhere Ausbeute ohne Ausbeutung—das ist typisch DDR," *Junge Welt* (May 27, 1988), p. 6.

87. *ADN International Service* (October 14, 1987).

88. See Stenographische Niederschrift des Treffens der Genossen des Politburos des Zentralkomitees der SED mit dem Generalsekretär des ZK der KPdSU und Vorsitzenden des Obersten Sowjets der UdSSR, Genossen Michail Sergejewitsch Gorbatschow (October 7, 1989), Berlin-Niederschoenhausen, pp. 13, 19, reproduced in Mittag, *Um jeden Preis,* pp. 371, 377. Accounts of this meeting circulated widely in the months following Honecker's fall. According to one rumor, Honecker went so far as to produce from his pocket a prototype copy of the 1-megabit chip as proof that the GDR was viable without any kind of *perestroika.* Gorbachev's reported reaction was a bemused "Tsss."

12. Between Autonomy and State Control

1. O. Reinhold, *Die Gestaltung unserer Gesellschaft* (Berlin, 1986); M. Lötsch, "Technological and Social Change in the GDR," in M. Gerber, ed., *Studies in GDR Culture and Society* 9 (Lanham: University Press of America, 1989), pp. 21–33.

2. The expert interviews took place as part of a research project funded by the BMFT (the Federal Ministry for Research and Technology), *Die Technologisierung der Biologie: Zur Durchsetzung eines neuen Wissenstyps in der Forschung* (R. Hasse, R. Hohlfeld, P. Nevers, W. Ch. Zimmerli; University of Bamberg, 1994). Some of the material comes from data gathering as part of the project *Wissenschaft und Wiedervereinigung* of the Berlin-Brandenburgische Akademie der Wissenschaften, 1995 and 1996. I would like to thank the academy for permission to use this material. The protocols are in the possession of the author. The nomenclature is explained at the first citation of the various interview series.

3. The institutes were given these names after the academy reform in 1968.

4. Nikolai Vavilov discovered that the diversity of strains of cultivated plants is not spread out equally over the entire world but is concentrated in certain mountain regions of the tropics and subtropics, the so-called diversity or gene centers. H. Stubbe, *Geschichte des Instituts für Kulturpflanzenforschung Gatersleben* (Berlin: Akademie-Verlag, 1982), p. 15.

5. H. Bielka, *Beiträge zur Geschichte der Medizinisch-Biologischen Institute Berlin Buch 1930–1995* (Berlin: Max-Delbrück-Centrum für Molekulare Medizin, 1995), p. 33.

6. Compare L. Graham, *Science and Philosophy in the Soviet Union* (New York: Alfred A. Knopf, 1972), pp. 3–68.

7. See P. Hanelt, "Tradition und Fortschritt einer Forschungseinrichtung," *Biologisches Zentralblatt* 113 (1994): 15–23. Stubbe himself also speaks of the good relations between the institute leadership in Gatersleben and the Soviet Occupation Authorities. Thanks to his connections, the Gatersleben grounds were exempted from land reform. H. Stubbe, *Geschichte des Instituts für Kulturpflanzenforschung Gatersleben* (Berlin: Akademie Verlag, 1982), p. 24.

8. See Chapters 1 and 2.

9. "The principle of self-regulation was invalidated by the SED leadership in the field of research." Deutscher Bundestag, *Materialien der Enquete-Kommission "Aufarbeitung von Geschichte und Folgen der SED-Diktatur in Deutschland"* (Frankfurt: Suhrkamp, 1995), vol. 1, p. 319.

10. H. Klare, "Zu Fragen der Akademiereform. Referat des Präsidenten der Deutschen Akademie der Wissenschaften zu Berlin auf der Beratung der Direktoren und leitenden Mitarbeiter der Institute und Einrichtungen. Berlin, 31, Juli." Archiv Bielka.

11. Bundesarchiv Potsdam, DQ 1 (Ministry of Public Health)/2519/Arbeitsmaterial zu Vorgaben für Vorhaben der sozialistischen Großforschung, 10.4.1969.

12. IEH GE (Interviews of Experts, Genetics) 1, 1993 (see note 2).

13. K. Müntz in "Wie beeinflußt Gentechnik unser Leben? Bericht der Neuen Fernseh Urania," DDR 1, March 12, 1986.

14. J. Schöneich in E. Geissler and H. Ley, eds., *Philosophische und ethische Probleme der modernen Genetik* (Berlin: 1972), p. 196.

15. Hanelt, "Tradition und Fortschrift einer Forschungseinrichtung," p. 21.

16. Schöneich, *Philosophische und ethische Probleme,* note 14.

17. According to E. Geißler, "it was hopeless to compete with Paul Berg and strong

foreign groups [because the department had to work] under the state's policy of *'Störfreimachung'* (making free of disturbances) especially since we didn't exactly enjoy a lot of support from the administration." E. Geißler, "Genetik zwischen Angst und Hoffnung, ethischen, ideologischen und ökonomischen Zwängen," in E. P. Fischer and E. Geißler, eds., *Wieviel Genetik braucht der Mensch? Die alten Träume der Genetiker und ihre heutigen Methoden* (Konstanz: Universitätsverlag, 1994), p. 10.

18. C. Coutelle, A. Speer, and H. D. Unger, "Humangenetik—humane Genetik," *Spectrum* 18, no. 3 (1987): 10–13.

19. K. Winter, *Das Gesundheitswesen in der Deutschen Demokratischen Republik* (Berlin: VEB Verlag Volk und Gesundheit, 1980), pp. 191–196.

20. See T. H. Matthes, L. Rohland, and H. Spaar, *Die medizinisch-wissenschaftlichen Gesellschaften der DDR* (East Berlin: Dietz-Verlag, 1981).

21. See the retrospective by E. Geißler, "Genetik zwischen Angst und Hoffnung, ethischer, ideologischen und ökonomischen Zwängen."

22. *Verfügungen und Mitteilungen des Ministeriums für Gesundheitswesen* (Berlin, October 2, 1986).

23. E. Geißler et al., eds., *Genetic engineering und der Mensch* (Berlin: Akademie-Verlag, 1981).

24. E. Geißler, H. E. Hörz, and H. Hörz, "Eingriffe in das Erbgut des Menschen," *Wissenschaft und Fortschritt*, no. 5 (1980): 188. The authors were not alone in advocating this position, nor was it a new one in the genetics debate in East Germany: compare R. Hohlfeld and H. B. Nordhoff, "Probleme gesellschaftlicher Entwicklung und die Rolle der humanwissenschaftlichen Forschung in der DDR," in I. Spittmann-Rühle and G. Helbig, eds., *Die DDR vor den Herausforderungen der achtziger Jahre* (Cologne, 1984), pp. 141–157.

25. See R. Hohlfeld and H. B. Nordhoff, "Probleme gesellschaftlicher Entwicklung und die Rolle der humanwissenschaftlichen Forschung in der DDR."

26. *Sinn und Form* 36 (1979): 1006.

27. K. Hager, "Marxismus-Leninismus und Gegenwart," *Neues Deutschland* (November 6, 1986), p. 4; see also E. Geißler and R. Mocek, "Gentechnik—Fluch oder Segen?" *Einheit* 5 (1989): 446–453, with a differentiated overview of the issue for the SED readership.

28. R. Hohlfeld and H. B. Nordhoff, "Organismen als Produktivkraft," *Deutschland-Archiv* 21 (1988): 182–196; A. M. Wobus and U. Wobus, eds., *Genetik zwischen Furcht und Hoffnung* (Leipzig: Urania-Verlag, 1991), especially the Foreword by the editors (pp. 7–10).

29. U. Wobus, "Gewinn für beide Seiten," *Spectrum* 18, no. 2 (1987): 25–28.

30. Report of the Neues Fernseh Urania (note 13).

31. K. von Lampe, "Die öffentliche Diskussion über die Gentechnik in der DDR," thesis, Free University of Berlin (1989), p. 82.

32. Leaving aside the period 1969 to 1971, when financing was tied to topics and results and was done via project groups.

33. EB ON (Interview of Experts, Oncology) 3, 1995.

34. E. Geißler "Genetik zwischen Angst und Hoffnung, ethischer, ideologischen und ökonomischen Zwängen," p. 9.

35. SAPMO DY 30/34836/2. Letter of the head of the Health Department of the Central Committee of the SED to the first secretary of the Academy of Sciences, H. Klemm. Almost identical letters—with the exception of the political request—were sent to the first secretaries of the SED regional leadership in Dresden, Leipzig, and Rostock.
36. EB ON 5, 1, 1995.
37. EB ON 1, 5, 1995.
38. EB ON 5, 1995.
39. EB GE 1, 1993.
40. EB BT (Interview of Experts, Biotechnology) 1, 1995.

13. Robert Havemann

1. A brief analysis of Havemann's speech in Leipzig in 1962, and of his lectures at Humboldt University in Berlin in 1963–1964, are provided in Dieter Hoffmann, ed., *R. Havemann: Dialektik ohne Dogma?* (Berlin: Deutscher Verlag der Wissenschaften, 1990), p. 230.
2. R. Havemann, "Was wir Stalin verdanken," *Tägliche Rundschau* 9 (March 20, 1953): 65, 1.
3. See M. Heinemann, "Der Wiederaufbau der Kaiser-Wilhelm-Gesellschaft und die Neugründung der Max-Planck-Gesellschaft (1945–1949)," in R. Vierhaus and B. v. Brocke, eds., *Forschung im Spannungsfeld von Politik und Gesellschaft. Geschichte und Struktur der Kaiser-Wilhelm-/Max-Planck-Gesellschaft* (Stuttgart: Deutsche Verlagsanstalt, 1990), p. 407ff. See also K. Macrakis, *Surviving the Swastika: Scientific Research in Nazi Germany* (New York: Oxford University Press, 1993), epilogue.
4. R. Havemann, "Trumans großer Theaterdonner," *Neues Deutschland* (February 5, 1950), in R. Havemann, *Warum ich Stalinist war und Antistalinist wurde. Texte*, ed. D. Hoffmann and H. Laitko (Berlin: Dietz Verlag, 1990), pp. 88–90.
5. R. Havemann, *Ein deutscher Kommunist* (Hamburg: Rowohlt, 1978), p. 72.
6. See H.-G. Bartel, "Robert Havemann—Naturwissenschaftler und Antifaschist," *Wissenschaft und Fortschritt*, 38, (1990): 8, 202; W.-D. Bilke and H. Pietsch, "Robert Havemann," *Zeitschrift für Physikalische Chemie* 98, (1990): 1078ff.
7. R. Havemann, *Einführung in die chemische Thermodynamik* (Berlin: Deutscher Verlag der Wissenschaften, 1957), p. 5.
8. See "Deckname 'Leitz,'" *Der Spiegel* 46, no. 21 (1995): 87–88.
9. R. Havemann, "Ja, ich hatte unrecht," in Havemann, *Warum ich Stalinist war,* p. 194.
10. R. Havemann, "Meinungsstreit fördert die Wissenschaften," *Neues Deutschland* (July 8, 1956), in Havemann, *Warum ich Stalinist war,* pp. 133–141.
11. See R. Havemann's letter to W. Ulbricht, Berlin, April 27, 1967, SAPMO IV 2/11/v. 4920, p. 8.
12. R. Havemann, "Rückantworten an die Hauptverwaltung 'Ewige Wahrheiten,'" *Sonntag* (October 28, 1956), in Havemann, *Warum ich Stalinist war,* pp. 149–156.

13. That is why Jürgen Rühle described the years 1956–57 as "June 17 of the intelligentsia," a reference to the workers' uprising on June 17, 1953.

14. SAPMO, IV A 2/9.04/101, p. 3.

15. SAPMO IV 2/1/01/388, folio 230.

16. R. Havemann, "Hat Philosophie den modernen Naturwissenschaften bei der Lösung ihrer Probleme geholfen?" (Leipzig, September 17, 1962), in Havemann, *Dialektik ohne Dogma?*.

17. Ibid., pp. 50ff.

18. The lecture was subsequently published in an obviously revised version as M. Steenbeck, "Über die Einheit von Natur- und Gesellschaftswissenschaften," *Deutsche Zeitschrift für Philosophie* 10 (1963): 1472–1488.

19. SAPMO NL 182/937, Nachlaß Ulbricht, pp. 74ff., 150ff.

20. SAPMO IV A2/9.04/304, Werner Neugebauer to Kurt Hager, Berlin, July 30, 1963.

21. Ibid.

22. See the select bibliography in Havemann, *Dialektik ohne Dogma?*, pp. 264–265.

23. SAPMO NL 281/56, p. 79.

24. "Der Jugend Vertrauen und Verantwortung. Kommunique des Politbüros des Zentralkommittees," in *Dokumente der SED*, Bd. IX (Berlin: Dietz Verlag, 1965), p. 691.

25. Ibid.

26. Walter Ulbricht, *Die Durchführung der ökonomischen Politik im Planjahr 1964. Rede auf der 5. Tagung des ZK der SED (3.-7.2.1964)* (Berlin: Dietz Verlag, 1964), p. 17.

27. SAPMO IV/2/11/4920. Letter from Havemann to the Central Party Control Commission, June 26, 1964.

28. R. Havemann, "Wir Deutschen machen alles besonders gründlich," *Echo am Abend* (March 11, 1964), p. 9, in Havemann, *Warum ich Stalinist war*, pp. 189–191.

29. R. Havemann, *Ein deutscher Kommunist. Rückblicke und Perspektiven aus der Isolation*, ed. M. Wilke (Reinbeck: Rowohlt-Verlag, 1978).

30. Interview with Kurt Hager (Kristie Macrakis).

31. Even colleagues who were sympathetic toward Havemann, like W. Heise and G. Klaus, could not understand why Havemann did this and criticized him not only in public statements but also in private letters.

32. SAPMO NL 281/56, p. 61.

33. SAPMO J IV 2/3–957.

34. SAPMO J IV 2/2–924.

35. SAPMO J IV 2/3–953.

36. SAPMO IV A2/9.04/103.

37. SAPMO J IV 2/2; IV A2/9.04/103.

38. Archiv der Berlin-Brandenburgischen Akademie der Wissenschaften, Akademieleitung (AAW, AL), no. 161, L. Pauling to W. Hartke, Pasadena, April 21, 1964.

39. R. Havemann, "Plädoyer für eine neue KPD (Die Partei ist kein Gespenst),"

Der Spiegel 19, no. 52 (1965): 30–32, in Havemann, *Warum ich Stalinist war*, pp. 197–204.

40. Ibid.

41. SAPMO J IV 2/2–1019.

42. *Neues Deutschland* 20, no. 350 (December 21, 1965): 2.

43. SAPMO IV A2/9.04/106.

44. SAPMO IV A2/9.04/106.

45. SAPMO J IV 2/2–1155.

46. See the documentary appendix in Havemann, *Dialektik ohne Dogma?* pp. 245–255.

47. *Die Zeit* 18, no. 12 (March 18, 1964): 3.

48. AAW AL no. 162.

49. Ibid., no. 164.

50. AAW AL, no. 163.

51. SAPMO J IV 2/3–1164.

52. "Erklärung des Präsidiums der Akademie der Wissenschaften," *Neues Deutschland* 21, no. 91 (April 1, 1966): 2.

53. See H. Laitko, *Robert Havemann—Stalinismuskritik und Sozialismusbild* (Berlin: Schriftenreihe der PDS, 1992); D. Hoffmann and H. Laitko, "Robert Havemann. Ein nichkonformer Kommunist in Deutschland," in Th. Bermann and M. Keßler, eds., *Ketzer im Kommunismus* (Mainz: Decaton-Verlag, 1993), pp. 320–338.

54. SAPMO J IV 2/2/1568.

55. H. Jäckel, ed., *Ein Marxist in der DDR. Für Robert Havemann* (Munich: Piper-Verlag, 1980), p. 7.

56. R. Havemann, "Schreiben für die DDR," in R. Havemann, *Berliner Schriften*, ed. A. W. Mytze (Berlin: Verlag europäische Ideen, 1977), p. 26.

14. Kurt Gottschaldt and Psychological Research in Nazi and Socialist Germany

1. Earlier versions of this chapter were presented at the colloquium for history of science at the Humboldt University of Berlin (Professor Rüdiger vom Bruch), June 21, 1994, and at the History of Science Society meeting in Minneapolis, Minnesota, October 29, 1995. Thanks to the participants in these events, and also to Kristie Macrakis and Dieter Hoffmann for their comments.

2. Mitchell G. Ash, "Verordnete Umbrüche, Konstruierte Kontinuitäten: Zur Entnazifizierung von Wissenschaftlern und Wissenschaften nach 1945," *Zeitschrift für Geschichtswissenschaft* 43 (1995): 203–223; idem, "Denazifying Scientists—and Science," in Matthias Judt and Burghard Ciesla, eds., *Technology Transfer out of Germany after 1945* (Chur: Howard Academic, 1995), pp. 61–80.

3. Bruno Latour, *Science in Action* (Cambridge, Mass.: Harvard University Press, 1987); Andy Pickering, "Big Science as a Form of Life," in M. de Maria et al., eds., *The Restructuring of Physical Sciences in Europe and the United States, 1945–1960* (Singapore: World Scientific, 1989), pp. 42–54; Andy Pickering,

"From Science as Knowledge to Science as Practice," in Andy Pickering, ed., *Science as Practice and Culture* (Chicago: University of Chicago Press, 1992).

4. For further examples of this approach, see Mitchell G. Ash, "Wissenschaftswandel in Zeiten politischer Umwälzungen—Entwicklungen, Verwicklungen, Abwicklungen," *NTM—Internationale Zeitschrift für Geschichte und Ethik der Naturwissenschaften, Technik und Medizin* 3 (1995): 1–21.

5. The circumstances of Gottschaldt's appointment cannot be described here. For more detailed accounts see Mitchell G. Ash, "Die Erbpsychologische Abteilung des Kaiser-Wilhelm-Instituts für Anthropologie, menschliche Erblehre und Eugenik 1935–1945," in Lothar Sprung and Wolfgang Schönpflug, eds., *Geschichte der Psychologie in Berlin* (Frankfurt am Main: Lang-Verlag, 1992), pp. 205–222; Mitchell G. Ash, *Gestalt Psychology in German Culture, 1890–1967: Holism and the Quest for Objectivity* (Cambridge, England, and New York: Cambridge University Press, 1995), pp. 354–361.

6. Gottschaldt had originally been proposed, at Eugen Fischer's suggestion, as the only candidate for a full professorship in the Philosophical Faculty, where he would have succeeded his teacher, the Gestalt psychologist Wolfgang Köhler. This was prevented by the head of the university chapter of the National Socialist Professors' League. AUB, K 221, Bd. 3, Bl. 66–68, 75.

7. Kurt Gottschaldt, "NSV und Zwillingsforschung," *Wochen-Dienst der NSDAP Gau Berlin, Amt für Volkswohlfahrt* 3:28 (July 11, 1939): 335ff.

8. Kurt Gottschaldt, "Phänogenetische Fragestellungen im Bereich der Erbpsychologie," *Zeitschrift für induktive Abstammungs- und Vererbungslehre* 76 (1939): 118–157.

9. For a detailed account of these methods, see Kurt Gottschaldt, *Die Methodik der Persönlichkeitsforschung in der Erbpsychologie* (Leipzig: Verlag Johann Ambrosius Barth, 1942), pp. 94ff. Contemporary twin research still employs a method of concordance and discordance; but this now involves measuring the correlation coefficients, rather than the averages, for the performances of identical and fraternal twin pairs.

10. See, e.g., Ulfried Geuter, *The Professionalization of Psychology under Nazism*, trans. Richard Holmes (New York and Cambridge, England: Cambridge University Press, 1992); Ute Deichmann, *Biologen unter Hitler* (Frankfurt am Main: Campus, 1992); Kristie Macrakis, *Surviving the Swastika: Scientific Research in Nazi Germany* (New York: Oxford University Press, 1993); Mark Walker and Monika Renneberg, eds., *Science, Technology and National Socialism* (Cambridge: Cambridge University Press, 1994).

11. Kurt Gottschaldt, "Erbpsychologie der Elementarfunktion der Begabung," in Günter Just, ed., *Handbuch der Erbbiologie des Menschen*, Bd. V/1 (Berlin: Springer-Verlag, 1939), 445–537, esp. p. 459.

12. Gottschaldt, "Phänogenetische Fragestellungen"; idem., "Zur Problematik der psychologischen Erbforschung (Eine Erwiderung auf den gleichnamigen Beitrag von Fritz Lenz)," *Archiv für Rassen- und Gesellschaftsbiologie* 36 (1942): 27–56. See also Franz Emanuel Weinert, Ulrich Geppert, Jürgen Dörfert, and Petra Viek, "Aufgaben, Ergebnisse und Probleme der Zwillingsforschung,

Dargestellt am Beispiel der Gottschaldtschen Längschnittsstudie," *Zeitschrift für Pädagogik* 40 (1994): 265–288.

13. See, e.g., Eugen Fischer, Tätigkeitsbericht, 15.6.1933, MPG-Archiv, KWI für Anthropologie, Bd. 2399.

14. Otmar Freiherr von Verschuer, *Leitfaden der Rassenhygiene,* 2nd ed. (Leipzig: Thieme, 1943), pp. 83ff.

15. For a brief summary of Gottschaldt's wide-ranging practical activities, see "Tätigkeitsbericht der Kaiser-Wilhelm-Gesellschaft zur Förderung der Wissenschaften für das Geschäftsjahr 1942/43," *Die Naturwissenschaften* 31 (1943): 520–541, on p. 522. Research on the practical side of Gottschaldt's activities under Nazism is still in progress.

16. For contacts with the Race Policy Office, see, e.g., Kurt Gottschaldt, "Zwillingsforschung," *Neues Volk* (February 1937): 9–13; Peter Weingart, *Doppel-Leben. Ludwig Ferdinand Clauss: Zwischen Rassenforschung und Widerstand* (Frankfurt a.M.: Campus-Verlag, 1995), esp. chapter 9.

17. Gottschaldt, "Erbpsychologie," p. 514.

18. Gottschaldt, Interview with Mitchell G. Ash, Göttingen, August 5, 1987.

19. Gottschaldt, Bericht, August 26, 1945; Verschuer to Gottschaldt, August 28, 1945; Ernst Telschow to Verschuer, September 3, 1945. Copies in MPG-Archiv, Abt. II, Rep. 1A, X97.

20. Havemann to Shulits, February 18, 1946, OMGUS-Hesse (OMGUS/OMGH) files, Hessisches Hauptstaatsarchiv Wiesbaden (HHW), Abt. 649, 8/22–2/8. On Havemann's career, see Chapter 13 of this book.

21. Gottschaldt to Shulits, May 8, 1946, HSW, OMGUS/OMGH, Abt. 649, 8/22–2/8.

22. "Handlanger des Verbrechens," *Tägliche Rundschau,* May 23, 1946. See also Peter Weingart, Jürgen Kroll, and Kurt Bayertz, *Rasse, Blut und Gene: Geschichte der Eugenik und Rassenhygiene in Deutschland* (Frankfurt am Main: Suhrkamp Verlag, 1988), pp. 574ff.

23. E.g., Verschuer to Otto Hahn, May 23, 1946, October 30, 1946, MPG-Archiv, Abt. II, Rep. 1A, X 97; Verschuer to Eugen Fischer, April 24, May 23, and June 24, 1946, Nachlaß Verschuer, Universitätsarchiv der Universität Münster.

24. Michael Wolfsohn to Benedikt B. Ferencz (Office of Chief Counsel for War Crimes, Berlin Branch), November 7, 1946, p. 3, "Als von Verschuer von Prof. Dr. Gottschaldt, dem Leiter der KWG (Institut für psychologische Forschung) befragt wurde, ob er über die Einzelheiten in Auschwitz informiert sei, bejahte er dies," HHW, Abt. 520/F, Nr. F2 5261.

25. Hartshorne to DeLong, July 2, 1946; Hartshorne to Gottschaldt, August 3, 1946, HHW, Abt. 649; Willy Viehweg to Fritz Karsen (E and RA Berlin), November 7, 1946, HHW, Abt. 520/F, Nr. F2 5261; Gottschaldt to Karsen, March 11, 1947; Gottschaldt to Rektor Stroux, May 17, 1947, AUB, Rektorat, Bnd. 550 (Schriftwechsel mit dem Institut für Psychologie, 1946–1962).

26. On the de-nazification of higher education in the Soviet zone, see Hans-Uwe Feige, "Aspekte der Hochschulpolitik der Sowjetischen Militäradministration in Deutschland (1945–1948)," *Deutschlandarchiv* 25 (1992): 1169–1180; idem., "Zur Entnazifizierung des Lehrkörpers an der Universität Leipzig," *Zeitschrift*

für Geschichtswissenschaft 42 (1994): 795–808; Mitchell Ash, "Verordnete Umbrüche, konstruierte Kontinuitäten" and "Denazifying Scientists—and Science."

27. AUB, G 395, Bl. 17.

28. Ibid., Bl. 33–46, 50, 55, 58.

29. Theodor Brugsch to Mathematics-Natural Science Faculty, December 1, 1947, AUB, G 395, Bl. 65; cf. Brugsch to Gottschaldt, March 20, 1946, AUB, G 395, Bl. 21.

30. Gottschaldt to Rektor Stroux, July 27, 1946, June 10, 1948, November 16, 1948, AUB, Rektorat, Bnd. 550.

31. For details on the institute and its growth, see Jürgen Meehl, "Das Institut für Psychologie der Humboldt-Universität zu Berlin," *Zeitschrift für Psychologie* 157 (1954): 149–153; Hans Hiebsch, "Die Ausbildung der wissenschaftlichen Aspiranten des Faches Psychologie," ibid., 157 (1954): 156–162. See also Hans-Dieter Schmidt, "Erinnerungen an Kurt Gottschaldt," *Psychologische Rundschau* 43 (1992): 252–260; Michaela Hausmann, "Psychologie an der Humboldt-Universität zu Berlin nach dem II. Weltkrieg: Das Wirken von Kurt Gottschaldt 1947–1961," in Siegfried Jaeger, Irmingard Staeuble, Lothar Sprung, and Horst-Peter Brauns, eds., *Psychologie im soziokulturellen Wandel— Kontinuitäten und Diskontinuitäten* (Frankfurt am Main: Peter Lang, 1995), pp. 281–285.

32. Gottschaldt was mentioned for the position of rector of Humboldt University as early as 1951. Wolfgang Harig to ZK, November 13, 1951, Bundesarchiv Abteilungen Potsdam (BAP), R-3, 210; Wohlgemuth to ZK (Central Committee Dept. of Science), July 5, 1951, AUB, G 395.

33. Vorschlag Heinrich Frank, November 26, 1952. Archiv der Berlin-Brandenburg Akademie der Wissenschaften (Akademie-Archiv), Bestand Akademieleitung/Personalia, Bnd. 128.

34. Mehl, "Das Institut für Psychologie," pp. 151–152; Gottschaldt, "Geleitwort," *Zeitschrift für Psychologie* 157 (1954): cf. Hausmann, "Psychologie an der Humboldt-Universität."

35. Gottschaldt, "Die Pädagogische Psychologie im Universitätsstudium der Lehrer," *Pädagogik* 1:5 (1946): 1–12.

36. Kurt Gottschaldt, *Probleme der Jugendverwahrlosung,* 2nd ed. (Leipzig: Verlag Johann Ambrosius Barth, 1950), pp. 5ff., 17–18; 1st ed. 1948.

37. Ibid., pp. 4, 39.

38. For Gottschaldt's funding applications, see Forschungsplan 1950, January 12, 1950, BAP R-2 1433, Bl. 49; Gottschaldt to Math.-Nat. Dekan Noack, May 23, 1949, AUB, Rektorat, Bnd. 550; Gottschaldt to Notgemeinschaft, November 15, 1949, Gottschaldt papers. For public acknowledgment of these funding sources, see Kurt Gottschaldt, "Zur Theorie der Persönlichkeit und ihrer Entwicklung," *Zeitschrift für Psychologie* 157 (1954): 1.

39. Gottschaldt, "Zur Theorie der Persönlichkeit," pp. 2–4.

40. Kurt Gottschaldt, "Über Persona-Phänomene," *Zeitschrift für Psychologie* 157 (1954): 163–200; Gottschaldt, "Zur Psychologie der Wir-Gruppe," *Zeitschrift für Psychologie* 163 (1959): 193–229.

41. Gottschaldt, "Zur Psychologie der Wir-Gruppe."

42. Kurt Gottschaldt, "Der Dank des Wissenschaftlers," *Berliner Zeitung,* no. 241, October 16, 1954. Colleagues from East Germany have suggested that such testimonials were often written by the newspaper's staff. In this case, the language appears to be Gottschaldt's own.

43. Gottschaldt to DAW Presidium, June 16, 1956, Akademie-Archiv, Bestand Akademieleitung—Naturwissenschaftliche Einrichtungen, Band 91.

44. Karras, Staatssekretariat für Hochschulwesen an Hauptreferent Hartmann, October 3, 1952; Personalbogen K. Gottschaldt, April 1, 1953 (new address), AUB, G 395, Bl. 82 resp. 1a.

45. Aktennotiz November 28, 1956, Gottschaldt papers, folder "Planung für Grundstück Oranienburger Str. 19."

46. Hans-Dieter Schmidt, "Bedingungsgrundlagen der sozialen Betriebsatmosphäre und Probleme der innerbetrieblichen Kooperation," *Zeitschrift für Psychologie* 159 (1959): 153–186, here, p. 157.

47. Alfred Katzenstein, "Gestalt- und Klinische Psychologie," *Neurologie, Psychiatrie und medizinische Psychologie,* Heft 7 (1956); Hans-Dieter Schmidt, "Über eine unsaubere Methode der Diskussion in der Psychologie. Eine notwendige Stellungnahme," *Zeitschrift für Psychologie* 160 (1956/57): 199–200.

48. Hans Hiebsch, "Aufgaben und Situation der Pädagogischen Psychologie in der DDR," *Pädagogik* 13:4 (1958): 251; Ulrich Ihlefeld, "Zur ideologischen Situation in der Pädagogischen Psychologie," *Pädagogik* 13:7 (1958): 508.

49. Stefan Busse, "Gab es eine DDR-Psychologie?" *Psychologie und Geschichte* 5 (1993): 40–62.

50. For this self-characterization, see Schmidt, "Erinnerungen an Kurt Gottschaldt."

51. For Gottschaldt's *Aussprache* with Hager, see, e.g., Walter Mäder to Hager, February 21, 1959, SAPMO IV 2.904.150, Bl. 28ff.

52. ZK memo to Hager, January 21, 1960, ibid., Bl. 153–154; see also report by Gretel Junge, March 11, 1960, ibid., Bl. 184.

53. Bericht über die Lage am Institut für Psychologie hinsichtlich der Institutsleitung, November 7, 1958, ibid., Bl. 11–16; Über die Lage an der Arbeitsstelle für experimentelle und angewandte Psychologie der DAW Berlin," December 22, 1958, ibid., Bl. 36–45.

54. Hans Gummel, quoted in Protokoll, Unterredung in Berlin-Buch, February 27, 1959, ibid., Bl. 69–71; Wolfgang Steinitz, quoted in ZK memo to Hager, February 21, 1959. See also Aktennotiz, Besprechung im ZK am 2.3.1960, BAP R-3, 226.

55. Gottschaldt to Hartke, August 4, 1959, ibid., Bl. 90–93.

56. ZK to Girnus (state secretary for higher education), December 11, 1959; ZK memo to Hager, December 22, 1959, ibid., Bl. 127–130.

57. Hartke to ZK, December 18, 1959; ZK to Hager, January 21, 1960, ibid., Bl. 120–122, 153–154.

58. On the "market economy" in scientists before 1961, see Ralph Jessen, "Zur Sozialgeschichte der ostdeutschen Gelehrtenschaft (1945–1970)," in Martin Sabrow and Peter Th. Walther, eds. *Historische Forschung und sozialistische Dik-*

tatur: Beiträge zur Geschichtswissenschaft der DDR (Leipzig: Leipziger Universitätsverlag, 1995), pp. 121–143: here, pp. 128ff.

59. ZK to Parteiorganisation Humboldt-Universität, March 29, 1960; Frommknecht to Hager, May 27, 1960, SAPMO IV 2.904.150, Bl. 181–182, 193ff.

60. Vertrauliche Aktennotiz, Parteiorganisation Humboldt- Universität to ZK, June 9, 1960, ibid., Bl. 228.

61. Thilo to Hager, February 25, 1961; Thilo to Dahlem, March 2, 1961; Dahlem to Tschersisch, February 27, 1961; Dahlem to Thilo, March 2, 1961, all in AUB, G 395. ZK memo, March 25, 1961, SAPMO IV 2.904.150, Bl. 307–310.

62. Girnus to Gottschaldt, September 8, 1961 and November 9, 1961, AUB, G 395; Mäder (ZK) memo, January 12, 1962, Academy of Sciences PO to ZK, January 16, 1962, SAPMO IV 2.904.150, Bl. 356–361.

63. Frohne, report to HA III 16/2, February 24, 1962, Bundesbeauftragte für die Unterlagen des ehemaligen Staatssicherheitsdienstes der DDR (BStU), MfS AP 17823/62; Frommknecht to Hager, February 23, 1962, SAPMO IV 2.904.150, Bl. 364.

64. Kurt Gottschaldt, "Begabung und Vererbung—Phänogenetische Befunde zum Begabungsproblem," in Heinrich Roth, ed., *Begabung und Lernen: Ergebnisse und Folgerungen neuerer Forschungen*, 2nd ed. (Stuttgart: Klett, 1969), pp. 132, 134. For further discussion, see Mitchell G. Ash, "From 'Positive Eugenics' to Behavioral Genetics: Psychological Twin Research under Nazism and Since," *Paedagogica Historica*, Supplementary Series, vol. 3 (1998), 335–358.

65. Kurt Gottschaldt, "Zwillingsforschung als Lebenslaufforschung: Längsschnittuntersuchungen über Entwicklungsverläufe von Zwillingen, aufgewachsen unter sich verändernden Zeitumständen," *Bericht über den 33. Kongress der Deutschen Gesellschaft für Psychologie* (Göttingen: Hogrefe, 1982), pp. 53–64.

66. Weinert et al., "Aufgaben, Ergebnisse und Probleme der Zwillingsforschung."

Biographies of Contributors

Mitchell Ash is Professor of Modern History at the University of Vienna in Austria. He lived in Berlin from 1977 to 1984, during which he received his doctorate from Harvard University. In 1990–91 he was a fellow at the Wissenschaftskolleg/Institute for Advanced Study in Berlin. He is the author of numerous articles and has co-edited three volumes on psychology in modern Germany. His book *Gestalt Psychology in German Culture, 1890–1967: Holism and the Quest for Objectivity* was published in 1995.

Dolores Augustine is Associate Professor of History at St. John's University in New York. After receiving her B.S. from Georgetown's School of Foreign Service, Augustine moved to Berlin in 1977 and completed her masters and doctoral degree (1991) at the Free University of Berlin. She conducted research in the GDR with an IREX grant in 1986. In addition to numerous articles, she is the author of *Patricians and Parvenues: Wealth and High Society in Wilhelmine Germany*, published in 1994.

John Connelly is Assistant Professor of History at the University of California, Berkeley. His interest in the GDR began during a year (1982–83) spent studying in Heidelberg, where he specialized in the GDR opposition. Since the summer of 1989 he has researched in the former GDR, the former Czechoslovakia, and Poland. He completed his dissertation, entitled "Creating the Socialist Elite: Communist Higher Education Policies in the Czech Lands, East Germany, and Poland, 1945–54," at Harvard University in 1994.

Eckart Förtsch is a political scientist who worked at the Institute for Science and Society in Erlangen-Nuremberg, the main West German Institute studying East Germany, until it closed in 1993. After that he began work at the Berlin branch of the Federal Ministry for Science and Technology. He is the author of several books, including *Forschung und Entwicklung in den neuen Bundesländern 1989–1991* (1991), and numerous published articles and reports on science policy in the GDR.

Gary Geipel is a senior fellow at the Hudson Institute, a public-policy research organization. He received his Ph.D. from Columbia University in 1993 with a dissertation entitled "Politics and Technology in the German Democratic Republic, 1977–1990." In addition to editing three major volumes for the Hudson Institute, including, most recently, *Rethinking the Transatlantic Partnership* (1996), he has written widely on Germany for scholarly presses as well as for newspapers and magazines. He spent 1984–1985 on an ITT international fellowship in Munich. In 1987–1988 he worked for the CIA as a graduate fellow writing reports on Eastern European affairs.

Dieter Hoffmann is a research scholar at the Max Planck Institute for the History of Science and a *Privatdozent* at Humboldt University. Between 1975 and 1991 he worked at the Institute for the Theory, History, and Organization of Science at the Academy of Sciences of the GDR. Subsequently he was a Humboldt Foundation fellow at the University of Stuttgart and Harvard University, as well as a research fellow at the Physikalische-Technische Bundesanstalt, in Berlin. He works on biographies and institutions in the history of modern physics; since 1989 he has increasingly dealt with the history of science and technology in the GDR. He is the author of numerous publications on the history of modern physics, including some books and anthologies.

Rainer Hohlfeld worked at the West German Institute for Science and Society in Erlangen-Nuremberg, an institute devoted to the study of East Germany, from 1981 until it closed in 1993. After receiving his Ph.D. in bacterial genetics at the University of Cologne in 1973, he worked until 1980 at the Max Planck Institute (Starnberg) for Research on the Conditions of Life in the Scientific-Technical World. His work has focused on the sociological analysis of biological and biomedical research. In addition to numerous analyses for the Erlangen institute, he has published articles on biotechnology and biomedicine in the GDR. He is currently at the Brandenburg Academy of Sciences in Berlin.

Hubert Laitko was Professor and Member of the GDR National Committee for the History and Philosophy of Science from 1979 until the fall of the Berlin Wall. He studied philosophy and journalism at the University of Leipzig, then received a doctorate in 1964 in the philosophy of science at Humboldt University, Berlin, under the direction of Hermann Ley. In 1969 he joined the Academy of Sciences' Institute for the Theory, History, and Organization of Science, where he headed a research group on the history of science from 1975 until the institute closed in the wake of German unification in 1991. He is the author of numerous articles on the history of scientific institutions and urban and regional history (especially in Berlin), as well as the methodology of history of science, and was the head of numerous collective projects. He is the author of *Wissenschaft als allgemeine Arbeit* (1979).

Kristie Macrakis is Associate Professor of the History of Science at Michigan State University. While a graduate student at Harvard University, she spent six months (1986) of a two-year German stay in East Germany on an IREX fellowship. She met most of the German participants in this project during that time. Macrakis is the recipient of numerous awards and fellowships, including, most recently, a Senior Fulbright Fellowship and a National Science Foundation grant. In addition to numerous articles on science in modern Germany, she is the author of *Surviving the Swastika: Scientific Research in Nazi Germany* (1993).

Peter Nötzoldt is a research associate with the study group on the history of academies of science in Berlin of the Berlin-Brandenburgische Academy of Sciences. He received his B.A. from the Technical University, Chemnitz, in the pedagogy of physics and mathematics. After a stint teaching physics to engineering students, he was a staff member at the Central Institute for Electron Physics at the Academy of Sciences of the GDR from 1976 to 1990. During that time he also worked closely with Robert Rompe, the GDR's doyen of physics. He recently received his doctoral degree from Humboldt University, completing a dissertation on the Academy of Sciences in the GDR.

Reinhard Siegmund-Schultze is a lecturer in the history of mathematics at Humboldt University, where he held a regular position in the Department of the History of Science until the department closed in the wake of unification in January 1991. Since then he has been a Humboldt Foundation fellow at the University of New Hampshire and Harvard University. Before studying the history of mathematics at the University of Leipzig, where he received his doctorate in 1979, he received an M.A. in mathematics from the University of Halle. He is the author of numerous publications in the history of mathematics (not including numerous popular articles), specializing in the social history of Nazi Germany and the history of functional analysis and function theory. His most recent book is *Mathematiker auf der Flucht vor Hitler*, 1998.

Raymond Stokes recently moved to the University of Glasgow, where he is Associate Professor in the Department for Economic and Social History. Before that he was a professor in the Rensselaer Polytechnic Institute's Department of Science and Technology Studies since 1987. He is the author of two books on the German chemical industry, the most recent, *Opting for Oil: The Political Economy of Technological Change in the West German Chemical Industry, 1945–1961*, was published in 1994.

Burghard Weiss is *Privatdozent* at the Technical University, Berlin, and Lecturer at the Medical University in Lübeck. He received his M.A. in Atomic Physics and studied the history of science and technology at the University of Hamburg, where he received his doctorate in 1987 (Habilitation, 1994). He is the author of numerous articles in the history of physics from the eighteenth to the twentieth century, including the history of experiments, atomic physics, and big research. His most recent book is *Forschungsstelle D*, published in 1997.

Index

ABC (Atomic-Biological-Chemical) weapons, 94, 214–215, 257
Abitur, 53–54
Abusch, Alexander, 164, 177
Academies and institutes; all-German tradition, 145, 147, 150–153, 158, 162, 168, 170, 176–178; reform, 58–61; self-government, 58, 156; Western members, 152–154, 156
Academy for Continuing Medical Education, 256
Academy Institute for Cybernetics and Information Processing, 117
Academy of Agricultural Sciences, 32, 37
Academy of Building Sciences, 32, 37
Academy of Sciences, 3, 7, 15, 31–32, 36–38, 43, 50, 58, 140, 142–157, 165, 178, 213, 215, 248–250, 252, 261, 264, 271, 282, 292, 293
Academy Reform, 153, 155, 157, 251–254
Acetylene, 207–208
Ackerman, Anton, 127
AEG Company, 191
AHB Electronics, 114, 118
Albrecht, Uwe, 109
von Allesch, Johannes, 299
Allied Control Council, 70, 212
Allied Military Command, 141
Anticommunism, 77
Antifascism, 3, 6, 20, 64, 67–80, 130–131, 163–166, 273
Anti-SDI program. *See* "Heide"; "Precision"
Apel, Erich, 12
Ardenne Institute, 31
von Ardenne, Manfred, 14
Arms dealing, 87
Arnold, Gerhard, 111
Association of Academy Institutes, 154
Associations of Publicly Owned Enterprises (VVBs), 49–50, 191, 193
"Atomic bank," 215

Atomic energy, 31. *See also* Nuclear power
Atomic weapons, 14
Aussennachrichtendienst (APN), 91
Austria, 5, 169
Aviation research and technology, 9–10, 26, 70, 96, 110, 212

Bareš, Gustav, 130–131
Barwich, Heinz, 216, 220
BASF, 205
Basic Treaty of 1972, 49
"Benchmarking," 202
Bereich Kommerzielle Koordinierung (KoKo). *See* Commercial Coordination Unit
Berlin Academy Institute for Mathematics, 80
Berlin Academy of Sciences, 78
Berlin-Buch Institute of the Academy of Sciences, 176
Berlin City Council, 141
Berlin Psychological Institute, 293
Berlin Wall, 34–35, 47, 49, 64, 78, 159, 161, 163–166, 168, 175–176, 178, 185, 193, 196, 205, 208, 218, 255, 274, 300; fall of, 16, 109
Berman, Jakub, 130–131
Bertag, Peter, 104
Bethge, Heinz, 160
Bieberbach, Ludwig, 69, 71, 73, 77
Biermann, Wolf, 277, 279–281, 285
Biermann, Wolfgang, 106–107, 114, 116
Bilke, Paul, 93, 96, 98
Biomedical research, 10, 143, 247–265
Blaschke, W., 70
Bloch, Ernest, 151
Bodenstein, Max, 271
Boeing, 231
Bourgeois science and scientists, 17, 64, 70–72, 127, 147–148, 153, 172, 177, 216, 248, 251, 296
Brandenburg Prison, 270